Climate Change and Water

International Perspectives on Mitigation and Adaptation

Editors

Carol Howe, CSIRO, Delft, the Netherlands

Joel B. Smith, Stratus Consulting, Boulder, Colo.

Jim Henderson, Stratus Consulting, Boulder, Colo.

American Water Works Association

IWA Publishing

Climate Change and Water: International Perspectives on Mitigation and Adaptation

Disclaimer

This book is provided for informational purposes only, with the understanding that the publisher, editors, and authors are not thereby engaged in rendering engineering or other professional services. The authors, editors, and publisher make no claim as to the accuracy of the book's contents, or their applicability to any particular circumstance. The editors, authors, and publisher accept no liability to any person for the information or advice provided in this book or for loss or damages incurred by any person as a result of reliance on its contents. The reader is urged to consult with an appropriate licensed professional before taking any action or making any interpretation that is within the realm of a licensed professional practice.

AWWA Publications Manager: Gay Porter De Nileon
AWWA Technical Editor/Project Manager: Melissa Valentine
IWA Publishing: Michelle Jones, Production Manager
Cover Art: Bright Idea Graphic Design, Gayle Hughes
Production Editor: Cheryl Armstrong

Library of Congress Cataloging-in-Publication Data

Smith, Joel B.
Climate change and water : international perspectives on mitigation and adaptation / Joel Smith, Carol Howe.
p. cm.
ISBN 978-1-58321-730-6
1. Water-supply--Environmental aspects. 2. Water resources development--Environmental aspects. 3. Water utilities--Management. 4. Climatic changes--Government policy--International cooperation. I. Howe, Carol. II. Title.
TC405.S566 2009
363.6'1--dc22

2009024259

Co-published by

American Water Works Association
ISBN: 9781583217306

and

IWA Publishing
ISBN: 1843393042
ISBN13: 9781843393047

About the American Water Works Association

AWWA is the authoritative resource for knowledge, information, and advocacy to improve the quality and supply of water in North America and beyond. AWWA is the largest organization of water professionals in the world, advancing public health, safety, and welfare by uniting the efforts of the full spectrum of the water community. Through our collective strength, we become better stewards of water for the greatest good of people and the environment.

For information on membership, publications, and conferences, contact

American Water Works Association
6666 West Quincy Avenue
Denver, CO 80235-3098 USA
800.926.7337
www.awwa.org

About IWA Publishing

IWA Publishing, the publishing arm of the International Water Association, provides information services on all aspects of water, sanitation, and related environmental fields. IWA Publishing is a specialist publisher focusing solely on the international marketing of water related publications to IWA members and the wider world. Spanning all areas of the water industry, IWA Publishing covers every aspect of water and wastewater treatment, management and distribution, including water quality and quantity management, and embracing scientific, technical engineering, appropriate technologies, legal, health and economic aspects.

IWA Publishing
Alliance House
12 Caxton Street, London SW1H 0QS, UK
Tel. +44 (0) 20 7654 5500, Fax +44 (0) 20 7654 5555
publications@iwap.co.uk
www.iwapublishing.com

Contents

INTRODUCTION 1

UNDERSTANDING THE IMPACT 3

1. No Doubt About Climate Change and Its Implications for Water Suppliers 5

John E. Cromwell III, national policy researcher for Stratus Consulting
Joel B. Smith, Vice-President, Stratus Consulting
Robert S. Raucher, Executive Vice-President, Stratus Consulting

2. How Should Water Utilities Prepare for Climate Change? 13

Richard G. Sykes, Manager of Natural Resources, EBMUD, Oakland, Calif.
Clifford C. Chan, Manager of Treatment and Distribution, EBMUD, Oakland, Calif.
Rey C. Encarnacion, Associate Civil Engineer, EBMUD, Oakland, Calif.
Kevin T. Richards, Associate Civil Engineer, EBMUD, Oakland, Calif.

3. Mountain Water and Climate Change 21

D. Vanham, University of Innsbruck, Innsbruck, Austria
W. Rauch, University of Innsbruck, Innsbruck, Austria

4. Prevailing Water Demand Forecasting Practices and Implications for Evaluating the Effects of Climate Change 41

Jack C. Kiefer, PhD, Principal Economist and Water Resources Practice Leader, CDM Federal Programs, Carbondale, Ill.

5. Impacts of Climate Change and Variability on Source Water Quality of Lake Cachuma, California 47

Joseph A. Drago, PE, PhD, Principal, Kennedy/Jenks Consultants, San Francisco, Calif.
Levi Brekke, PhD, Water Resources Modeler, US Bureau of Reclamation, San Francisco, Calif.
Catriona E. Rogers, PhD, Environmental Scientist, USEPA Global Change Research Program, Washington, D.C.
Kevin D. Walsh, PE, General Manager, Goleta Water District, Goleta, Calif.

6. The Climate Footprint and the Practical Application at Water Companies in the Netherlands 55

Peter Weirs, Climate Partners
Henk de Boer, Brabant Water
Theo Janse, Operations Research, Waternet
Leo Joosten, Climate Chance BV
Jan Peter van der Hoek, Strategic Centre, Waternet
Dirk van der Woerdt, Waterbedrijf Groningen

7. **Climate Footprint and Mitigation Measures in the Dutch Water Sector 73**
Jos Frijns, KWR, Watercycle Research Institute
Mirabella Mulder, Grontmij
Jelle Roorda, MWH (formerly Grontmij)

8. **The Water Footprint of Bio-Energy 81**
Prof.dr.ir. A.Y. Hoekstra, University of Twente, the Netherlands
Dr. P.W. Gerbens-Leenes, University of Twente, the Netherlands
Prof.dr.ir. T.H. van der Meer, University of Twente, the Netherlands

MITIGATION 97

9. **The Water–Energy–Climate Nexus—Systems Thinking and Virtuous Circles 99**
Monique Retamel, Institute for Sustainable Futures, University of Technology, Sydney, Australia
Andrea Turner, Institute for Sustainable Futures, University of Technology, Sydney, Australia
Stuart White, Institute for Sustainable Futures, University of Technology, Sydney, Australia

10. **Energy Use in Urban Water 111**
S.J.Kenway, Research Scholar, Advanced Water Management Centre, University of Queensland
A. Priestley, Principal Research Scientist, CSIRO, Australia
S. Cook, Research Scientist, CSIRO, Australia
A. Gregory, Theme Leader, Water for a Healthy Country Urban, CSIRO, Australia
A. Lovell, Manager, Science and Sustainability, Water Services Association of Australia
N. Smith, Project Analyst, Water Services Association of Australia

11. **WATERGY: Energy and Water Efficiency in Municipal Water Supply and Wastewater Treatment 123**
Sudha Setty, Alliance to Save Energy

12. **Station Efficiency Reduces Greenhouse Gas Emissions 137**
Robert Eckard, founder, Robert Eckard Consulting, Woodland, Calif.

13. **Climate Change—Mitigation Policy Issues 145**
Cynthia Lane, Regulatory Specialist, American Water Works Association

14. **Climate Change Mitigation Strategies in the Water Sector in Developing Countries 157**
Juan van den Broeck, NOLASCO & Associates, Buenos Aires, Argentina
Daniel A. Nolasco, NOLASCO & Associates, Buenos Aires, Argentina

ADAPTATION 169

15. Incorporating Climate Change in Water Planning 173
Kathy Freas, Water Resources and Ecosystem Management Services, CH2M HILL, Albuquerque, N.M.
Bob Bailey, President of Water Business Group, CH2M HILL, Albuquerque, N.M.
Armin Munevar, Senior Water Resources Engineer and Climate Technologist, CH2M HILL, Albuquerque, N.M.
Susan Butler, Senior Water Resources Planner, CH2M HILL, Albuquerque, N.M.

16. Climate Change and Water Utilities 183
Marloes H.N. Bakker, PhD, Co-operative Programme on Water and Climate, the Netherlands
Henk van Scheik, Deputy Programme Coordinator, Co-operative Programme on Water and Climate, the Netherlands

17. Half Full or Half Empty? Either Way It's Time to Plan 193
Perri Standish-Lee, Director of Integrated Watershed Planning, Black & Veatch Corp., Sacramento, Calif.
Erik Loboschefsky, Junior Engineer, Black & Veatch Corp., Sacramento, Calif.
Kristina Lecina, Water Resources Engineer, Black & Veatch Corp., Sacramento, Calif.

18. Climate Change Is Real: How Can Utilities Cope with Potential Risks? 199
Joel B. Smith, Vice-President, Stratus Consulting, Boulder, Colo.

19. Planning Strategy in a Changing Climate 205
Sarah Deslauriers, Carollo Engineers, Walnut Creek, Calif.
Lydia Holmes, PE, Carollo Engineers, Walnut Creek, Calif.
Steve McDonald, PE, Carollo Engineers, Walnut Creek, Calif.
Jon Loiacono, PE, San Francisco Public Utilities Commission, San Francisco, Calif.

20. Climate Change: Charting a Water Course in an Uncertain Future 215
Michael J. Wallis, Director of Operations and Maintenance, EBMUD, Oakland, Calif.
Michael R. Ambrose, Manager of Regulatory Compliance, EBMUD, Oakland, Calif.
Clifford C. Chan, Manager of Water Treatment and Distribution, EBMUD, Oakland, Calif.

21. Implementation of Climate Adaptation and Mitigation Strategies for Drinking Water Production in the Netherlands 227
Gerard A. van der Berg, KWR Water Cycle Research Institute, Nieuwegein, the Netherlands
Jos Frijins, KWR Water Cycle Research Institute, Nieuwegein, the Netherlands
J.J. Gertjan Zwolsman, KWR Water Cycle Research Institute, Nieuwegein, the Netherlands

22. Meeting the Challenges of Climate Change: Singapore 241
Chang Chian Wui, Policy Planning, PUB, Singapore

23. Climate Change and Adaptation in Southern California 251
Maryline C. Laugier, Malcolm Pirnie Inc., Irvine, Calif.
Edward G. Means III, Malcolm Pirnie Inc., Irvine, Calif.
Jennifer A. Daw, Malcolm Pirnie Inc., Denver, Colo.
Michael Hurley, Malcolm Pirnie Inc., Los Angeles, Calif.

24. Melbourne Water Climate Change Study 263
Carol Howe, Director Future Cities Programme, CSIRO, Australia
Bruce Rhodes, Manager, Urban Water Planning, Melbourne Water, Melbourne, Australia

25. Climate Change Impacts on Urban Drainage Systems in Scandinavia 277
O. Mark, DHI Water, Environment & Health, Hørsholm, Denmark
G. Svensson, DHI, Göteborg, Sweden
A. Konig, DHI, Trondheim, Norway
J.J. Linde, PH Consultant, Søborg, Denmark

INDEX 289

Introduction

"According to many experts, water and its availability and quality will be the main pressure on, and issues for, societies and the environment under climate change."
—IPCC Technical Paper VI

ADDRESSING CLIMATE CHANGE AND UNCERTAINTY

For many years, it has been widely accepted that climate change will have significant implications for management of water resources. The effects of climate change are starting to emerge. Increasing temperatures are increasing evaporation rates, changing precipitation patterns, affecting runoff and water quality, and changing water use demands. Impacts of changing climate will likely affect many water utilities, from increased salinity of coastal aquifers affected by sea level rise to earlier peak runoff in snowmelt dominated areas to reduced water supplies in the Mediterranean, Australia, and other regions.

Other aspects of urban water management including stormwater and wastewater collection and treatment systems may also feel impacts, such as increased incidences of flooding and sewer overflows. Knowledge of these impacts and possible adaptation measures are still limited. Impacts from climate change are felt in both developed and developing countries, but the ability to respond and the approaches needed to respond to this challenge can be quite different.

Even though urban water agencies will be affected by climate change, water utilities also contribute to climate change. They are an obvious source of greenhouse gas (GHG) emissions through their consumption of energy and emission of gases such as methane. Urban water managers need to be at the forefront to address the challenges from climate change, by mitigating their own contribution to greenhouse gas emissions and by building resilience into their systems to adapt to potential future climate scenarios.

The water management community, however, has been a leader in recognizing that it must respond to climate change. Indeed, the American Water Works Association (AWWA) and International Water Association (IWA) have been studying climate change and advising water resource managers about how to reduce their greenhouse gas emissions, how they may be affected by climate change, and what they can do to plan for the consequences.

In 1997, AWWA's public advisory forum recommended that water agencies consider climate change in their planning. In 2005, the Awwa Research Foundation (currently the Water Research Foundation) published *Climate Change and Water Resources: A Primer for Municipal Water Providers,* written by leading researchers on climate change and water resources at the National Center for Atmospheric Resources. AWWA has also been sponsoring training seminars on climate change across the United States for water managers.

IWA is assisting the global water community in meeting intensified challenges and uncertainties to build systems that are more robust, flexible, and resilient. IWA's Climate Change programme focuses on the need to model, predict, and assess future impacts and costs for society and the environment, and to support decision making under uncertainty. Over the past 10 years, IWA's journal, *Water21,* highlighted aspects of climate change such as the increase in waterborne diseases (*Water21* 2001). IWA's specialist group on Climate

Change and Adaptation brought together practitioners and researchers from around the world through a series of climate impact and adaptation conferences and discussion groups. IWA has influenced global policy through official statements, documentation, and providing expert advice at different forums.

For years, AWWA and IWA have published many articles and performed thorough studies on various aspects of climate change. The articles have addressed how utilities may be affected by climate change, as well as how to reduce their greenhouse gas emissions and begin adapting to climate change. This book is a collaboration of some of the best and most recent articles (some new, some republished) on climate change and its implications for water utilities.

The articles are arranged in three sections. The first section discusses the impact of climate change and of water utilities on the climate. This section includes articles explaining how the climate is expected to change and the consequences such changes will likely have for water utilities. It also addresses how impacts can be measured and communicated through climate and water footprints.

The second section examines mitigation strategies for utilities to reduce their carbon footprint. The first set of articles in this section examines the connection between urban water use and energy. This includes how energy is consumed in the production, use, and treatment of urban water. The second set of articles addresses mitigation issues and strategies for water and wastewater utilities to reduce their greenhouse gas emissions.

The final section addresses adaptation. The first set of articles in this section give general guidance on how utilities can build climate change adaptation into their planning processes. Utilities, however, are already starting to adapt to climate change. The second set of papers presents case studies on how utilities worldwide have taken action to understand and reduce their vulnerability to climate change.

Building on the long involvement of AWWA and IWA in helping urban water, wastewater, and drainage utilities to address their contribution to climate change and their risks from it, this book is a good start for urban water managers wishing to understand climate change and how to address it.

Carol Howe

Joel B. Smith

Jim Henderson

Understanding the Impact

INTRODUCTION

Understanding the Problem

An understanding of climate change begins with an understanding of the science and potential implications of the changes in climate for water resources management. This section succinctly summarizes the science of how climate is projected to change and what the most important consequences may be for utilities.

Understanding the impacts begins with an understanding of the science of climate change. Cromwell, Smith, and Raucher briefly describe the science of climate change and review projected impacts likely to concern water utilities. The authors analyze how effects will differ in major regions of the United States. Cromwell et al. conclude by describing approaches such as portfolio management and triple bottom line that can be used by utilities to reduce their vulnerability.

A critical question for utilities is whether the science is advanced enough for utilities take action on adaptation. Sykes, Chan, Encarnacion, and Richards examine this question and do so by focusing on the situation in the East Bay Municipal Utility District (EBMUD). They conclude that in general, there is enough information for utilities to begin adapting to climate change.

Mountains are a water reservoir for many regions of the world. Vanham and Rauch give a global overview of how climate change will impact high altitude hydrogeology, the storage of water in the form of ice and snow, and the potential consequences of water availability in both upstream and downstream parts of river basins including the Indus, Rhine, Nile, and Colorado. They conclude that regions dependent on glacier melt water will be strongly affected by climate change.

An important aspect of climate change management that is often overlooked is the impact of climate change on demand for water. Keifer assesses different methods that water managers can use to estimate change in demand.

Another important topic that has received insufficient attention in the climate change literature is the impacts of climate change on water quality. Drago, Brekke, Rogers, and Walsh explore this topic by modeling the effects of climate change and El Niño/Southern Oscillation (ENSO) on several indicators of water quality in Lake Cachuma near Goleta, California.

Paper 1. **No Doubt About Climate Change and Its Implications for Water Suppliers**

Paper 2. **How Should Water Utilities Prepare for Climate Change?**

Paper 3. **Mountain Water and Climate Change**

Paper 4. **Prevailing Water Demand Forecasting Practices and Implications for Evaluating the Effects of Climate Change**

Paper 5. **Impacts of Climate Change and Variability on Source Water Quality of Lake Cachuma, California**

Measuring the Impact—Climate and Water Footprints

Understanding the problem includes understanding how water utilities are contributing to climate change. Climate footprints provide water utilities with a practical tool for a good overview of the greenhouse gas emissions that an individual utility can influence.

The climate footprint of three Dutch water utilities is explored by Wiers et al. The paper illustrates the significant emission differences that can occur between companies even within a small geographic area and illustrates how this practical tool can be used by a utility to target and to benchmark its own emission performance over time.

A broader overview of the source of emissions from water, sewerage, and wastewater treatment and their contribution to the Dutch water sector climate footprint is discussed by Frijns, Mulder, and Roorda. The relative impact of direct and indirect emissions along with potential mitigation measures is discussed. The paper concludes that mitigation measures across the whole water cycle will be required to achieve a substantial reduction of the climate footprint.

As the energy sector shifts to biofuels, the consequent water demands will cause increasing pressure on freshwater resources. Hoekstra, Gerbens-Leenes, and van der Meer provide a different perspective on water and climate by giving a global overview of the demands of bio-energy through the use of the water footprint. They conclude that biomass production includes large water requirements and suggest that there should be an ethical discussion on whether limited water resources should be used for food or for energy.

Paper 6. **The Climate Footprint and the Practical Application at Water Companies in The Netherlands**

Paper 7. **Climate Footprint and Mitigation Measures in the Dutch Water Sector**

Paper 8. **The Water Footprint of Bio-Energy**

– 1 –

No Doubt About Climate Change and Its Implications for Water Suppliers

John E. Cromwell III, Joel B. Smith, and Robert S. Raucher

With the release of its Fourth Assessment Report: "Climate Change 2007," the Intergovernmental Panel on Climate Change (IPCC) has removed many doubts that previously shrouded both scientific and policy discussions of climate change. Mounting evidence about climate change and its effects made the situation much clearer to the scientists and government policy analysts from around the world who participated in a six-year process and arrived at the consensus presented in the report.

The World Meteorological Organization and the United Nations Environment Program (UNEP) established the IPCC in 1988. IPCC's role is to assess on a comprehensive, objective, open, and transparent basis the scientific, technical, and socioeconomic information relevant to understanding the scientific basis for the risk of human-induced climate change, its potential effects, and the options for adaptation and mitigation. The IPCC does not carry out research nor does it monitor climate-related data or other relevant parameters. Its assessment is based primarily on peer-reviewed and published scientific/technical literature.

A main activity of the IPCC is to provide at regular intervals an assessment of the state of knowledge on climate change. The First Assessment Report was completed in 1990, the second in 1995, and the third in 2001. Background reports are written by hundreds of scientists from around the world and are subject to extensive peer review. The "Summary for Policy Makers," typically a 20-page summary of the entire report that usually receives extensive attention in the press, is drafted by scientists but subject to approval by governments. In approving, adopting, and accepting reports, the IPCC makes every effort to reach consensus. If consensus cannot be reached, differing views are explained, and scientific and policy differences are distinguished. The IPCC's conclusions are not official until they have been accepted in a plenary meeting by the governments. Thus governments cannot ignore the conclusions of the IPCC because they have endorsed them.

The IPCC's adherence to scientific rigor and openness, coupled with the objective of articulating the best possible consensus views, might result in a slower and more cautious process of deliberation than some would prefer, but it also gives tremendous weight to IPCC findings when they are finally put forward as consensus statements.

During the six years since the release of the Third Assessment Report, the scientific evidence regarding climate change has become much more compelling. Accordingly, the Fourth Assessment Report (IPCC 2007a–c), released in 2007, contains consensus statements that are much more profound than those in previous reports. This article provides a brief review of some of the statements that pertain to water resources. These statements serve as a foundation from which to examine the implications of climate change for water suppliers.

IPCC REPORT REMOVES DOUBTS ABOUT CLIMATE CHANGE

The Fourth Assessment Report made headlines because the IPCC made very strong statements that left no room for doubt that global warming is producing long-term effects on natural systems and that anthropogenic sources are a likely cause. The summary statements that captured attention were:

- Warming of the climate system is unequivocal, as is now evident from observations of increases in global average air and ocean temperatures, widespread melting of snow and ice, and rising global average sea level.
- Most of the observed increase in global average temperatures since the mid-twentieth century is very likely due to the observed increase in anthropogenic greenhouse gas concentrations.
- Observational evidence from all continents and most oceans shows that many natural systems are being affected by regional climate changes, particularly temperature increases.
- A global assessment of data since 1970 has shown it is likely that anthropogenic warming has had a discernable influence on many physical and biological systems.
- Anthropogenic warming and sea level rise would continue for centuries due to the time scales associated with climate processes and feedbacks, even if greenhouse gas concentrations were stabilized.

The first three of the preceding conclusions had been made in previous IPCC assessments, although with less confidence. This time the authors felt there was much stronger evidence. In addition, the conclusions tying human emissions to these impacts are new.

This IPCC assessment made it clear that continued warming is inevitable. Six scenarios of greenhouse gas emissions were considered. All would result in an acceleration of the rate at which global temperatures and sea levels are rising.

CONSEQUENCES FOR WATER RESOURCES

The Fourth Assessment Report states with "high confidence" that among the effects of global warming on hydrologic systems the following are presently occurring:

- Increased run-off and earlier spring peak discharge in many glacier- and snow-fed rivers.
- Warming of lakes and rivers in many regions, with affects (sic) on thermal structure and water quality.

Turning to the future, the Fourth Assessment Report makes the following predictions regarding water resources:

- By mid-century, annual average river runoff and water availability are projected to increase by 10–40% at high latitudes and in some wet tropical areas, and to decrease by 10–30% over some dry regions at mid-latitudes and in the dry tropics, some of which are presently water stressed areas. In some places and in particular seasons, the changes differ from these annual figures.
- Drought-affected areas will likely increase in extent. Heavy precipitation events, which are very likely to increase in frequency, will augment flood risk.
- In the course of the century, water supplies stored in glaciers and snow cover are projected to decline, reducing water availability in regions supplied by meltwater from major mountain ranges, where more than one-sixth of the world population currently lives.

- Coasts are projected to be exposed to increasing risks, including coastal erosion, due to climate change and sea level rise (0.2–0.6 m by 2100, and probably more).

The IPCC's more specific projections regarding future effects of global warming in North America include the following statements:

- Warming in western mountains is projected to cause decreased snowpack, more winter flooding, and reduced summer flows, exacerbating competition for over-allocated water resources.
- Disturbances from pests, diseases, and fire are projected to have increasing impacts on forests, with an extended period of high fire risk and large increases in area burned.
- Moderate climate change in the early decades of the century is projected to increase aggregate yields of rain-fed agriculture by 5–20% (in mid- and high latitudes), but with important variability among regions. Major challenges are projected for crops that are near the warm end of their suitable range or depend on highly utilized water resources.
- Cities that currently experience heat waves are expected to be further challenged by an increased number, intensity, and duration of heat waves during the course of the century, with potential for adverse health effects.
- Coastal communities and habitats will be increasingly stressed by climate change impacts interacting with development and pollution. Population growth and the rising value of infrastructure in coastal areas increase vulnerability to climate variability and future climate change, with losses projected to increase if the intensity of tropical storms increases.

BOTTOM LINE IMPLICATIONS FOR WATER SUPPLIERS: ADAPTATION

The IPCC concludes adaptation will be necessary to address the effects of warming that are already unavoidable because of past emissions. A further 0.6°C (1°F) increase in global mean temperature—relative to temperatures during the period from 1980 to 1999—is projected to occur by the end of the twenty-first century even if greenhouse gas concentrations remained at 2000 levels. The IPCC projects that the average rise in global temperatures could range from 1.1 to 6.4°C (about 2 to 12°F) by 2100 when examining a range of greenhouse gas emissions scenarios. Temperatures in the lower 48 states are expected to rise about a third more than the global average (Wigley 1999).

One of the simplest ways to envision many of the implications of global warming for water resources is to follow the logic of what happens when water is heated; global warming will basically accelerate the pace of the hydrologic cycle. Consistent with this, forecasters predict effects for water resources in the arid western regions of North America that are different from those in the humid eastern regions. Recent trends indicate that many of these effects are already happening.

In the West, warming effects may be seen most prominently in reduced water supply capacity. Snowpack will be smaller and melt earlier, altering the recharge of surface water and groundwater sources. In addition to less rainfall in this region, droughts are expected to be more extensive, with more heat waves and dry days, accompanied by increases in evaporation and greater irrigation demands. When precipitation does come, it is likely to be more intense. The combination of earlier snowmelt and more intense precipitation will likely increase turbidity, sedimentation, and the risk of flooding in many areas.

In the East, evidence of warming will primarily come in the form of increased rainfall frequency and intensity. The increased rainfall intensity will likely produce increased runoff and erosion with resulting increases in turbidity and sedimentation. Related effects include direct flood damage to water and wastewater facilities, loss of reservoir storage capacity for flood control and water supply, and increased sewage overflows during wet weather events.

In both the East and the West, the changes in temperature and hydrology will produce changes in watershed vegetation and aquatic ecosystems, which in turn will have implications for water suppliers. One result will be changes in watershed conditions because of wildfires and pest infestations, both of which are likely to increase. With increased water temperatures and shallower reservoirs (because of lower base flows and sedimentation), eutrophic conditions will be more prevalent. Rising sea levels will pose the risk of flood damage to coastal water and wastewater facilities in both the East and the West, especially as a result of more intense coastal storms. Rising sea level, coupled with lower base flows in freshwater sources because of altered recharge, will also increase the salinity of both coastal aquifers and brackish surface water sources.

The individual effects of climate change are staggering enough, but they will also have compound effects on water resources because of the interactions of natural and human systems. The bottom line in water supply planning has always been a matter of coping with variability. With the coming changes in climate, there will be a heightened need to respond to increased variability. Global warming will change the variability of many key parameters affecting the quantity and quality of water that would normally be available at specific times and places. In addition, the capability to store water in various forms and the demand for water will be changed.

Portfolio Approach to Planning Encouraged

Given the difficulty of predicting the magnitude of individual effects, let alone compound effects, IPCC and others have recommended that adaptation planning should employ a portfolio approach. This approach would maintain a maximum degree of flexibility within the portfolio by devising coping strategies to address an array of possible climate change scenarios that may affect the quantity and quality of supply sources, and the demands placed on them. Reevaluating the entire portfolio from source to tap with respect to adaptation strategies may seem like a daunting task; however, there are several guiding philosophies that can offer a pragmatic way to start.

A first practical step in evaluating the adaptation strategies is to conduct a "bottom-up" vulnerability assessment. Rather than starting with complex climate modeling, it is more tractable to begin with what a utility already knows about its own systems and water sources to determine the points at which water quality or water quantity changes would present a major challenge. This type of "threshold analysis"—based on a utility's current models and knowledge—can identify key climate-related tipping points (e.g., what reduction of in-stream flows would be most problematic given current supply strategies). Where possible, these thresholds could be verified through simulation modeling of altered operating regimes or through actual system tests. Once these thresholds are identified, climate expertise can be brought to bear in a focused manner to examine the potential process changes that could lead to these key tipping points being exceeded.

Once a water utility is equipped with a sense of its vulnerabilities, the IPCC and others recommend development of adaptation strategies by focusing first on ways to improve system flexibility and resiliency across the entire portfolio. In concert with this emphasis on flexibility, it may be prudent to adopt a step-wise approach that recognizes that irreversible choices or capital commitments might be easier to optimize within a portfolio as information improves over time.

TRIPLE BOTTOM LINE IMPLICATIONS: MITIGATION

In addressing water supply planning in the context of portfolio optimization, it is necessary to consider all of what should go into a sustainable solution. As has already happened in many water-short areas in both the East and West, it is possible to conceive of many ways to enhance the reliability of water resource management outcomes by essentially investing more energy to produce more water. But in evaluating these options, it must also be acknowledged that water utilities account for a significant share of total electric consumption and that power plant emissions account for a significant share of greenhouse gases. There is no doubt, therefore, that water utilities need to apply a broader "triple bottom line" discipline to the design of adaptation strategies in order to balance the cost and reliability of water supply against social and environmental consequences. The key to implementing this approach in practice will be to apply it to overall portfolio outcomes and not just to the individual project elements. There are several areas in which the need for this broader thinking arises most visibly.

As urban and suburban areas continue to grow, water resource managers have already devised elaborate portfolio strategies to tap into multiple sources of supply and to make strategic investments in capabilities to move and store water. To wring every last drop from available supplies, flexibility in transmission and storage operations has been taken to great extremes in some water-short areas in order to stay within complex constraints imposed by environmental withdrawal limits and seasonal and annual swings in water availability. Although flexibility in transmission and storage will be valuable in re-optimizing current schemes to meet future challenges, some system features designed for the current understanding of climatic variability may not be reversible or easily adaptable under altered operating regimes that were never envisioned. Operating flexibility needs to be an even more important design consideration for future system improvements.

Balancing Fiscal and Environmental Responsibility

The rising cost of electric power has caused many water utilities to reexamine their transmission and distribution operating strategies in search of ways to curtail electricity use during peak periods and conserve electric power. Some cost-saving strategies—such as using backup generators to serve peak-period loads (where not prohibited by air quality regulations)—may save money, but they do not reduce greenhouse gas emissions. On the other hand, renewable energy supply strategies such as solar- or wind-powered pumping, or in-line hydropower generation, may be more expensive initially but provide fuel-cost savings and avoid greenhouse gas emissions. The challenge is to integrate such strategies at the level of the overall portfolio to produce the best systemwide operating outcomes in terms of cost, reliability, and social/environmental consequences. For example, the reliability profile of a solar- or wind-powered option may be that it is only viable if a backup generator is available to fill critical gaps. Although this would produce some greenhouse gas emissions, overall it would be less than a conventional power supply strategy.

As many water suppliers are already aware, the cost and energy investment required to reliably meet water supply needs can rise exponentially when hard, unmovable constraints arise in one part of the portfolio and cause everything else in the strategy to be shifted to more extreme levels. Global warming has the potential to cause simultaneous adverse changes in both the quantity and quality of available water. These compounded effects could greatly weaken the foundation of some existing integrated supply strategies. It is hard to compensate for such things as a smaller snow pack, an earlier spring runoff, or increased salinity.

Many water suppliers in overconstrained settings have already begun to turn to energy-intensive membrane treatment processes to enable desalination of saline water sources and reuse of highly treated wastewater effluent. These processes make it possible to overcome a deterioration in the reliability of normal supplies by meeting part of the demand from sources that will be abundant under most climate change scenarios (i.e., yields from water reuse and desalination supplies are drought-resistant). Although the costs—especially the energy cost—of these technologies are significant, the triple bottom line must be evaluated in the context of the overall portfolio. If these technologies can plug a gap or shore up a vulnerability produced by climate change processes in a way that enables a broader scope for optimization across the entire portfolio, then they could play a critical role in making the broader strategy more sustainable. The economies of scale involved in such facilities also offer opportunities to address greenhouse gas emissions during the design phase of related power generation facilities and incorporate renewable energy sources into the strategy. Perhaps tomorrow's water can be decoupled from yesterday's energy, making the water sector a leader in the transition to carbon-neutral energy sources.

Demand Management Plays Numerous Roles

No portfolio of adaptation strategies to meet global warming challenges would be complete without including a comprehensive review of the water demand side of the equation. Warming processes will lead to altered demand patterns as a result of seasonal shifts in precipitation, more evaporation, more frequent heat waves, and more extensive droughts. From a triple bottom line perspective, conservation programs offer a bonus in reducing both water supply needs and energy use. Bolstering conservation incentives (and disincentives to outdoor water use) may become more essential as warming processes increase water demands, especially during peak demand periods when both water supply and electric power capacities are stretched to their limits. Any margin of relief from the more extreme water and energy supply strategies required to meet peak demands can significantly improve the triple bottom line of the overall supply strategy.

In many regions, demand management is also related to population growth. Regional strategies for optimizing water supplies often meet with opposition in an indirect effort to control population growth. When broader regional portfolio optimization is blocked in this way, it can force individual jurisdictions to adopt more narrow and therefore less sustainable options because warming processes make it continually harder to meet demands with a limited range of adaptation options. At the other end of the spectrum, the energy intensity of moving water across distances and terrain probably limits the potential of broader regional approaches in terms of the triple bottom line criterion.

IPCC REPORT ALSO AFFIRMS HOPE

Despite the IPCC's sobering confirmations that global warming processes are well under way and will produce effects that are unavoidable over the next several decades, the Fourth Assessment Report also holds out hope in showing that actions taken to reduce greenhouse gas emissions over the same time period are well within our technical and economic capacity and should be capable of stabilizing the situation.

- Studies indicate that there is substantial economic potential for the mitigation of global greenhouse gas emissions over the coming decades that could offset the projected growth of global emissions or reduce emissions below current levels.
- In order to stabilize the concentration of greenhouse gases in the atmosphere, emissions would need to peak and decline thereafter. The lower the stabilization level, the more quickly this peak and decline would need to occur. Mitigation efforts over the

next two to three decades will have a large impact on opportunities to achieve lower stabilization levels.

Adaptation by water utilities could ameliorate many adverse effects, although there will be costs to adaptations, and maintaining reliability may be more challenging. Many water utilities are already engaged in considering adaptation strategies and mitigation of greenhouse gas emissions. The state of California and municipalities from New York, N.Y., to Miami, Fla., to Los Angeles, Calif., to Seattle, Wash., are assessing their risks and developing strategies to lessen the effects of climate change. There is no doubt that it is time for all to rise to this call.

ABOUT THE AUTHOR

John Cromwell consults for utilities and performs national policy research for Stratus Consulting Inc. in Washington, D.C. He has bachelor's degrees in biology and economics as well as a master's degree in public policy analysis, all from the University of Maryland. He can be contacted at jcromwell@stratusconsulting.com. Joel Smith is vice-president and Bob Raucher is executive vice-president, both at Stratus Consulting.

REFERENCES

IPCC (Intergovernmental Panel on Climate Change), 2007a (S. Solomon, D. Qin, M. Manning, Z. Chen, M. Marquis, K.B. Averyt, M.Tignor, and H.L. Miller, editors). Summary for Policy Makers. Climate Change 2007: The Physical Science Basis. Contribution of Working Group I to the Fourth Assessment Report of the Intergovernmental Panel on Climate Change. Cambridge University Press, Cambridge, UK and New York. http://ipcc-wg1.ucar.edu/wg1/Report/ AR4WG1_Pub_SPM-v2.pdf/.

IPCC, 2007b (M.L. Parry, O.F. Canziani, J.P. Palutikof, P.J. van der Linden, and C.E. Hanson, editors). Summary for Policy Makers. Climate Change 2007: Impacts, Adaptation, and Vulnerability. Contribution of Working Group II to the Fourth Assessment Report of the Intergovernmental Panel on Climate Change. Cambridge University Press, Cambridge, UK, and New York. http://www.ipcc-wg2.org.

IPCC, 2007c (B. Metz, O.R. Davidson, P.R. Bosch, R. Dave, and L.A. Meyer, editors). Climate Change 2007: Mitigation. Contribution of Working Group III to the Fourth Assessment Report of the Intergovernmental Panel on Climate Change, Cambridge University Press, Cambridge, UK and New York. http://www.ipcc.ch/SPM040507.pdf.

Wigley, T.M.L., 1999. The Science of Climate Change: Global and U.S. Perspectives. Arlington, Virginia: The Pew Center on Global Climate Change.

Originally printed in Journal AWWA *September 2007.*

— 2 —

How Should Water Utilities Prepare for Climate Change?

Richard G. Sykes, Clifford C. Chan, Rey C. Encarnacion, Kevin T. Richards

INTRODUCTION

East Bay Municipal Utility District (EBMUD) provides potable water supply to 1.3 million customers on the east side of San Francisco Bay in California. Average daily water demand in the EBMUD service area is slightly above 200 million gallons per day. About 90 percent of EBMUD's water supply originates in the 577-square-mile watershed of the Mokelumne River in the Sierra Nevada Mountains located approximately 100 miles east of the service area. EBMUD owns and operates two water supply reservoirs on the Mokelumne River, Pardee, and Camanche Reservoirs, which provide water supply, flood protection, recreation, and hydropower. The Mokelumne River water supply is conveyed to the EBMUD service area via three 90-mile long aqueducts stretching from Pardee Reservoir in the Sierra Nevada foothills to the city of Walnut Creek in the East Bay. EBMUD has been actively monitoring the progress of climate change research to understand and predict potential future impacts to its water supply and operations.

Scientists and researchers are in general agreement that over the next century, temperatures around the globe will increase. For the western United States, a 3°C warming over the next century is likely, which may result in significant impacts to the region's water supply systems. Moreover, without significant reductions in greenhouse gas emissions, temperatures may rise even higher than predicted. Models also predict greater geographical variability of runoff, which may be problematic for water utilities relying on small watersheds such as EBMUD's Mokelumne Watershed. There is still debate among the scientific community whether global warming will lead to more intense and frequent storms, or whether the region may experience an accelerated hydrologic cycle between drought and wet years.

In the western United States, climate change has the potential to affect water supply availability, flood protection, water quality, and water demand. The difficulty for most water utilities is in determining how to evaluate the information on climate change and its impacts in light of more pressing and predictable priorities. EBMUD has found that the information currently available is useful for water resources and operational planning.

POTENTIAL CLIMATE CHANGE IMPACTS ON WATER SYSTEMS

Water Supply Availability Impacts

Research by the Lawrence Livermore National Laboratory indicates the Central Sierra Nevada region will experience significant changes to hydrological conditions, including a higher elevation snow line, earlier and increased peak runoff, and less late spring runoff. In the last 50 years, the western United States has indeed seen a decrease in the size of the snowpack, and the percentage of runoff occurring between April and July has also decreased by approximately 10 percent over the last century. Additional warming of 3°C could shift 28 percent of the April–July runoff to the November–March period in the Mokelumne River Watershed.

Most water utilities in the Western United States enjoy the benefits of a snow-fed system where the snowpack provides "storage" that can be used in the late spring and early summer when precipitation is low. With smaller snowpack and a decrease in the April–July runoff, water utilities may not have the reservoir capacities sufficient to store the earlier runoff. Compounding this problem, many water utilities must evacuate reservoir storage in late fall for flood control, relying on the spring runoff to refill the system. Thus, current operations procedures may result in releasing water that should have been stored.

Flood Control/Storage Impacts

Earlier runoff could result in an increase in flood control releases for systems that are operated with the dual purposes of water supply and flood control. These reservoirs typically have well-defined flood control space requirements. According to the climate change scenario, there will be less spring runoff to refill these reservoirs, and summer water deliveries, power generation, and carryover storage could be affected. Water supply managers face a delicate balance between water supply needs and flood control needs. If too much water is released too early, there may not be enough water to supply customers. Conversely, if too little water is released, the risk of flooding downstream increases.

Operational Reliability Impacts

Another potential consequence of climate change is the predicted rise in the sea level. Researchers conservatively estimate that over the next century, sea levels may rise by as much as three feet. Under more extreme scenarios, it is estimated that the sea level may rise up to 20 feet over the next century. Higher sea levels will inundate coastal areas around the globe. They will also seriously affect water supplies in California including the Sacramento/San Joaquin River Delta, which provides drinking water to over 20 million people. For EBMUD, failure of the levees adjacent to its aqueducts that cross the Delta could result in catastrophic damage that would interrupt deliveries of raw water to the service area.

For the California Delta, rising sea levels will significantly affect water supplies and levee integrity. The aging levee system is already vulnerable to floods and earthquakes, and rising sea levels will only exacerbate these risks. Levee failure will not only damage low-lying infrastructure but will lead to massive intrusion of poor quality water into water supply intakes serving much of California.

Water Quality Impacts

As noted in the previous example, potential infrastructure damage as a result of climate change may affect water quality, but climate change can also directly impact water quality. Increased severity of storms will likely increase turbidity levels in raw water supplies and consequently reduces treatability of those supplies. An increase in water temperature during the spring and summer may also affect water quality by promoting algae growth and result in increased algal by-products, such as taste and odor compounds.

In addition to drinking water quality concerns, many water systems with fishery management responsibilities already experience temperature-related ecosystem management problems. Following recent dry winters, the summers of 2007 and 2008 illustrated the difficulty in sustaining cold-water reserves for fishery management. Water utilities have a limited toolbox to address this problem, and climate change will only make temperature management more challenging.

VULNERABILITY ANALYSES

EBMUD has used the results from climate change studies to analyze and understand the impacts on its water supply system and operations. In general, the results have indicated only modest change to water supply reliability. EBMUD is fortunate in this respect. Many other water utilities may find that the impact on their water supplies will be more severe as a result of climate change. The study results showed a more dramatic impairment of EBMUD's ability to contain floods in a warmer climate. The following is a summary of the studies that EBMUD performed to review water supply reliability and flood control.

Water Supply Reliability

The effects of climate change on water supply reliability were evaluated by running EBMUD's water supply model simulating the 80-year historical hydrologic record under a changed climate. These simulations assumed no change in the total annual precipitation, but they did alter the runoff hydrograph, shifting runoff patterns earlier in the year to match the expected effects of a warmer climate. Figure 2-1 illustrates the century-averaged unimpaired hydrographs for EBMUD's historical record and for a climate change scenario with a 28 percent shift of the April–July runoff to earlier in the year.

The study indicated that a 28 percent shift would have a relatively small impact on water deliveries. The reason for this is threefold—the amount and timing of demands, the reservoir storage-to-runoff ratio, and the relatively large percentage of spring runoff in the Mokelumne Basin.

The total demand on the Mokelumne River is less than the average annual runoff, which enables EBMUD to refill its water supply system with less-than-average runoff. Moreover, the timing of year-round municipal demands, as compared to seasonal irrigation demands, allows EBMUD to make direct use of early-winter runoff. Even with climate change increases in early-winter runoff, total storage between EBMUD's Camanche and Pardee Reservoirs and the Pacific Gas & Electric Company reservoirs upstream exceeds 800 thousand acre-feet (TAF).

This storage volume is greater than the average annual runoff of 750 TAF, which means there is sufficient capacity to capture large volumes of early runoff in most years. This is a big benefit in dry and critically dry years when the only significant runoff may be early in the season. Moreover, the Mokelumne Watershed historically has had more April–July runoff than can be stored; during years of below-normal and normal runoff, the April–July component has exceeded storage capacities in 8 out of 10 years, requiring flood control releases. Under an altered climate with decreased spring runoff, there would be fewer flood control releases, which would reduce power generation but rarely affect carryover storage (the amount of water in storage that is available to carry forward into the new water year starting each October 1).

The analyses previously described were performed at a 2040 level of development. While there were some impacts to EBMUD's carryover storage, these were not substantial. The analyses showed that there were carryover storage impacts in 10 percent of the 80 simulation years, with a maximum carryover storage reduction of 49 TAF. EBMUD's carryover storage capacity is 600 TAF, and water deliveries remain unaffected unless carryover storage is

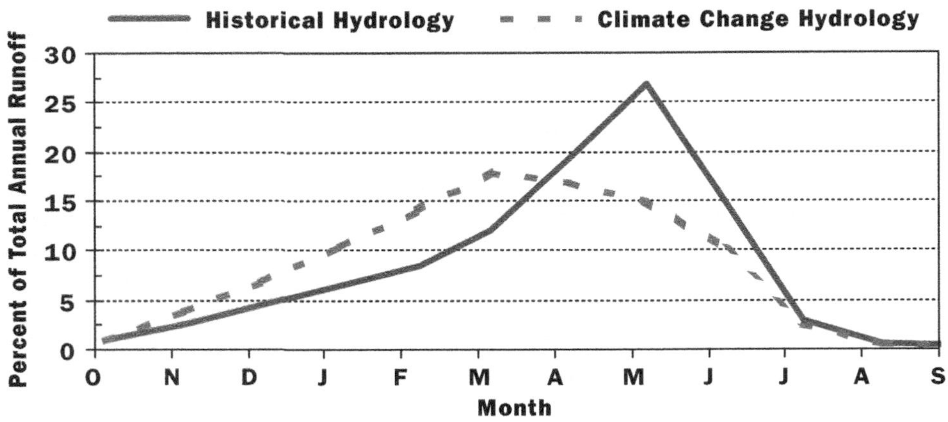

Figure 2-1. Comparison of long-term average Mokelumne Hill unimpaired runoff under historical and climate change conditions

less than 500 TAF at the beginning of the water year. By the year 2100, EBMUD's water demands may be larger and carryover storage impacts would be larger. Moreover, if the total annual precipitation changes, or if a more significant change in the timing of the runoff occurs, such as a near-complete loss of the spring snowpack, greater impacts would be expected.

Flood Control

Increased runoff during the winter months as a result of a shift in the runoff patterns over the next century would present a challenge for flood control management at EBMUD. Although flood control requirements are well-designed and have served their purpose in the past, the channel capacity downstream of Camanche Reservoir is limited to 5,000 cubic feet per second (cfs), which restricts the ability to respond to extreme flood events, such as the 1997 El Niño floods in California.

Studies were performed to assess the flood control implications for a warmer-than-historical flood using a simulation of the events of the 1997 El Niño storms. The simulation was run assuming temperatures were 3°C warmer. The results of this study showed that at its peak, the total release from Camanche Reservoir would have been approximately 15,500 cfs, well over the 5,000 cfs capacity of the Mokelumne River downstream of Camanche Reservoir. Under this scenario, significant flooding would occur in the lower Mokelumne River area.

Figure 2-2 shows the onset of flooding potential for the five worst floods on record.

The red line indicates critical storm-runoff volumes over time as defined by EBMUD's flood control requirements. Storm runoff volumes in excess of the red line would lead to uncontrolled spill from Camanche Reservoir and the potential for downstream flooding. It should be noted that the system has adequately handled every storm since the construction of Camanche Dam without exceeding the 5,000 cfs capacity of the channel downstream. However, the two worst storms on record have occurred in the last 20 years. Assuming a 5,000-cfs release rate is initiated at the beginning of an impending storm, 10 TAF of water per day can be safely evacuated from the upstream reservoirs. Although the frequency and intensity of future storms under climate change is unknown, this graph illustrates the importance of more accurate forecasting—for each day of lead time, 10 TAF of additional flood control space

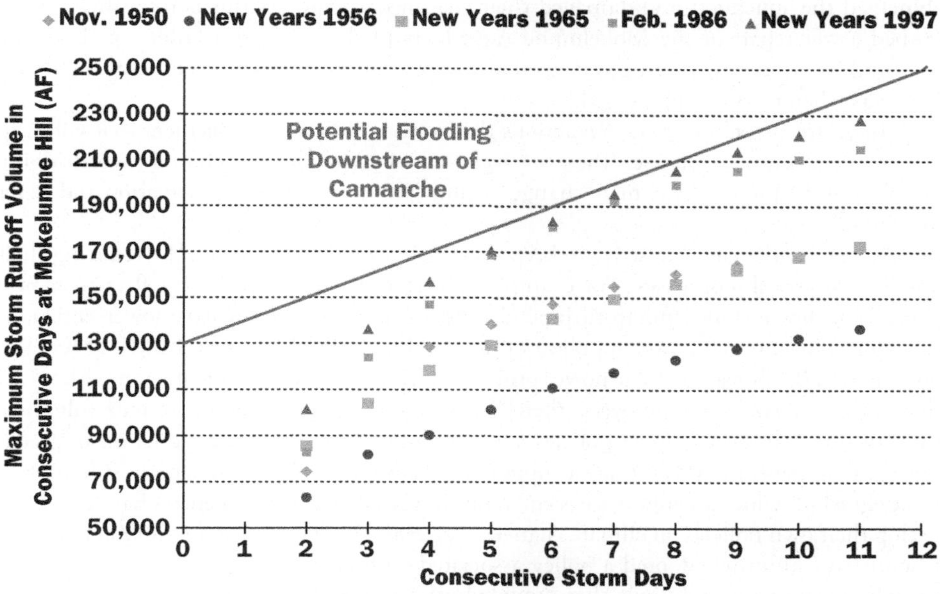

Figure 2-2. Mokelumne Basin storm-runoff development and flooding potential of the five largest storms on record

can be gained. The figure also shows that a modest increase in intensity, volume, or duration of the 1986 or 1997 storms would have caused flooding downstream of Camanche Reservoir.

PREPARING FOR CLIMATE CHANGE

As utilities evaluate the published information on climate change and its potential impacts, the question remains—How should water utilities prepare for climate change? The following offers some suggestions and provides examples for utility planners and resource managers.

Utilities should consider how climate change could affect their sources of supply. For example, if a utility's primary source of raw water is runoff from melted snowpack, consideration should be given to effects on storage and late season water availability. As appropriate, utilities should evaluate the availability of supplemental supplies that may not be as affected by climate change. Diversifying a utility's water supply portfolio will improve the ability to manage the impact that climate change may have during any given year. For example, EBMUD is adding supplemental water supplies including sources from the Sacramento River, a watershed several hundred miles north of the Mokelumne Watershed, to ensure water availability should climate change result in greater geographical variation in precipitation. EBMUD is also investigating groundwater banking projects and desalination. In addition, EBMUD has active recycled water and conservation programs to help reduce future demands on increasingly limited supplies.

Water agencies should consider the repercussions of rising sea levels on their water systems and operations, and find cost effective solutions to address those issues. For example, the Mokelumne Aqueducts, which convey water from EBMUD's reservoirs in the Sierra to the Bay Area, cross the California Delta. The Delta levees are extremely vulnerable to sea-level rise and have been breached on numerous occasions. A breach in June 2004 submerged the aqueducts and damaged their coatings. To address this issue, EBMUD has planned a wet repair of the Mokelumne Aqueducts in the event of a failure of the Delta levees and is considering coating systems that will provide corrosion protection for these steel aqueducts when they are submerged.

There are also many ***no regrets*** actions that water utilities can implement that will help them adapt to climate change. These ***no regret*** actions, or solutions that are worth doing regardless of the impacts of climate change, include promoting water conservation and water reclamation.

Water utilities can also strive to keep their operations *climate neutral* by reducing their emissions of greenhouse gases. For example, EBMUD has changed its car fleet to hybrid vehicles and has installed micro-turbine electrical generators (which have lower emissions than most conventional power supplies) on its administration buildings. EBMUD has also constructed a 0.4 megawatt solar-power project at one of its water treatment plants. EBMUD offers transit subsidies to employees, thereby encouraging low-emission commute solutions. Water utilities can also support policies and legislation that encourage reduction in greenhouse gas emissions. EBMUD has a support position on California Assembly Bill 1058, a bill designed to reduce greenhouse gas emissions in California. Some agencies have chosen to develop their own policies on climate changes. For example, the Metropolitan Water District of Southern California adopted a policy associated with climate change to support further research and no-regret solutions that provide both immediate benefits to water supply and quality as well as future ability to manage climate change impacts.

Water utilities should support additional research to better plan for climate change. Watershed specific models capable of providing streamflow conditions under an altered climate could help make more accurate predictions. Currently, EBMUD's studies are based on a 3°C increase in temperature, a 28 percent shift in the spring runoff, and a 3-ft rise in sea level. Under these conditions, EBMUD's water supply system will be able to operate without severe impairment, but with slightly less carryover storage, slightly greater risk of flooding, and slightly increased risk of infrastructure damage. However, if the forecasts are incorrect or if the regional temperature increases lead to unexpected variations in precipitation or storm frequency and intensity, impacts may be significantly underestimated.

CONCLUSION

The effects of climate change are already evident and are expected to continue through this century even if greenhouse gas emissions stabilize. Current research and data show that on average, the planet is getting warmer, and that the impact on the Western United States may be greater than other areas because of the region's reliance on snowpack as late-spring and early-summer storage. The research is useful for making preliminary studies to determine what effects climate change will have on water supply systems. Fortunately, it is relatively simple to perform sensitivity analyses related to earlier runoff, rising temperatures, and rising sea levels.

Water utility managers have an obligation to consider potential climate change impacts and to assess mitigations for those impacts as a part of ongoing planning efforts. Planning efforts can include evaluating how future infrastructure investments will perform in a changed climate. Such evaluations may lead to changes for some projects. This is most evident for new or supplemental water supply projects such as storage, conveyance, inter-agency connections, groundwater development, etc. where water availability may not be as predicted by historical records. Other climate change outcomes, such as rising sea levels, can easily be evaluated as affected infrastructure is readily identified. Long-range planning may allow for future impacts to be mitigated at relatively low cost.

The climate change information currently available is suitable for evaluating water system performance and vulnerabilities, and it can be useful when evaluating current and future infrastructure investments. This information will help lead to more cost-effective, long-term solutions that will help utilities avoid or minimize climate change effects in the future.

Originally printed in Opflow *January 2009.*

– 3 –

Mountain Water and Climate Change

D. Vanham and W. Rauch

MOUNTAINS: WATER TOWERS OF THE WORLD

General

In many regions of the world, mountains act as water towers. Figure 3-1 gives an overview of the major mountain ranges in the world. By definition of Meybeck et al. (2001), mountains and higher-altitude areas make up about 25% of global continental surface excluding Antarctica and Greenland. This definition distinguishes a total of 15 relief categories clustered into nine main types, derived from average cell altitude (eight classes), and an average roughness indicator (seven classes). These 25% of the Earth's total land area account for 32% of surface runoff. However, depending on the climate zone, mountain discharge can represent temporarily up to 95% of total river basin flow, when surrounding lowlands are characterised by a dry climate (Viviroli et al. 2003, 2007). Examples of such wet islands in dry climate zones include the Hindu–Kush–Himalayas mountain range for the Indus river basin, the Rocky Mountains for the Colorado river basin, the Taurus–Elburs–Zagros mountain ranges for the Tigris–Euphrates river basin, or the Ethiopian Highlands for the Nile river basin. A global map of Earth system mountain types (Figure 3-2) was generated by Viviroli et al. (2007). Within the authors' definition, mountains and higher-altitude areas (hills, mid-altitude plains, and platforms of medium to very high altitude) make up 39% of global continental surface excluding Antarctica and Greenland. This map shows different Earth system mountain types, hydrologically distinguished according to the contributing potential of a mountain cell (resolution of 0.5° × 0.5°) to the respective analysis basin's lowland runoff. The authors show that more than 50% of mountain areas have an essential or supportive role for downstream regions.

The Earth system mountain type I in Figure 3-2 (Viviroli et al. 2007)—high contributing potential to dry lowland—matches the concept of mountains being elevated humid islands within a dry region. Examples have been given in the previous paragraph. Earth system mountain type II is defined by a medium or low contributing potential to a dry lowland area. Examples of these occasional (either seasonally or in single years) water towers are the Hoggar and Tibesti mountains in the Sahara of Northern Africa. The European Alps are an example for Earth system mountain type III—medium or high contributing potential to a wet lowland area—for the Rhine, Rhone, Danube, and Po river basins. The Eastern Himalayas are also of this type for the Yangtze River in China. Although located in a region where water supply in the lowlands is sufficient on average, these mountain cells have an important supporting function and make important seasonal contributions to lowland runoff. The Earth system mountain type IV has low contributing potential to a wet lowland area and is

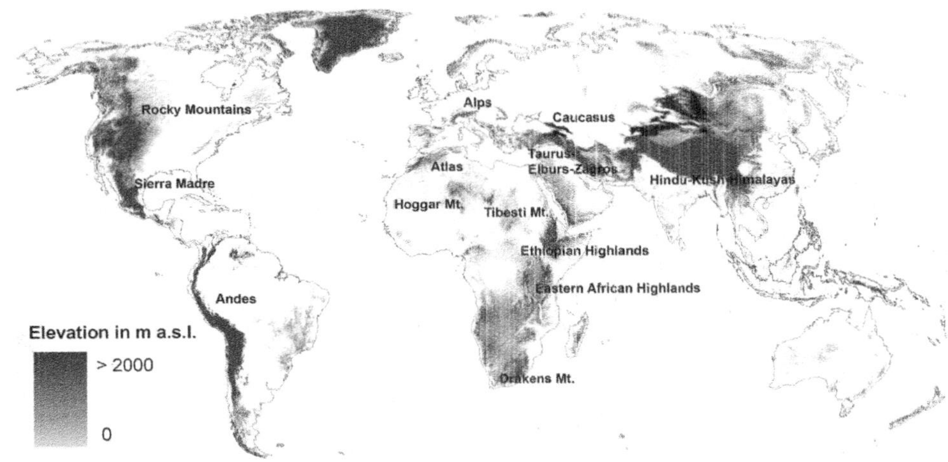

Figure 3-1. Major mountain ranges of the world

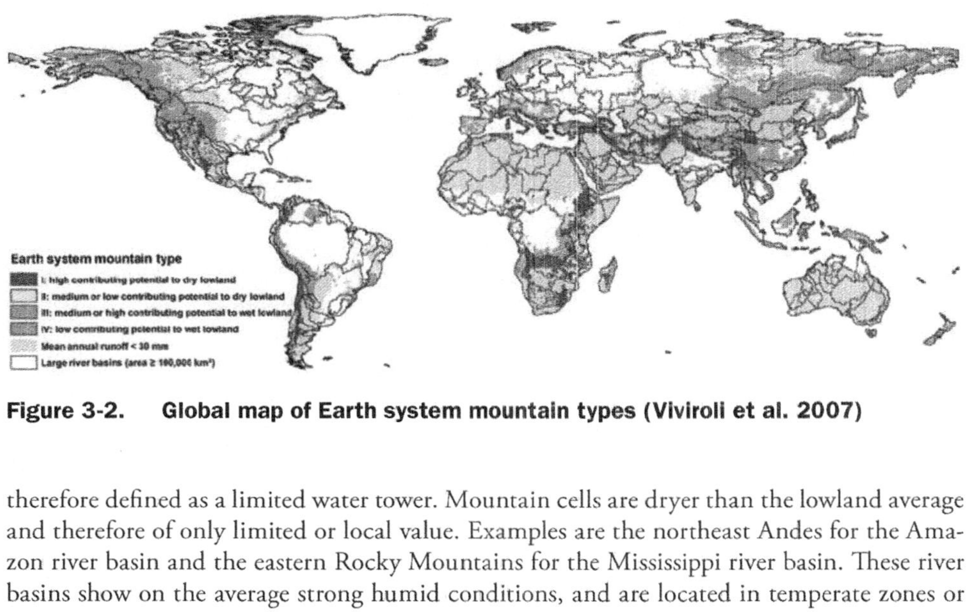

Figure 3-2. Global map of Earth system mountain types (Viviroli et al. 2007)

therefore defined as a limited water tower. Mountain cells are dryer than the lowland average and therefore of only limited or local value. Examples are the northeast Andes for the Amazon river basin and the eastern Rocky Mountains for the Mississippi river basin. These river basins show on the average strong humid conditions, and are located in temperate zones or the tropics.

In many mountain ranges of the world, snow and ice significantly affect catchment hydrology by temporarily storing and releasing water on various time-scales. The seasonal snow cover causes a time lag between the precipitation event and runoff of typically several months. According to Smakhtin (2001), in cold or mountainous regions, in addition to the usual catchment parameters, low flows are subject to the special influences of ice and snow melting. Rivers often have their lowest flows in winter due to the temporal storage of precipitation as snow. Contrary, in the European Alps, this period of minimal water availability is met by the peak water demand from different stakeholders, due to winter tourism (more municipal water demand) and water demand for technical snow production (Vanham et al. 2008c). Therefore, the effect of global warming on the behaviour of seasonal snowpacks and glaciers is a key factor for future water availability within these areas and their river basin lowlands.

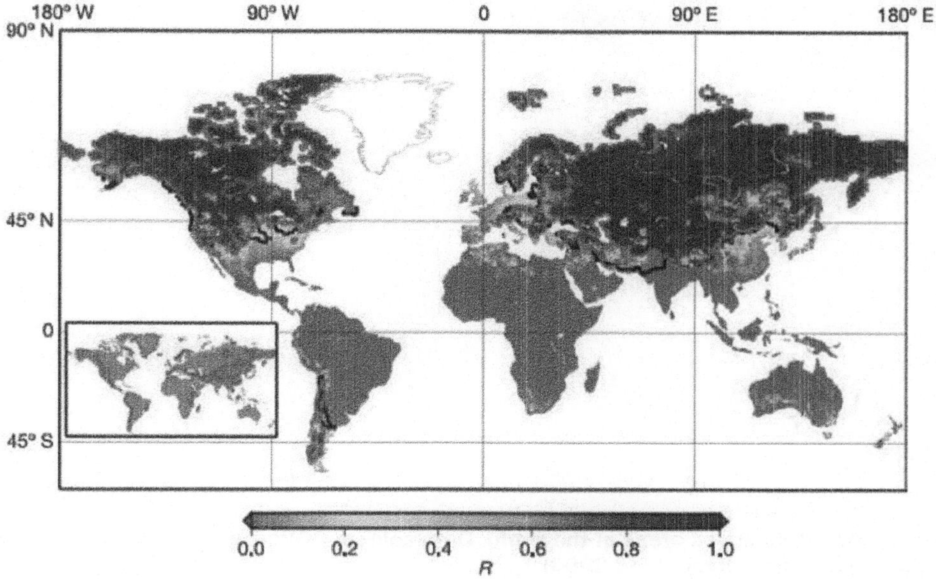

Figure 3-3. Accumulated annual snowfall divided by annual runoff over the global land regions for the period 1980–1999 (Barnett et al. 2005)

Barnett et al. (2005) identified the regions in the world where seasonal snowpacks have a significant influence on the hydrological cycle (Figure 3-3). The authors used a spatially distributed macroscale hydrology model to identify the regions of the globe where snowmelt plays a dominant role in the seasonal patterns of stream flow. They approximated the importance of snow to annual runoff by using the ratio R of the accumulated annual snowfall to annual runoff. A snowmelt-dominated gridcell is defined by the criterion that $R > 0.5$. The red outline in Figure 3-3 shows the domain where runoff is snowmelt-dominated minus four basins identified as having large storage capacities relative to runoff. They state that within this domain, water resources are arguably susceptible to warming-induced shifts in streamflow seasonality. Mountainous regions—the inset of Figure 3-3—(except those nearest the Equator) are generally snowmelt-dominated. The domain of influence within the red line of Figure 3-3 does not account for areas where water availability is predominantly influenced by snowmelt that is generated upstream. Therefore, the authors extended the domain of influence into sub-basins where the annual runoff originating in the snowmelt dominated cells accounts for at least 50% of the runoff for the entire sub-basin (black lines in Figure 3-3). According to a year 2000 population map, approximately one-sixth of the world's population lives within this combined snowmelt-dominated, low-reservoir-storage domain, not taking into account for populations that derive their water resources from outside the basins in which they dwell.

Glaciers, ice caps and continental ice sheets cover about 10% of the earth's land surface at the present time, whereas during the ice ages, they covered about three times this amount (Paterson 1994). The glacier moraines formed between the 17th century and the second half of the 19th century mark Holocene maximum extents of glaciers in most of the world's mountain ranges (UNEP and WGMS 2008). Since then, glaciers around the globe have been shrinking significantly, with strong glacier retreats in the 1940s, stable or growing conditions around the 1970s, and again increasing rates of ice loss since the mid-1980s. On a shorter

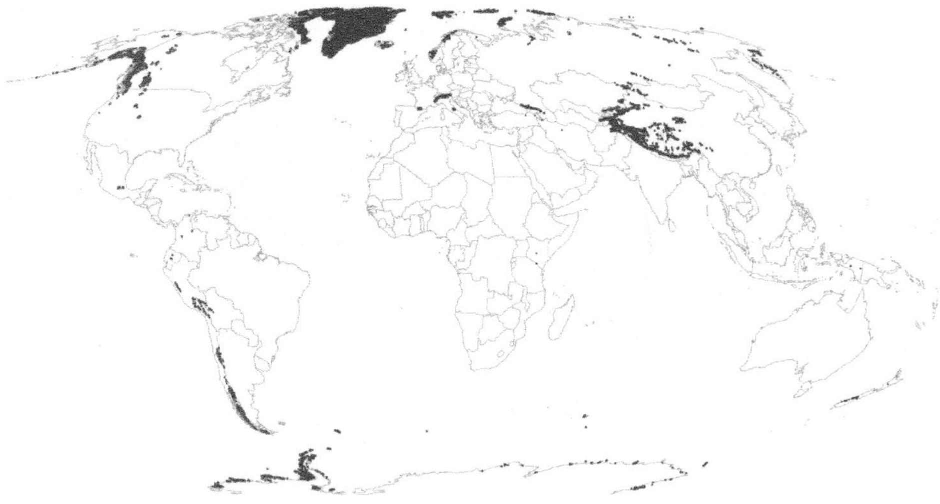

Figure 3-4. Glacerised areas of the world. Data source: DCW (Digital Chart of the World) glacier layer (Raup et al. 2000)

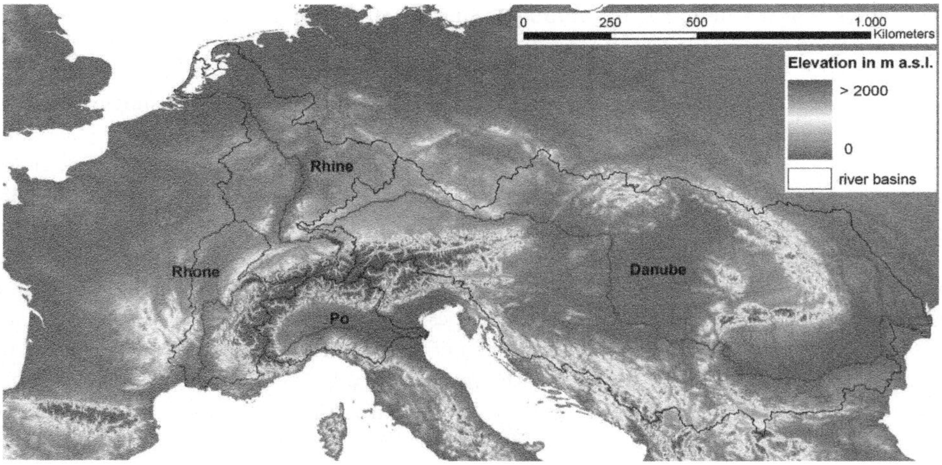

Figure 3-5. Major rivers originating in the European Alps—Rhine, Rhone, Po and Danube

time scale, glaciers in various mountain ranges have shown intermittent re-advances. Figure 3-4 shows the distribution of glacerised areas in the world. A comparison with Figure 3-1 shows that these areas—apart from the poles—are all located in the major mountain ranges of the world.

Case Study: Rhine River Basin

As a case study, the Rhine river basin is discussed. Discharges in the alpine tributaries of the Rhine river basin are dominated by snow and glacier melt in spring and summer (Verbunt et al. 2003). The Rhine is one of four major rivers originating in the European Alps (Figure 3-5).

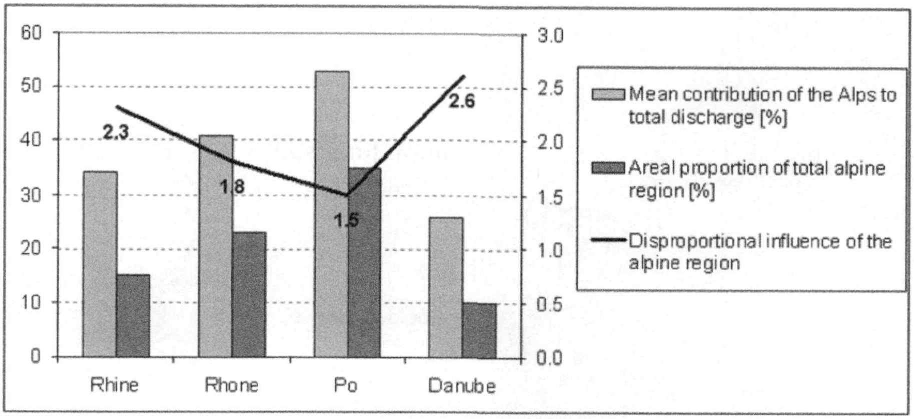

Figure 3-6. Contribution of the Alps to total discharge of the four major alpine streams, adapted from Weingartner et al. (2007)

As indicated before, for these four river basins the Alpine mountain range has medium or high contributing potential to the wet lowland area. With a mean contribution of 34% of the total discharge (Figure 3-6), the mountain region of the Rhine river system supplies 2.3 times more water than might be expected on the basis of surface area alone (disproportional influence) (Weingartner et al. 2007). This disproportional influence ranges from 1.5 (Po) to 2.6 (Danube). However, during summer months this value can be much higher. The mountainous part of the Rhine River thus plays a distinctive role in the hydrology of the whole river basin, because of higher precipitation and lower evapotranspiration due to the effect of temperature decreases with altitude. This results in a more effective runoff generation in the mountainous catchments as compared to in the lowlands. The components of a hydrological water balance are precipitation (*P*), runoff (*R*), evapotranspiration (*ET*), and change in water storage (d*S*).

Under specific hydrogeological conditions, in particular in karst areas, natural underground inflow and outflow (*I*) must also be taken into account. The equation for the water balance is therefore:

$$P = R + ET + dS - I \qquad \text{(Eq. 3-1)}$$

An analysis of the mean monthly discharge of two subcatchments of the Rhine river basin (Figures 3-7 and 3-8), the Lahn river subcatchment (gauge Leun, area 3,571 km^2) in the lowlands and the glacerised Ill River (gauge Gisingen, area 1,281 km^2) in the Alps, show a large difference in hydrological characteristics. The discharge in the alpine subcatchment is considerably higher (annually 1,628 mm) than in the lowland subcatchment (annually 314 mm), due to higher precipitation and lower evapotranspiration values. The discharge difference is particularly large in summer, when snowmelt occurs (April–June) and glacier melt (July–August). In winter, most precipitation in the Alps is stored as snow. The summer runoff from the Alps arrives at a time of low flows in the lowlands, where the largest water demands (municipalities, small trade and industry, agriculture, cooling water, …) are concentrated.

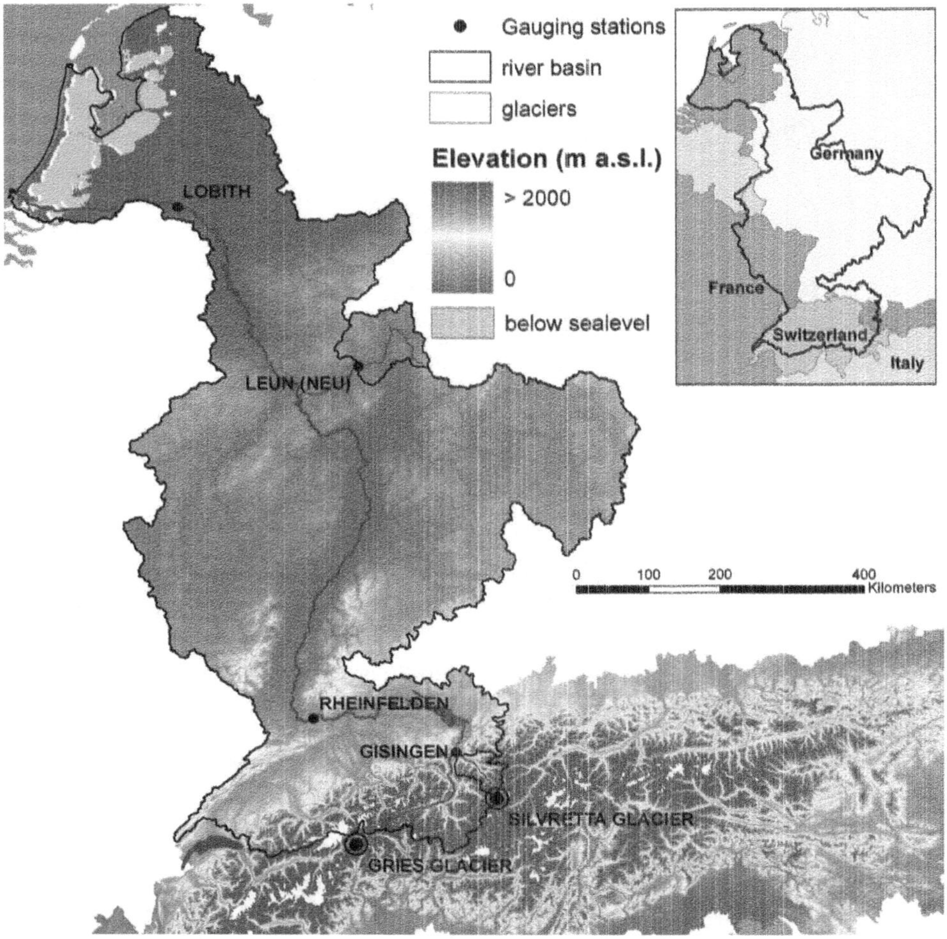

Figure 3-7. Rhine river basin with two discharge measuring stations on the Rhine (Rheinfelden, 260 m a.s.l., and Lobith, 9 m a.s.l.), one on its tributary the river Ill (Gisingen, 434 m a.s.l.), and one on its tributary the river Lahn (Leun, 135 m a.s.l.)

CLIMATE CHANGE AND MOUNTAIN HYDROLOGY

General

Mountainous areas are particularly vulnerable to temperature and precipitation changes (Beniston 2003). In the European Alps for example, the 20th-century warming has been roughly three times the global average (Jungo and Beniston 2001). Mountains are also early indicators of climate change. Many climatologists believe that the changes occurring in mountain ecosystems provide an early glimpse of what may come to pass in lowland environments. Over the last century, glaciers in the European Alps and the Caucasus Mountains have shrunk to half their size. In Africa, only 8% of Mount Kenya's largest glacier remains. If current trends continue, by the end of this century many of the world's mountain glaciers will have vanished entirely.

Because the hydrological cycle will be enhanced under warmer climatic conditions, the current distribution, seasonality, and amount of precipitation may undergo significant changes in various geographical regions. The consequences for river runoff are likely to affect not only the watersheds within the mountains themselves, but also in the lowland regions that are heavily dependent on this mountain resource (Kundzewitz et al. 2007). A very robust finding of hydrological impact studies is that warming leads to changes in the seasonality of river flows where much winter precipitation currently falls as snow. In most temperate mountain regions, the snowpack is close to its melting point, so it is very sensitive to changes in temperature. As warming progresses in the future, current regions of snow precipitation increasingly will experience precipitation in the form of rain. In these catchments, earlier peak flows in spring are forecasted, as well as higher flows in winter and lower flows in summer (resulting in longer dry periods) (Eckhardt and Ulbrich 2003; Kundzewitz et al. 2007; Vanham et al. 2008a, b). A shift in peak river runoff to winter, away from summer and autumn when demand is highest, is observed (Barnett et al. 2005). This has been found in the European Alps, Scandinavia, the Himalayas, and western, central, and eastern North America. The effect is greatest at lower elevations, where snowfall is more marginal, and in many cases peak flows by the middle of the 21st century would occur at least a month earlier (Kundzewitz et al. 2007). Where storage capacities are not sufficient, much of the winter runoff will be lost to the oceans. In many mountain areas, e.g., in the tropical Andes and many Asian mountain ranges, where glaciers provide the main runoff during pronounced dry seasons, water volumes stored in glaciers and snow cover are projected to decline. Runoff during warm and dry seasons is enhanced while glaciers are shrinking, but will dramatically drop after they have disappeared. Millions of people depend directly or indirectly on these natural water storage facilities for drinking water, agriculture, industry, and power generation during critical periods of the year. Climate change can also have a significant impact on water quality, for example by concentrating pollutant loads from wastewater treatment plants during prolonged dry periods.

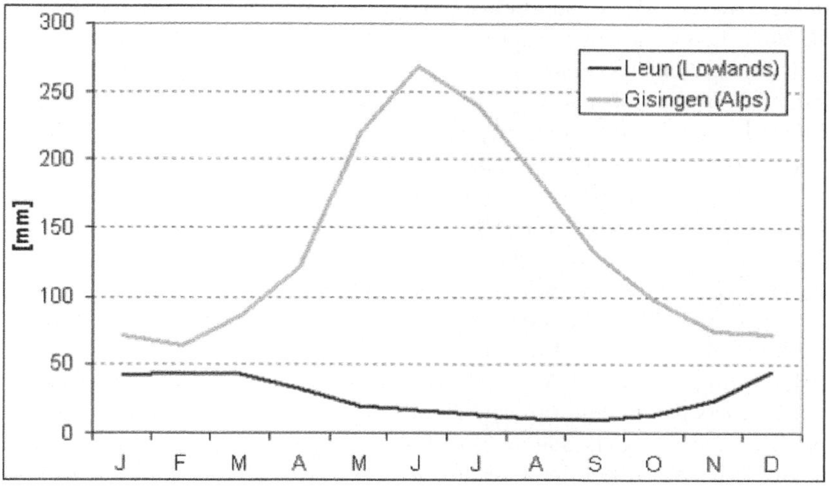

Figure 3-8. Mean monthly discharge (mm) for the period 1961–1990 for two subcatchments of the Rhine river basin, the Lahn river subcatchment upto the discharge measuring station Leun (lowlands Germany, area 3,571 km^2) and the Ill river subcatchment upto the discharge measuring station Gisingen (Austrian Alps, area 1,281 km^2). Data Source: EHYD and the Global Runoff Data Centre, 56058 Koblenz, Germany.

In the following text, the effect of global warming on water resources management in the major mountain ranges of the world is discussed. As a detailed case study, the Rhine river basin was chosen.

Detailed Case Study: Rhine River Basin

The Rhine river basin is one of the most densely populated and industrialised in Europe. It occupies a central position in the continent, both economically and geographically. The International Commission for the Hydrology of the Rhine Basin (CHR) carried out a research project to assess the impact of climate change on the river flow conditions in the Rhine Basin. The results are described in Middelkoop et al. (2001). The authors conclude an increase in winter discharge as a result of intensified snowmelt and increased winter precipitation, and lower summer discharge due to the reduced winter snow storage and an increase of evapotranspiration. As a matter of fact, the Rhine river basin will shift from a combined rainfall and snowmelt regime to a more rainfall-dominated regime. This leads to increases in the frequency and height of peak flows during winter. Longer and more frequent periods of low flow during the summer lead to a reduction in water availability for domestic use, industry, agriculture, and hydropower generation during the season of peak demand (further stressed by an increase in summertime demand due to higher temperatures). Low flows during summer will adversely affect inland navigation. During this period, there will be an increase in water temperatures and a possible decrease in water quality. Longer periods of low flow exacerbate many forms of water pollution. In the Rhine estuary in the Netherlands, the sea may penetrate further inland, leading to groundwater and surface water salinisation. For a densely populated country like the Netherlands, this poses major threats to water supply. Power plants are not allowed to produce at the necessary level if the water temperature has reached a certain high level. This may lead to energy shortage or, in the worst case, electricity breakdowns.

An extreme heatwave was experienced over the whole European continent in the summer of 2003. Numerous publications have been written about the effect of this summer on hydrology, water resources, and water supply (Beniston 2004; Beniston and Diaz 2004; BUWAL 2004; Eybl et al. 2005; Zappa and Kan 2007; Vanham et al. 2008a). This extreme climatic situation resembles simulations by regional climate models of summer temperatures in the latter part of the 21st century under the A2 scenario (Beniston and Diaz 2004). Schär et al. (2004) also state in the renowned journal *Nature* that at the end of this century (2071–2100), every second summer can be as hot (or even hotter) and as dry (or even drier) as the summer of 2003 in Europe. This heatwave is thus seen by many as a shape of things to come, reflecting the extremes of temperature that summers are projected to have in the latter decade of the 21st century. This event therefore provides a unique opportunity to observe the effects of global warming on water availability throughout much of Europe.

During the year 2003, a record melting of the glaciers of the Alps was recorded (BUWAL 2004). Based on data collected through an international network of glaciologists, the World Glacier Monitoring Service (WGMS) of the International Association of Cryospheric Sciences with their partners provide information about glacial changes. The glacier mass balance of two of their reference glaciers (Figure 3-9) shows a record negative value during 2003. The Silvretta Glacier is located on the watershed border between Rhine and Danube, the Gries Glacier south of the Rhine river basin (Figure 3-10). For the 14 reference glaciers of WGMS in the European Alps, the average mass balance loss equalled 2.6 meter water equivalent in 2003 (WGMS 2008), more than four times the average loss rate during the period 1980–2000. This melt rate would imply a reduction of 50% of the ice volume by 2025 as compared to the situation in 1970–1980 (BUWAL 2004). In 2100, only 5% of glacier ice would remain.

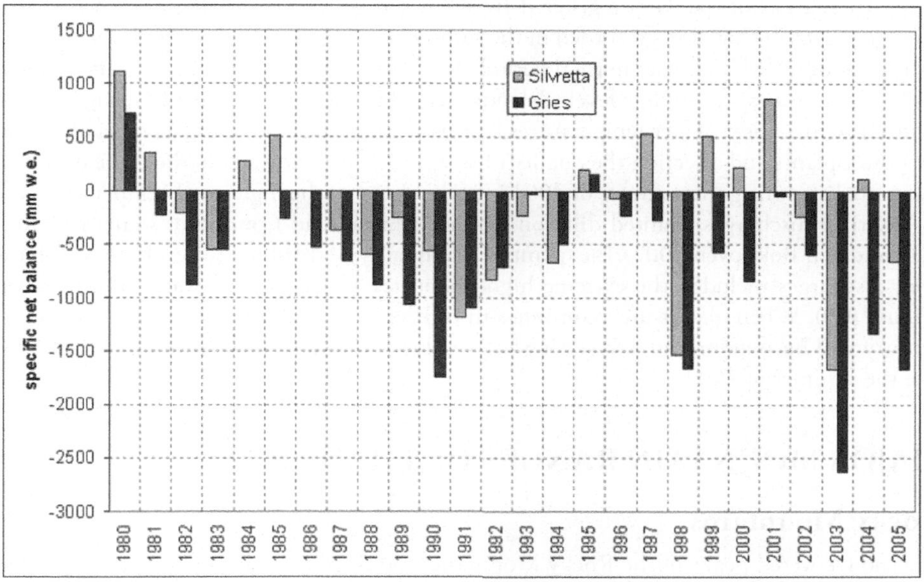

Figure 3-9. **The glacier mass balance of two of their reference glaciers shows a record negative value during 2003. Source: (IAHS AND UNESCO 1998; IUGG (CCS) et al. 2005; WGMS 2008)**

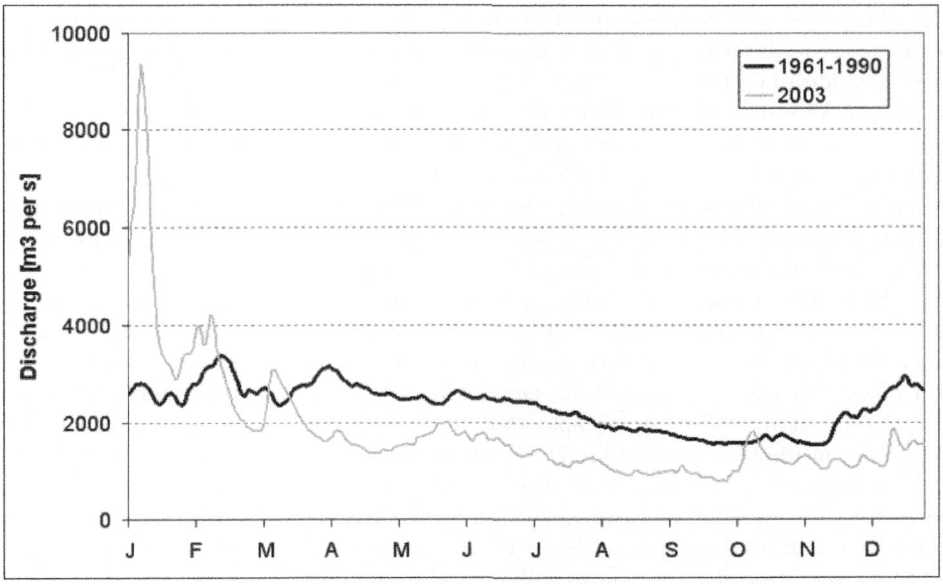

Figure 3-10. **Gauge Lobith on the Rhine river (lowlands, 9 m a.s.l.), daily average of the discharge 2003 and averaged daily values of the discharge of the time series 1961–1990. Data Source: Global Runoff Data Centre, 56058 Koblenz, Germany.**

An analysis of the daily registered river flows at gauge Lobith (Figure 3-9 and 3-10) shows the extreme low flows recorded in the summer of 2003 as compared with the long year average of the WMO climate normal period 1961–1990. During July, August, and September, the river was at record low levels. Islands formed in the river and inland navigation was reduced significantly. Water temperatures increased, therefore restricting its use as a coolant for power plants and affecting the quality of surface and groundwater bodies. The decline in water quality during the summer of 2003 is both related to the high water temperatures and to low river discharges (limited dilution of the chemical load from point sources) (Zwolsman and van Bokhoven 2007).The quantity and quality of drinking water decreased. However, discharges included the extreme high quantities of glacier melt water from the Alps (Figure 3-8). When glacier melt continues in future, its contribution at the end of the 21st century will become minimal, resulting in even lower discharges for future summer events like the summer of 2003.

MAJOR MOUNTAIN RANGES OF THE WORLD

Rocky Mountains

The snow-covered peaks of the Rocky Mountains and Sierra Nevada act as water towers for the North American West. Each annual snowmelt recharges the region's river basins and reservoirs. Global warming will bring more rain, less snow, and earlier melting. The most significant impact of a general warming in the Western USA was found to be a large reduction in mountain snow pack and a substantial shift in stream-flow seasonality, so that by 2050, the spring stream-flow maximum will come weeks earlier in the year (Barnett et al. 2005). Not all river systems throughout the West would be equally affected. For example, those with higher ratios of storage to average annual runoff would be impacted less than those with lower ratios (Standish-Lee and Lecina 2008). There is not enough reservoir storage capacity over most of the West to handle this shift in maximum runoff and so most of the "early water" will be passed on to the oceans. Winter and early spring flows increase (raising flooding potential), and summer flows decrease substantially (Kundzewitz et al. 2007). These hydrological changes have considerable impacts on water availability. Over-allocated water systems of the Western USA and Canada relying on capturing snowmelt runoff could be especially vulnerable, as those are the systems that rely upon runoff from glaciers.

In British Colombia for example, projected impacts include increased winter precipitation, more severe spring floods on the coast and the interior, and more summer droughts along the south coast and southern interior. This would decrease stream flow in the region, and affect both fish survival and water supplies in the summer, when demand is the highest (Kundzewitz et al. 2007). For example, in the Columbia river system, less winter snowfall and earlier melting has profound implications for the multiple objectives within the basin, and particularly for those objectives that are most sensitive to declines in summer stream flows (e.g., irrigation, urban water supply, fish flows) (Hamlet and Lettenmaier 2000). Seasonal flows shift markedly toward larger winter and spring flows and smaller summer and autumn flows. These changes in flows will be likely to coincide with increased water demand, principally from regional growth but also induced by climate change. Climate change is also projected to impact urban water supplies within the basin. For example, a 2°C warming projected for the 2040s would increase demand for water in Portland, Ore, by 5.7 million m^3/year with an additional demand of 20.8 million m^3/year due to population growth, while decreasing supply by 4.9 million m^3/year (Mote et al. 2003).

Potential impacts of global warming on water resources management in Southern California is discussed by Beuhler (2003). This semi-arid region is partly supplied by mountain water from northern California (the State Water Project) and from the Colorado River, originating in the Rocky Mountains. The mountainous section of the Colorado River represents an extremely important water tower for the arid lowlands downstream. The proportion of discharge generated in the mountain areas exceeds 90% of total discharge in the majority of cases (Viviroli et al. 2003). Southern California is a region where a population increase is forecasted (Standish-Lee and Lecina 2008), and water demand is expected to grow by 26% up to 2025 (Beuhler 2003). Vulnerabilities for the region include conversion of snowfall to rain, which effectively reduces the storage provided by snowpack in the mountains. This poses difficult challenges facing water resource managers in the western United States, who rely heavily on artificial storage of freshwater supplies (Standish-Lee and Lecina 2008). A reduced snowpack will lead to increased runoff during winter and less snowmelt during spring, making it more difficult to refill reservoirs in late spring and early summer. The overall availability of surface water supplies will be reduced and potential conflicts among agricultural, municipal, and environmental water users will increase. Reduced reservoir levels would also limit lake recreation and hydropower generation.

Andes

A large fraction of the population living west of the South American Andes relies on the glacial melt from those mountains to feed the area's rivers to supply municipal water and hydropower (Barnett et al. 2005). A large part of the Andes range is identified by (Viviroli et al. 2007) to have a high contributing potential to the surrounding dry lowlands (Figure 3-11). In a region where population growth continues, with consequences for food demand and irrigation, water demands increase and water availability will decrease in the long run due to glacier melt. Therefore, a (largely poor) dense population will be confronted with increased water stress in the near future.

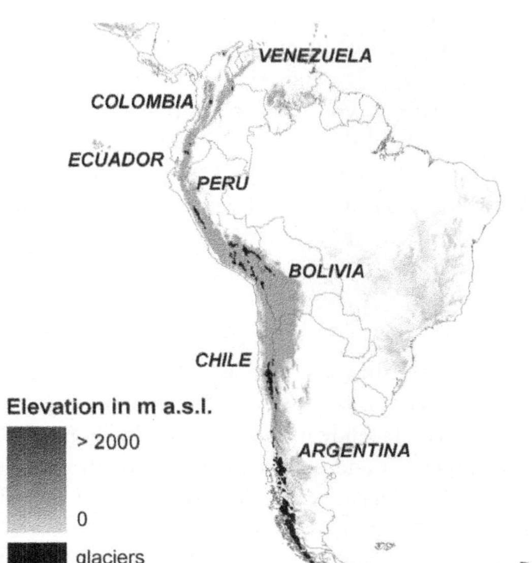

Figure 3-11. Glacerised areas of South America. Data source: DCW (Digital Chart of the World) glacier layer (Raup et al. 2000)

Coudrain et al. (2005) describe the impact of shrinking of glaciers in Peru, Bolivia, and Ecuador. In contrast to the Alpine glaciers, which undergo a long accumulation period in winter and a short ablation season in summer, the glaciers of the tropical Andes experience, in their lower part, an ablation regime throughout the year. Tropical Andean glaciers have a seasonal precipitation distribution (dry and wet season), while seasonal temperature variations are small, compared to a strong diurnal (day–night) amplitude. However, glaciers in the tropical Andes are undergoing rapid retreat with potentially devastating consequences for populations who rely on them for water resources (Mark and Mckenzie 2007). During the last decades, a substantial mass reduction for many glaciers from Colombia and Venezuela to Chile was observed (Casassa et al. 2007). The smallest glaciers have been affected the most (Ramírez et al. 2001).

As climate models predict that greenhouse warming will cause temperatures to rise faster at higher than at lower altitudes, glaciers may soon disappear in the tropical Andes (Bradley et al. 2006). Many basins, including large areas of glaciers in the cordilleras, have experienced an increase in runoff in recent decades, while precipitation has not changed, or tended to decrease. This is due to glacier retreat, leading to increased river flows in the short term, but the contribution of glacier melt will gradually decline over the next few decades (Coudrain et al. 2005).

In the dry Andes, glacial meltwater contributes substantially to river flow (Coudrain et al. 2005). The melting of glaciers may lead to water shortages for millions of people, especially during the long dry period. Among the regions most at risk from glacier melting due to climate change are Bolivia, Ecuador, and Peru, where glaciers feed rivers all year round. In both densely and sparsely populated areas, millions of people depend directly or indirectly on these natural water storage facilities for drinking water, agriculture, industry, and power generation during key parts of the year. Many large cities in the Andes are located above 2,500 m and thus depend almost entirely on high-altitude water stocks to complement rainfall during the dry season. For example, Ecuador's capital Quito currently receives part of its drinking water from a rapidly retreating glacier on Volcano Antizana. Other cities, like La Paz in Bolivia and many smaller cities and villages, likewise partially depend on glacier sources for drinking water (Bradley et al. 2006). In Peru, 10 million residents of Lima depend on freshwater from the Quelcaya Glacier. In many dry, inter-Andean valleys, agriculture relies on glacier runoff. Furthermore, in most Andean countries, hydropower is the major source of energy for electricity generation. Bradley et al. (2006) state as possible practical measures to prepare for, and adapt to, these changes: the conservation of (or price controls on) water supplies in urban areas, a shift to less water-intensive agriculture, the creation of highland reservoirs to stabilize the cycle of seasonal runoff, and a shift to power generation from resources other than hydropower.

European Alps

The case study of the Rhine river basin was discussed in detail as representative for the Alps.

A smaller snowfall dominated case study in the Austrian Alps showed a decrease in snow cover duration with the rate dependent on elevation (Vanham et al. 2008b). This leads to a shortening of the winter season and a shift in the seasonality of river flows and water availability (springs, groundwater, surface water) with more water availability in winter and less in spring to autumn. In the current situation, winter is the most critical period of the year, due to high demands and low water availability (Vanham et al. 2008c). However, with global warming, water availability in winter increases. On the other hand, water demand for snowmaking increases (Vanham et al. 2008b). Summer dry periods become more critical.

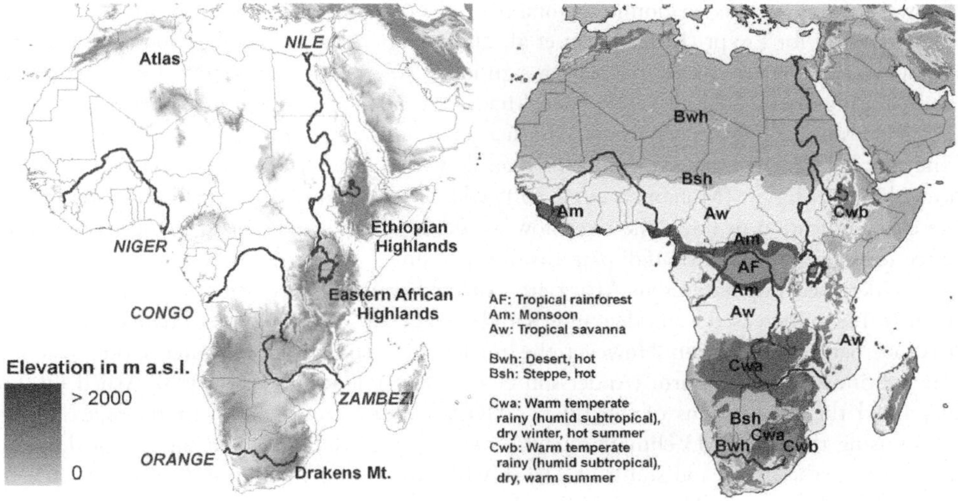

Figure 3-12. Major mountain ranges and rivers in Africa (left); Major climatic zones 1950–2000 according to the Köppen–Geiger system of climate classification (Köppen, 1936) adapted according to Hijmans et al. (2005) by C.A.J.M. de Bie (2007) (right).

African Mountain Ranges

In Schulz and de Jong (2004), the importance of the snow cover within the High Atlas Mountains as a major source for freshwater renewal and water availability for the semi-arid lowlands of southeastern Morocco is emphasized. As a matter of fact, the Atlas is the only major mountain range in Africa where snow plays an important role in the overall hydrology of the region. Viviroli et al. (2007) define parts of the Atlas mountain range as a high contributing potential to the surrounding dry lowlands (Figure 3-1). The larger Mediterranean area is projected to experience increasing average temperatures and decreasing average precipitation during the 21st century (Kundzewitz et al. 2007). The resulting loss in snow cover duration and volumes in the Atlas will decrease the natural storage of water significantly.

The mountain ranges where the three African rivers Nile, Orange, and Niger originate act as an essential water tower for these rivers. Viviroli et al. (2007) describe the Nile, Orange, and Niger as three of the rivers in the world that are by far the most dependent on mountain discharge. The Nile and Orange rivers have part of their middle courses and their lower courses in dry lowland, more particularly dry hot desert and steppe land (Figure 3-12) according to the Köppen–Geiger scheme (Köppen 1936; Viviroli and Weingartner 2004). Köppen's classification divides roughly into tropical (A), dry (B), temperate (C), cold (D), and snow and ice (E) climate types. Further distinctions are made according to variations and seasonal patterns in precipitation and temperature. The Orange has its source in the Drakens Mountains in South Africa, whereas the Nile has its source in both the Ethiopian Highlands (Blue Nile) and the Eastern African Highlands (White Nile) (Figure 3-12). The upper and lower courses of the Niger are located in a tropical climatic zone, but its middle course in desert.

Egypt, for example, is very dependent on the mountain waters feeding the Nile River. About 80% of annually available per-capita water resources in the country comes from the Nile river influx from abroad (Zehnder 1999). A major challenge for the country is to close the rapidly increasing gap between the limited water availability and the escalating demand for water from various economic sectors. The rate of water utilisation has already reached its maximum for Egypt (Kundzewitz et al. 2007). Agriculture consumes about 85% of the annual total water resource and plays a significant role in the Egyptian national economy, contributing about 20% of GDP. More than 70% of the cultivated area depends on low-efficiency surface irrigation systems. Another major concern is the rapid population growth. The effect of climate change on the Nile river flow is therefore essential to assess future water availability in Egypt (Mohamed et al. 2005). However, Conway (2005) found that there is no clear indication of how Nile river flow would be affected by climate change, because of uncertainty in projected rainfall patterns in the basin.

Increased populations in Africa are expected to experience water stress before 2025, mainly due to increased water demand (Kundzewitz et al. 2007). Climate change is expected to exacerbate this condition. However, the impact of climate change on water resources across the continent is not uniform (Andersson et al. 2006; Giannini et al. 2008). Arnell (2004) described the implications of the IPCC's SRES scenarios for a river runoff projection for 2050 using the HadCM3 climate model. These experiments indicate a significant decrease in runoff in the north and south of Africa, while the runoff in eastern Africa and parts of semi-arid sub-Saharan Africa is projected to increase. However, multi-model results show considerable variation among models, with the decrease in northern and southern Africa and the increase in eastern Africa emerging as the most robust responses. Parts of the East African region may expect an increase in rainfall that could even put it into the wet regime (de Wit and Stankiewicz 2006). Under these scenarios, mean annual precipitation in the Nile's sources in the humid Ethiopian and East African highlands is likely to increase. Precipitation in the Drakens Mountains would slightly increase (Hewitson and Crane 2006). Within the lowlands of the Orange River, precipitation will decrease dramatically (Hewitson and Crane 2006).

Taurus–Elburs–Zagros

As case study for this mountain range, the Tigris–Euphrates river basin is discussed. For many thousands of years, civilizations have relied on this basin. Today, Turkey, Iraq, and Syria (Figure 3-13) depend on the rivers' water and hydroelectric potential. Tension is rising between the first two countries, as Turkey is implementing its southeast Anatolian development project (GAP), an ongoing plan to install 22 dams and 19 hydroelectric plants along the Euphrates and the Tigris. On the other hand, the rivers feed large irrigation systems in the dry steppe and desert lowlands and are used for drinking water purposes.

Viviroli et al. (2003) describe the Tigris–Euphrates system as some of the rivers in the world that are by far the most dependent on mountain discharge. The Euphrates and Tigris rivers have part of their middle courses and their lower courses in dry lowland, more particularly dry hot desert and steppe land (Figure 3-13), according to the Köppen–Geiger scheme (Köppen 1936; Viviroli and Weingartner 2004). The lowest annual precipitation (< 300 mm) occurs in the Mesopotamian plain and the highest precipitation (> 600–1,200 mm) falls over the mountainous belt to the north (e.g., Taurus Mountains) and east (Zagros Mountains) of the plains (Jones et al. 2008). These authors state that snow melt is the dominant source for the flow in the rivers. Particularly in spring, the discharge pattern of the rivers is dominated by large volumes of meltwater from the mountains (Viviroli et al. 2003). This melting snow in spring in the mountains of Turkey, Iran, and Kurdistan causes the rivers Tigris and Euphrates to flood between March and May at a time when the long, hot, dry summer in Iraq is beginning (Sorman et al. 2007). During summer months, the mountain water represents up to 100% of the discharge in the rivers in the lowlands. Snow accumulation therefore acts as a natural storage of water for dry periods.

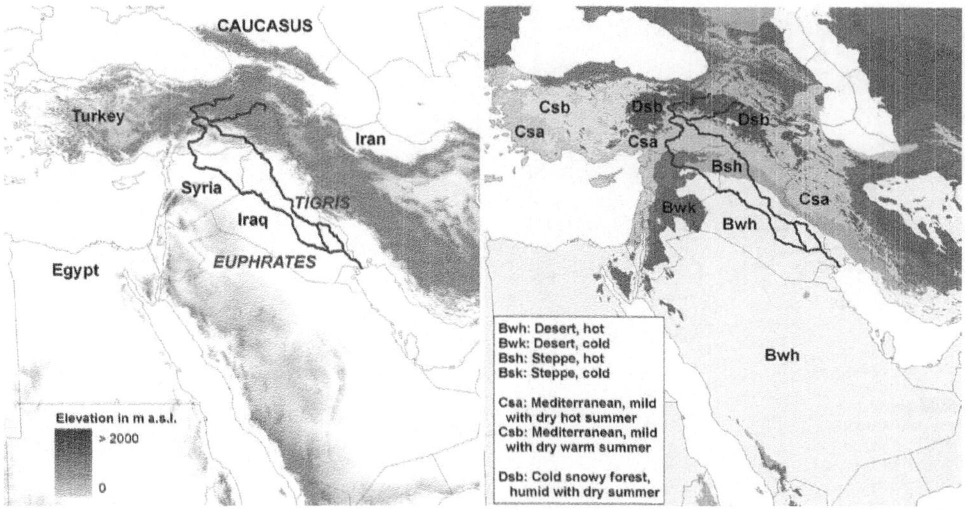

Figure 3-13. River system of the Tigris–Euphrates with mountain range (left); Major climatic zones 1950–2000 according to the Köppen–Geiger system of climate classification (Köppen 1936) adapted according to Hijmans et al. (2005) by C.A.J.M. de Bie (2007) (right).

Due to climate change, changes in seasonality and amount of water flow in the Tigris and Euphrates are expected (Kundzewitz et al. 2007). The larger Mediterranean area is projected to experience increasing average temperatures and decreasing average precipitation during the 21st century (Kundzewitz et al. 2007). As more precipitation will fall as rain in stead of snow, the natural water storage effect of the mountain range feeding the Tigris and Euphrates will decrease.

Hindu–Kush–Himalayas (HKH)

In India, Pakistan, Nepal, Bangladesh, and China, water shortages have been attributed to issues such as rapid urbanisation and industrialisation, population growth, and inefficient water use, which are all aggravated by changing climate and its adverse impacts on demand, supply, and water quality. In the countries situated in the Brahmaputra–Ganges and Indus basins, water shortages are also the result of the actions of upstream riverside-dwellers in storing water.

The major rivers draining on the south and southeast side of the Hindu–Kush–Himalayas mountain range are the Indus, Ganges–Brahmaputra, Irrawaddy, Mekong, and Yangtze (Figure 3-14). The extensive glacerised river basin of the Indus is identified by Viviroli et al. (2003) as one of the major rivers in the world that are by far the most dependent on mountain discharge. The Indus has part of its middle course and its lower course in dry hot desert and steppe lowland (Figure 3-14). The mountain range serves as a water tower. During the low flow season (April–May) before summer monsoon (June–September), glacier melt contributes a very large part of the total river flow. Snow cover and glacier melt also contributes to the discharge in the Ganges River. The Eastern Himalayas have an important supporting function and make important seasonal contributions to lowland runoff in the Yangtze River in China. A noticeably different pattern of specific runoff is observed for the Mekong (monsoon climate). Here, specific runoff increases in the lower reaches of the river, which means that discharge generation in the lowlands is clearly more important than in

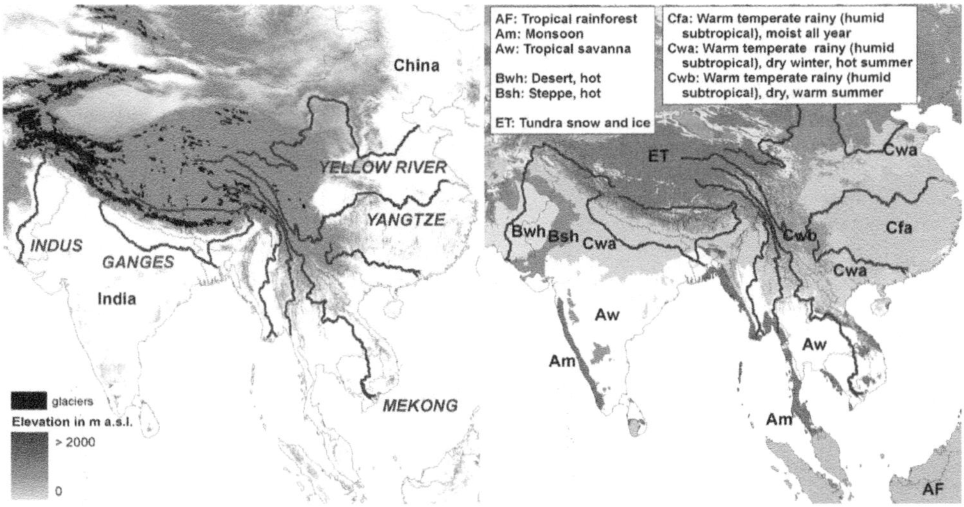

Figure 3-14. Major rivers originating in the Hindu-Kush-Himalayas mountain range with location of glaciers according to DCW (Digital Chart of the World) glacier layer (Raup et al. 2000) (left); Major climatic zones 1950–2000 according to the Köppen–Geiger system of climate classification (Köppen 1936) adapted according to (Hijmans et al. 2005) by C.A.J.M. de Bie (2007) (right).

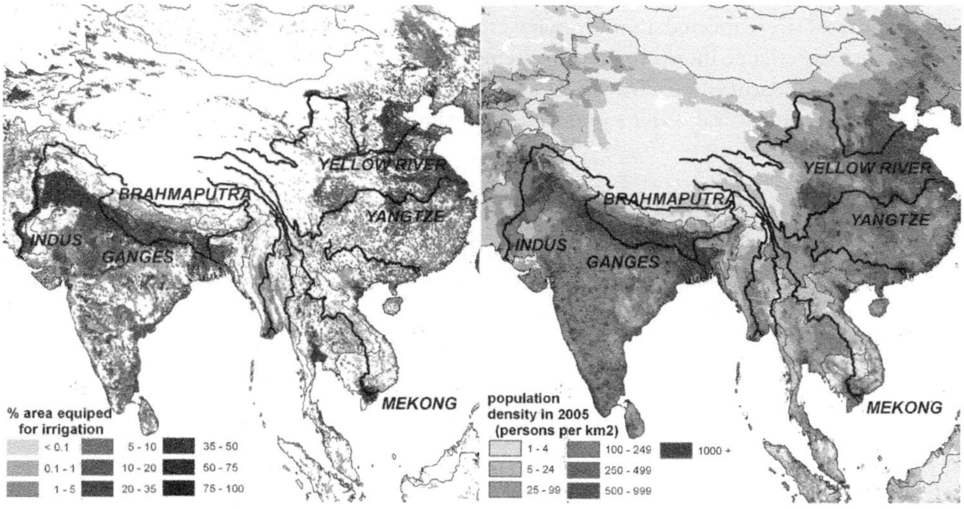

Figure 3-15. Percentage of area equipped for irrigation (left), Data Source: FAO; Population density in 2005 in persons per km² (right), Data Source: CIESIN.

the Himalayas (Viviroli et al. 2003). This illustrates the completely different significance of lowland areas in the humid tropics as opposed to those in arid regions.

Barnett et al. (2005) state that perhaps the most critical region in which vanishing glaciers and diminishing snowpacks will negatively affect water supply in the next few decades will be China and parts of Asia, including India, because of the region's huge population (about 50–60% of the world's population) (Figure 3-15). The increasing population and climate change effects on water availability result in increasing water demands for irrigated agriculture (Figure 3-15), industry, and energy. Extreme events like flooding and drought are likely to occur more often in future. For the south Asian monsoon in summer (June–August) an increase in precipitation is predicted. But with increasing temperatures, glaciers in the HKH mountain range are melting and snow cover accumulation decreases. River runoff will initially increase in winter or spring but eventually will decrease as a result of loss of ice resources. In the Hindu–Kush–Himalayas region, there may (for the next several decades) appear to be normal, even increased, amounts of available melt water to satisfy dry season needs. The shortage, when it comes, will likely arrive much more abruptly in time (Barnett et al. 2005). This will lead to much reduced water availability downstream for the extensive irrigation schemes in the Indus and Ganges–Brahmaputra river basins (Figure 3-15), but also for drinking water for a huge population, and water for industries and hydropower.

CONCLUSIONS

In many regions of the world, mountain water acts as an important and reliable source of fresh water for both the mountain regions themselves and the surrounding lowlands. This is, on the one hand, due to a higher specific runoff as a result of higher precipitation (orographic effect) coupled with lower evapotranspiration rates. On the other hand, mountain runoff shows a stable pattern with retarding of discharge due to (1) the storage of water in geological aquifers in the mountain core and (2) the accumulation and storage of winter precipitation in the form of snow and ice. This natural storage effect can compensate for water shortages both in the mountains and lowlands in critical dry periods like summer. In many river basins in the world (Indus, Rhine, Nile, Colorado, …), summer runoff variability in the lowlands is reduced, particularly through snow- and glacier melt (Viviroli and Weingartner 2004).

However, mountainous areas are particularly vulnerable to temperature and precipitation changes. Over the last century, glaciers throughout the world have shrunk considerably in size. On a seasonal level, both precipitation increases and decreases have been predicted for the different mountain ranges in the world. In the European Alps, for example, an increase in winter precipitation but a decrease in summer precipitation is forecasted. Temperature increase, however, is likely to occur throughout the year in most regions of the world, resulting in increased snow and glacier melt. The consequences for river runoff will affect both the mountainous watersheds and the lowland regions that depend on them.

Due to temperature increases, more precipitation will fall as rain and less as snow. This results in a shift in the seasonality of river flows. Higher flows in winter, earlier peak flows in spring, and lower flows in summer and autumn are expected. This will result in more water availability in winter but longer dry periods in summer. The European Alps and the Western USA, for example, are regions where these effects will occur. This shift in river runoff away from summer and autumn when water demands are highest, reduces the natural storage of water that can be released to the lowland during summer periods. Where man-made storage is not available, much of winter flows will be lost.

Regions that are very dependent on glacier melt water will be strongly affected. In different regions of the Andes and the Hindu–Kush–Himalayas, glacier melt contributes significantly to river discharge, especially during the dry season. Glacier retreat will lead to increased river flows in the short term, but the contribution of glacier melt will gradually decline over the next few decades. Millions of people depend directly or indirectly on these natural water storage facilities for drinking water, agriculture, industry, and power generation during critical periods of the year. This is, for example, observed for the Indus river basin, where glacier melt water feeds the largest irrigation system in the world and serves an increasing population number with drinking water. In many river basins like the Nile, Indus, or Ganges, water shortages have been attributed to population growth, rapid urbanisation and industrialization, inefficient water use, increasing demands for hydropower, and irrigation. Climate change and its adverse impacts on demand, supply, and water quality, will only aggravate these water shortages.

ACKNOWLEDGMENTS

The authors would like to thank the Global Runoff Data Centre (GRDC), 56058 Koblenz, Germany, for the provision of discharge data. Part of this work was presented during a workshop organized by the IWA specialist group on climate change at the 6th World Water Congress, Vienna, Austria, 8–12 September 2008.

REFERENCES

Andersson, L., Wilk J., Todd, M.C., Hughes, D.A., Earle, A., Kniveton, D., Layberry, R., and Savenije, H.H.G. (2006). Impact of climate change and development scenarios on flow patterns in the Okavango River. *Journal of Hydrology*, 331 (1–2), 43–57.

Arnell, N.W. (2004). Climate change and global water resources: SRES emissions and socio-economic scenarios. *Glob. Environ. Change*, 14 31–52.

Barnett, T.P., Adam, J.C., and Lettenmaier, D.P. (2005). Potential impacts of a warming climate on water availability in snow-dominated regions. *Nature*, 438 (7066), 303–309.

Beniston, M. (2003). Climatic change in mountain regions: A review of possible impacts. *Climatic Change*, 59 5–31.

Beniston, M. (2004). The 2003 heat wave in Europe. A shape of things to come? *Geophysical Research Letters*, 31 L02022.

Beniston, M., and Diaz, H.F. (2004). The 2003 heat wave as an example of summers in a greenhouse climate? Observations and climate model simulations for Basel, Switzerland. *Global and Planetary Change*, 44 73–81.

Beuhler, M. (2003). Potential impacts of global warming on water resources in southern California. *Water Science and Technology*, 47 (6–8), 165–168.

Bradley, R.S., Vuille, M., Diaz, H.F., and Vergara, W. (2006). Threats to water supplies in the tropical Andes. *Science*, 312 (5781), 1755–1756.

BUWAL. (2004). *Auswirkungen des Hitzesommers 2003 auf die Gewässer (Impact of the hot summer 2003 on water resources)*. Bundesamt für Umwelt, Wald und Landschaft, Bern, Switzerland.

Casassa, G., Haeberli, W., Jones, G., Kaser, G., Ribstein, P., Rivera, A., and Schneider, C. (2007). Current status of Andean glaciers. *Global and Planetary Change*, 59 (1–4), 1–9.

Conway, D. (2005). From headwater tributaries to international river: observing and adapting to climate variability and change in the Nile basin. *Global Environmental Change*, 15 99–114.

Coudrain, A., Francou, B., and Kundzewicz, Z.W. (2005). Glacier shrinkage in the Andes and consequences for water resources. *Hydrological Sciences Journal*, 50 (6), 925–932.

de Wit, M., and Stankiewicz, J. (2006). Changes in surface water supply across Africa with predicted climate change. *Science*, 311 (5769), 1917–1921.

Eckhardt, K., and Ulbrich, U. (2003). Potential impacts of climate change on groundwater recharge and streamflow in a central European low mountain range. *Journal of Hydrology*, 284 (1–4), 244–252.

Eybl, J., Godina, R., Lalk, P., Lorenz, P., Mueller, G., and Weilguni, V. (2005). Das Trockenjahr 2003 (The dry year 2003). *Mitteilungsblatt des Hydrografischen Dienstes in Österreich*, 83 1–38.

Giannini, A., Biasutti, M., Held, I.M., and Sobel, A.H. (2008). A global perspective on African climate. *Climatic Change*, 90 359–383.
Hamlet, A.F., and Lettenmaier, D.P. (2000). Long-range climate forecasting and its use for water management in the Pacific Northwest region of North America. *Journal of Hydroinformatics*, 2 (3), 163–182.
Hewitson, B.C., and Crane, R.G. (2006). Consensus between GCM climate change projections with empirical downscaling: precipitation downscaling over South Africa. *International Journal of Climatology*, 26 1315–1337.
Hijmans, R.J., Cameron, S.E., Parra, J.L., Jones, P.G., and Jarvis, A. (2005). Very high resolution interpolated climate surfaces for global land areas. *International Journal of Climatology*, 25 1965–1978.
IAHS and UNESCO. (1998). *Fluctuations of Glaciers 1990–1995. Volume VII.* World Glacier Monitoring Service, Zürich.
IUGG (CCS), UNEP, and UNESCO. (2005). *Fluctuations of Glaciers 1995–2000, Volume VIII.* World Glacier Monitoring Service, Zürich.
Jones, C., Sultan, M., Yan, E., Milewski, A., Hussein, M., Al-Dousari, A., Al-Kaisy, S., and Becker, R. (2008). Hydrologic impacts of engineering projects on the Tigris–Euphrates system and its marshlands. *Journal of Hydrology*, 353 (1–2), 59–75.
Jungo, P., and Beniston, M. (2001). Changes in the anomalies of extreme temperature anomalies in the 20th century at Swiss climatological stations located at different latitudes and altitudes. *Theoretical and Applied Climatology*, 69 1–12.
Köppen, W. (1936). *Das geographisca System der Klimate.* In Handbuch der Klimatologie, Köppen, W., and Geiger, G. (ed.), 1. C. Gebr, Borntraeger, 1936.
Kundzewitz, Z.W., Mata, L.J., Arnell, N.W., Doell, P., Kabat, P., Jiménez, B., Miller, K.A., Oki, T., Sen, Z., and Shiklomanov, I.A. (2007). *Freshwater resources and their management.* In Climate Change 2007: Impacts, Adaptation and Vulnerability. Contribution of Working Group II to the Fourth Assessment Report of the Intergovernmental Panel on Climate Change, Parry, M.L., Canziani, O.F., Palutikof, J.P., van der Linden, P.J., and Hanson, C.E. (ed.), Cambridge University Press, 2007.
Mark, B.G., and Mckenzie, J.M. (2007). Tracing increasing tropical Andean glacier melt with stable isotopes in water. *Environmental Science & Technology*, 41 (20), 6955–6960.
Meybeck, M., Green, P., and Vörösmarty, C.J. (2001). A new typology for mountains and other relief classes: An application to global continental water resources and population distribution. *Mt. Res. Dev*, 21 (1), 34–45.
Middelkoop, H., Daamen, K., Gellens, D., Grabs, W., Kwadijk, J.C.J., Lang, H., Parmet, B.W.A.H., Schädler, B., Schulla, J., and Wilke, K. (2001). Impact of climate change on hydrological regimes and water resources management in the Rhine Basin. *Climatic Change*, 49 (1–2), 105–128.
Mohamed, Y.A., van den Hurk, B.J.J.M., Savenije, H.H.G., and Bastiaanssen, W.G.M. (2005). Hydroclimatology of the Nile: results from a regional climate model. *Hydrol. Earth Syst. Sci.*, 9 263–278.
Mote, P.W., Parson, E.A., Hamlet, A.F., Keeton, W.S., Lettenmaier, D., Mantua, N., Miles, E.L., Peterson, D.W., Peterson, D.L., Slaughter, R., and Snover, A.K. (2003). Preparing for climatic change: the water, salmon, and forests of the Pacific Northwest. *Climatic Change*, 61 45–88.
Paterson, W.S.B. (1994). *The Physics of Glaciers.* Pergamon Press, Oxford.
Ramírez, E., Francou, B., Ribstein, P., Descloitres, M., Guérin, R., Mendoza, J., Gallaire, R., Pouyaud, B., and Jordan, E. (2001). Small glaciers disappearing in the tropical Andes: a case study in Bolivia: the Chacaltaya glacier, 16°S. *Journal of Glaciology*, 47 187–194.

Raup, B.H., Kieffer, H.H., Hare, T.M., and Kargel, J.S. (2000). Generation of Data Acquisition Requests for the ASTER Satellite Instrument for Monitoring a Globally Distributed Target: Glaciers. *IEEE Transactions On Geoscience and Remote Sensing*, 38 (2), 1105–1112.

Schär, C., Vidale, P.L., Lüthi, D., Frei, C., Häberli, C., Liniger, M.A., and Appenzeller, C. (2004). The role of increasing temperature variability in European summer heatwaves. *Nature*, 427 332–336.

Schulz, O., and de Jong, C. (2004). Snowmelt and sublimation: field experiments and modelling in the High Atlas Mountains of Morocco. *Hydrology and Earth System Sciences*, 8 (6), 1076–1089.

Smakhtin, V.U. (2001). Low flow hydrology: a review. *Journal of Hydrology*, 240 147–186.

Sorman, A.Ü., Akyürek, Z., Sensoy, A., Sorman, A.A., and Tekeli, A.E. (2007). Commentary on comparison of MODIS snow cover and albedo products with ground observations over the mountainous terrain of Turkey. *Hydrol. Earth Syst. Sci.*, 11 1353–1360.

Standish-Lee, P., and Lecina, K. (2008). Getting ready for climate change implications for the western USA. *Water Science & Technology*, 58 (3), 727–733.

UNEP and WGMS. (2008). *Global Glacier Changes: facts and figures.*

Vanham, D., Fleischhacker, E., and Rauch, W. (2008a). Impact of an extreme dry and hot summer on water supply security in an alpine region. *11th international specialized conference on watershed and river basin management*, 4–5 September 2008, Budapest, Hungary.

Vanham, D., Fleischhacker, E., and Rauch, W. (2008b). Impact of snowmaking on alpine water resources management under present and climate change conditions. *4th Biennial International Young Water Professionals Conference*, 16–18 July 2008, Berkeley, USA.

Vanham, D., Fleischhacker, E., and Rauch, W. (2008c). Technical Note: Seasonality in alpine water resources management—a regional assessment. *Hydrology and Earth System Sciences*, 12 (1), 91–100.

Verbunt, M., Gurtz, J., Jasper, K., Lang, H., Warmerdam, P., and Zappa, M. (2003). The hydrological role of snow and glaciers in alpine river basins and their distributed modeling. *Journal of Hydrology*, 282 36–55.

Viviroli, D., and Weingartner, R. (2004). The hydrological significance of mountains: from regional to global scale. *Hydrology and Earth System Sciences*, 8 (6), 1016–1029.

Viviroli, D., Weingartner, R., and Messerli, B. (2003). Assessing the hydrological significance of the world's mountains. *Mountain Research and Development*, 23 32–40.

Viviroli, D., Dürr, H.H., Messerli, B., Meybeck, M., and Weingartner, R. (2007). Mountains of the world, water towers for humanity: Typology, mapping, and global significance. *Water Resources Research*, 43 W07447.

Weingartner, R., Viviroli, D., and Schädler, B. (2007). Water resources in mountain regions: a methodological approach to assess the water balance in a highland–lowland–system. *Hydrological processes*, 21 578–585.

WGMS. (2008). *Fluctuations of Glaciers 2000–2005, Volume IX.* Haeberli, W., Zemp, M., Kääb, A., Paul, F., and Hoelzle, M. (Eds.) ICSU(FAGS)/IUGG(IACS)/UNEP/UNESCO/WMO. World Glacier Monitoring Service, Zürich.

Zappa, M., and Kan, C. (2007). Extreme heat and runoff extremes in the Swiss Alps. *Natural Hazards and Earth System Sciences*, 7 375–389.

Zehnder, A.J.B. (1999). Wassernutzung und Nahrungsmittelproduktion—eine internationale Arbeitsteilung? *EAWAG news*, 46 18–20.

Zwolsman, J.J.G., and van Bokhoven, A.J. (2007). Impact of summer droughts on water quality of the Rhine River—a preview of climate change? *Water Science & Technology* 56 (4), 45–55.

– 4 –

Prevailing Water Demand Forecasting Practices and Implications for Evaluating the Effects of Climate Change

Jack C. Kiefer, Ph.D.

INTRODUCTION

Climate change may alter the reliability of water supply systems via direct impacts on hydrology and source supplies. Because providing reliable supplies of adequate quality to meet demands is a fundamental mission of municipal water providers, the evaluation of climate change scenarios is becoming more important for long-term planning. While there have been several studies of the potential impacts of climate change scenarios on water resources, there has been very little study of the potential impacts of climate change on the demand for municipal and industrial water supply.

Because long-term water demand forecasts provide the basis for evaluating the reliability of water services—by presenting expectations for the amount of water a water provider should be prepared to deliver in the future—the choice of water demand modeling and projections methodologies is critical to assessing potential climate change impacts.

The Water Demand Function and Climate Change

The demand for water is a function of several variables whose values vary over time and space. Generally speaking, water demand is influenced both by socioeconomic factors and weather conditions. The range of socioeconomic factors that influence demand include, but are not limited to, the price of water, income, land use patterns, the mix and scale of employment and production processes, and other population characteristics.

Meanwhile, air temperature and precipitation represent the chief weather variables that affect water use. To an economist, a change in price implies a *movement along* the *demand curve*, while changes in other factors cause *shifts* in the demand curve. Estimation of an empirical relationship to model and forecast demand will typically not distinguish between these effects.

It is logical to conclude that climate change would affect water demand through its potential influence on precipitation and temperature and on corresponding seasonal cycles of weather conditions. However, climate change may affect water use over longer time periods via more indirect means associated with other socioeconomic determinants of water demand. In other words, *adaptation* by people and businesses, as well as by water providers, may enter or be evaluated via the demand function (Figure 4-1).

Depending on the degree of climate change and how this affects the location of people and economic activity, customer adaptation in broad terms may involve geographical in- and

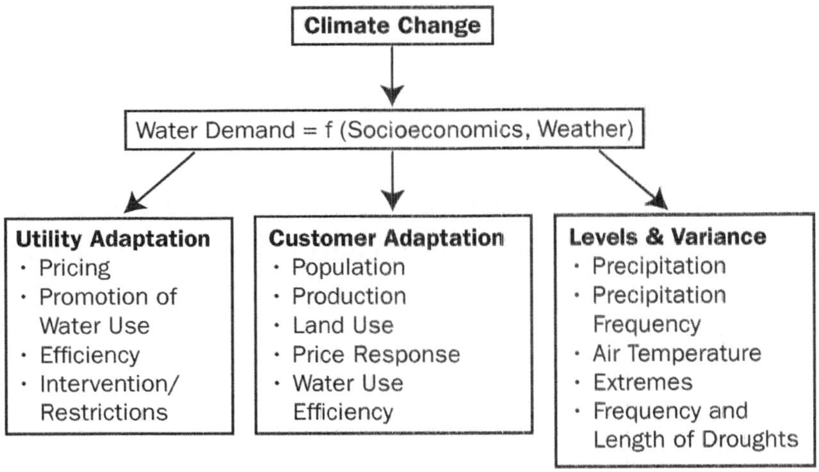

Figure 4-1. Climate change may affect the demand for water directly due to the effects of weather and indirectly through socioeconomic changes associated with human adaptation

out-migration, shifts in production or production processes, and changes in land use. If climate change affects the cost of water service, utility adaptation may result in changes in price or price structure, bringing corresponding changes in the demand for water.

Furthermore, certain scenarios may be hypothesized to result in more frequent water supply shortages. More frequent shortages may lead to demand-side adaptations such as increases in water-use efficiency or direct intervention on the part of water providers to restrict demand. A robust water demand model for evaluating and forecasting climate change scenarios would incorporate the ability to estimate changes in demand stemming from changes in these principal indicators. The direct indicators for municipal and industrial water demand are changes in the levels of and the variance of air temperature and precipitation, as well as the frequency in time of measurable precipitation events and the duration of extremes, whether these extremes represent hot–dry or cool–wet ends of the spectrum. Specification of indirect indicators would require variables that are capable of measuring or portraying the process of human responses through consumer and utility adjustments. Such a set of variables should include the price of water and some measure(s) that drive changes in demand or demand growth, such as population, housing, land use, and economic activity.

SUMMARY OF INDUSTRY PRACTICES AND CONSTRAINTS

An informal review of U.S. water utility industry practices demonstrates that there are wide differences in how urban water providers specify, model, and forecast the demand for water. Industry forecasting techniques vary in their sophistication and methodological abilities to account for determinants of water use, and therefore in their capacity to evaluate climate change scenarios. Specifically

- Traditional (per capita) long-term forecasting methods that tie projections of water use exclusively to population are insufficient for evaluating weather impacts;
- Disaggregate unit use approaches offer a slight improvement on the per capita method, because they are capable of differentiating water use sectors and climatic seasons, but fall short of explicitly defining cause-effect relationships;

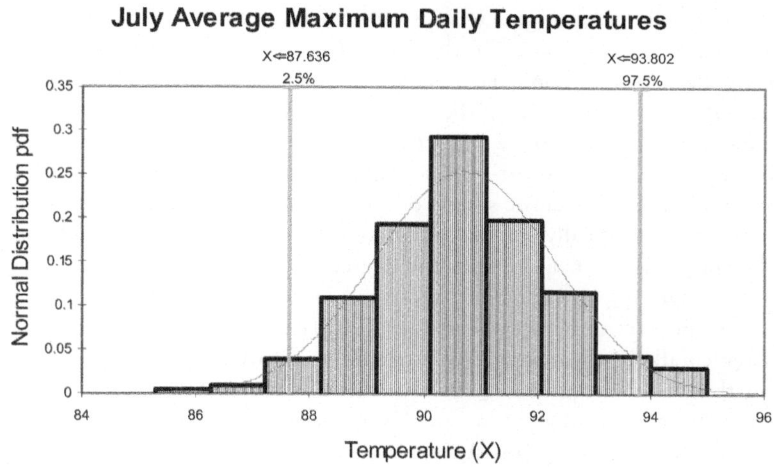

Figure 4-2. Uncertainty about climate change can be analyzed by examining potential effects on the distributions of weather variables.

- Predictive model-based approaches to water demand forecasting (sometimes referred to as *econometric models*) are the most ideal for evaluating the potential effects of climate change, because they are capable of directly incorporating the principal indicators that help trace impacts of climate change;
- Model-based approaches require additional data to implement, which often cause predictive models to be underspecified in practice (meaning important variables are left out of the equation due to data limitations); and
- Model-based approaches that disaggregate data into water use sectors and time periods further augment the capability to analyze climate change by isolating impacts on the underlying components of water use.

Simulations that account for uncertainty in future values of important variables more readily evaluate climate change scenarios, though these approaches are data intensive and have seen limited application in industry practice. The choice of a particular forecasting methodology is affected by the data that can be used to model relationships among demand determinants and sector water use. The availability and quality of data to support the development of models typically serves as a practical constraint on the options that are applicable for forecasting. Furthermore, the specific goals and objectives of any particular water demand forecasting effort may not require the enhancement of the prediction and informational capabilities of more complex methods that are most ideal for evaluating climate change scenarios.

POWER OF DEMAND SIMULATION METHODS

Simulation methods for examining forecast uncertainty are potentially the most powerful techniques for evaluating climate change scenarios. In particular, these methods permit an assessment of the entire distribution of any given weather variable that is influential in predicting water use. In other words, these techniques can account for scenarios that involve changes in both the central tendency and range of possible weather conditions. Thus, any particular climate change scenario may have a direct effect on the quantity of water demanded to the extent that it is realized in a way that influences normal (or mean) weather conditions, the variation around the mean, or both (Figure 4-2).

Simulation of climate change scenarios on demand need not be limited to the assessment of potential weather impacts. To the extent that any climate change scenario may be anticipated to affect socioeconomic factors that influence water use (via the adaptation path described in Figure 4-1), or increase the uncertainty about future values of these factors, simulation methods can incorporate a fuller accounting of the *system* that causes and creates variation in water demand (Figure 4-3). Also referred to as the *risk-based approach* to demand forecasting, simulation of the entire system of causal factors has multiple benefits for evaluating climate change. Specifically, and in conjunction with information or prediction of water supplies, the approach can help identify the degree of exposure in a future time period to conditions that might imply water shortages or surplus capacity.

Analysis of these scenarios may be used to identify risk reduction strategies that may be affected incrementally by the uncertainty surrounding future climate and may reveal coping mechanisms that do not require large investments in structural solutions. For example, in this context, water conservation and/or pricing measures may provide alternatives that can effectively offset climate change impacts on water resources availability.

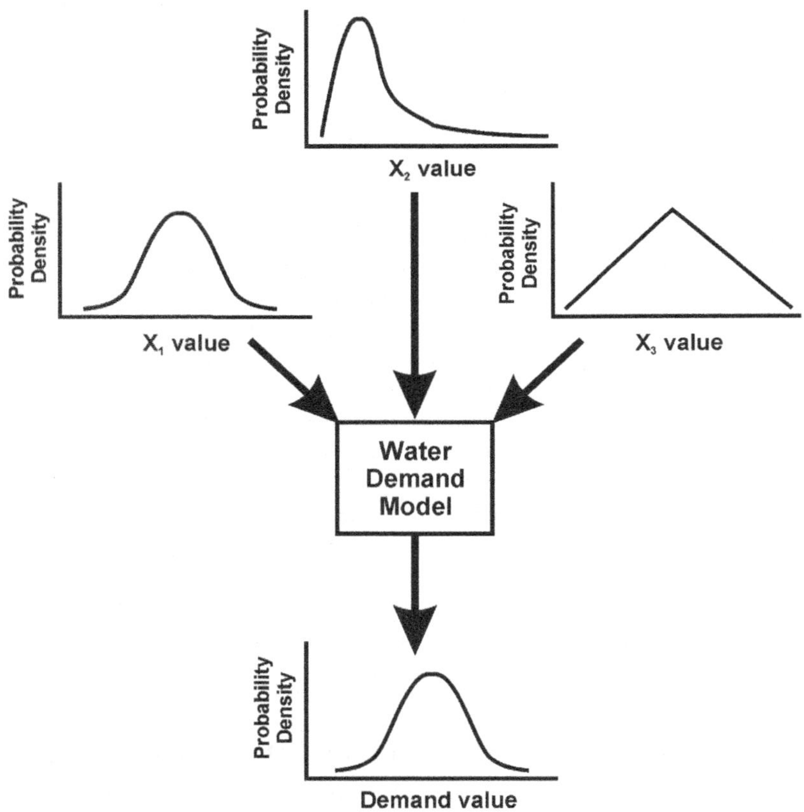

Figure 4-3. Promulgation of uncertainty can be analyzed with assumptions about the distributions of several variables that affect water demand, including those that are affected by climate change.

CONCLUSIONS

Climate change may affect in any particular geographic region both the physical quantity of available water and the demand for water services. Thus, analyzing the threats of climate change will require evaluation methods that address long-term water supply reliability. Water demand forecasts represent a key component in long-term reliability planning, as they characterize expectations about the amount of water that a water provider should be prepared to deliver in the future.

The degree to which any particular model of water demand or demand forecast is capable of evaluating climate change scenarios depends on a variety of practical considerations. There are a variety of ways in which water providers analyze and forecast water demands, regardless of whether climate change is or could be a planning consideration. The main reasons for this include

- The cost and availability of data that can be used to measure weather and socioeconomic factors;
- In-house expertise or the cost of external assistance in statistical estimation methods and interpretation; and
- Goals and objectives of existing water demand forecasting and long-range water supply planning efforts.

The implication of variation in water demand modeling and forecasting practices is that only certain forecasting models are sufficient for evaluating climate change—namely those that at least incorporate variables for weather and more preferably for socioeconomic factors as well. The degree to which these models exist in actual practice dictates the extent to which municipal and industrial water demand impacts can be analyzed for any given geographic area.

Predictive, model-based approaches to water demand forecasting are well-suited for climate change evaluation. Notwithstanding data constraints, they are capable of directly incorporating the principal indicators that help trace impacts of climate change, which essentially represent the same weather and socioeconomic factors that would serve to influence demand in the absence of climate change. Disaggregation of data into water use sectors and time periods further augments the capability to analyze climate change by providing an opportunity to isolate impacts on the underlying components of water use. Risk-based forecasting approaches, which are designed to simulate the effects of uncertainty and variability on projections of water use, offer additional advantages for evaluating the potential impacts of climate change.

Uncertainty about the future drives the need for planning. Models using risk-based approaches provide a means through which uncertainty can also drive the formulation of alternative planning strategies for coping with the potential risks associated with climate change.

ACKNOWLEDGMENTS

This work is supported by USEPA under contract GS-10F-0227J. Any opinions expressed in this paper are those of the author and do not necessarily reflect the official positions and policies of the USEPA. Any mention of products or trade names does not constitute recommendation for use by the USEPA.

Originally a 2006 AWWA Water Sources Conference proceeding.

– 5 –

Impacts of Climate Change and Variability on Source Water Quality of Lake Cachuma, California

Joseph A. Drago, P.E., Ph.D, Levi Brekke, Ph.D., Catriona E. Rogers, Ph.D., Kevin D. Walsh, P.E.

INTRODUCTION

The Goleta Water District serves over 75,000 people in Santa Barbara County, Calif. The District receives most of its water supply from Lake Cachuma, a 190,000 acre-foot reservoir formed by Bradbury Dam, which was constructed by the United States Bureau of Reclamation on the Santa Ynez River. Water from Lake Cachuma is diverted through a multiport outlet tower and conveyed through the Tecolote Tunnel, a 6.4-mile, 84-in. diameter tunnel through the Santa Ynez Mountains. Groundwater infiltration into the tunnel has some influence on water quality. The 28-mile South Coast Conduit delivers water to the district, the city of Santa Barbara, and several other small water districts. The district treats its surface water supply at the Corona del Mar Water Treatment Plant. Treatment concerns are primarily turbidity removal, disinfection, control of disinfection by-products, and taste and odor control. Thus, the district has an interest in water quality variability in Lake Cachuma and impacts caused by climate change.

Lake Water Quality

Table 5-1 provides a summary of water quality data maintained by the Cachuma Operations and Maintenance Board (COMB) that are collected at the reservoir outlet structure. Parameters measured include water temperature, pH, dissolved oxygen (DO), threshold odor number (TON), turbidity, apparent color, iron, manganese, electrical conductivity (EC), and total dissolved solids (TDS). For this study, water temperature and DO values were used to compute percent DO saturation.

As shown in Table 5-1, lake water quality is quite variable. For this study, the influence of seasonal and annual hydroclimate conditions on lake water quality was examined.

Lake Hydroclimate

Typical of California coastal regions, the lake hydroclimate consists of a wet season from October through March and a dry season from April through September. The lake stratifies in the spring (April–May) and turns over in late autumn (October–November). The National Oceanographic and Atmospheric Administration (NOAA)'s National Climate Data Center maintains weather data for Lake Cachuma. For this study, the lake hydroclimate was defined using six precipitation variables, five air temperature variables, and monthly evaporation rates.

In addition, a monthly water balance was constructed for Lake Cachuma using lake precipitation and evaporation, Santa Ynez River flow above Lake Cachuma, and lake water elevations.

California Region Climate Change Models

Global projections of climate variables are typically developed using general circulation models (GCMs) of the atmosphere. Two studies on California-region climate change (Miller et al. 2003, Snyder et al. 2002) were selected to supply climate projection information with respect to precipitation and air temperature. Miller et al. (2003) used global projection data from two GCMs, HadCM2 developed by the United Kingdom Hadley Center, and PCM developed by the U.S. National Center for Atmospheric Research, in which atmospheric carbon dioxide was allowed to increase at 1 percent per year relative to present conditions. HadCM2 projected a warmer and wetter climate while PCM projected a warmer and drier climate. Snyder et al. (2002) used the Community Climate Model CCM3 (Kiehl et al. 1998), which considered a doubling of atmospheric carbon dioxide scenario. CCM3 projected a warmer climate with no change in mean annual precipitation.

Indicators of Climate Variability

Other studies (Redmond and Koch 1991, Piechota 1997) have found significant relationships between southern California hydroclimate (e.g., precipitation and air temperature) and climate phenomena happening several months earlier in the equatorial Pacific. These coupled patterns of climate variability between two geographically separated regions, or *teleconnections*, are associated with El Nino/Southern Oscillation (ENSO) indicators. ENSO is an ocean and atmospheric climate phenomenon occurring in the equatorial Pacific Ocean between South America and Southeast Asia. The phenomenon's state is always in flux and tends to oscillate between *warm phase* and *cool phase* conditions.

ENSO is diagnosed by sea surface monitoring and sea-level atmospheric monitoring. Figure 5-1 shows the location of ENSO monitoring stations and regions in relation to Lake Cachuma. Sea surface monitoring is indexed by regionally averaged sea surface temperature anomalies in various equatorial regions (i.e. ,"Nino SST" regions, commonly referred to as Nino4, Nino34, Nino3, and Nino12 moving from west to east on Figure 5-1). Sea level atmospheric conditions are monitored by sea level pressure "difference anomalies" between Tahiti Island and Darwin, Australia, which are expressed in various forms to describe a Southern Oscillation Index (SOI). Historical ENSO monitoring data are available from NOAA (http://www.cdc.noaa.gov/ENSO/).

The use of ENSO indicators to forecast future water quality was assessed in this study. An assessment was made of how climate change might impact Lake Cachuma source water quality caused by shifts in mean precipitation and mean air temperature. The general approach was to relate lake water quality conditions (e.g., water temperature, turbidity, color, TON, pH, etc.) to hydroclimate parameters (e.g., precipitation, evaporation, lake inflow, and air temperature) and then use climate change models to predict how the hydroclimate will change in the future. Figure 5-2 provides a general overview of the conceptual lake water quality model.

Hydroclimatic conditions were defined as hydrologic and weather conditions averaged over a period of three months or longer. The periods of interest used in this study included water year (October through September), dry season (October through March), autumn (October through December), winter (January through March), dry season (April through September), spring (April through June), and summer (July through September). Available water quality parameters were paired with hydroclimate conditions for these periods to

Table 5-1. Summary of Lake Cachuma water quality

Water Quality Parameter	Units	Minimum Value*	Maximum Value*
Water temperature	°C	6.75	26
Odor (TON)	Odor units	2	50
Dissolved oxygen (DO)	mg/L	0.6	14
pH	Std. units	7.44	8.87
Turbidity	ntu	0.2	37
Apparent color	Color units	3	183
Iron	mg/L	<0.01	1.2
Manganese	mg/L	<0.01	0.12
Electrical conductivity	µmho/cm	520	1090
Total dissolved solids (TDS)	mg/L	360	978

*Monthly samples, with data gaps, for 1978 through 2002.

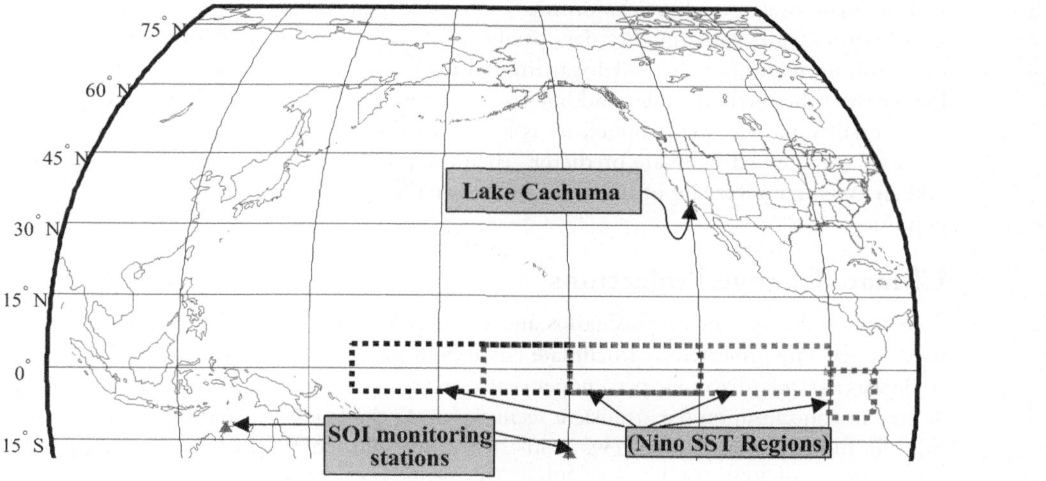

Figure 5-1. Location of ENSO SOI monitoring stations and Nino SST monitoring regions

develop hydroclimate water-quality correlations. Various global climate projections are applied to Lake Cachuma hydroclimate to assess how various climate change scenarios might impact Lake Cachuma source water quality.

Lake Water Quality Model

Forecast models of lake water quality conditions based on local hydroclimate predictors were developed using discriminant analysis. Discriminant analysis is often used as a classification technique, e.g., dividing a water quality parameter into groups above (high values) and below (low values) the median value. When applied to water quality forecasting models in this assessment, discriminant analysis relies on two main concepts: defining possible *water quality outcome* categories prior to model development, and applying *Bayesian forecasting* methods relative to these categories to fit historical data into *likelihood distributions*. An example

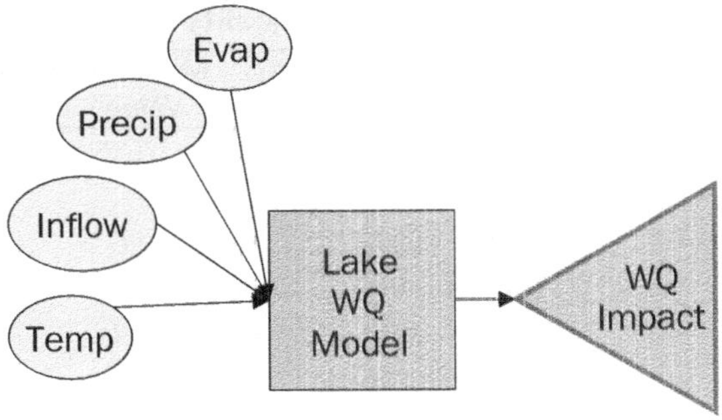

Figure 5-2. Lake water quality model

is predicting the probabilities of lake water quality conditions given area hydroclimate conditions.

Discriminant analysis models were developed for this assessment to describe various hydroclimate–water quality relationships at Lake Cachuma. Because the record of coincident hydroclimate–water quality data produced sparsely populated likelihood distributions, each discriminant analysis model was limited to two water quality outcomes (i.e., lower or higher than the median value) and one hydroclimate predictor. Once developed, these lake water quality models were applied to assign water quality outcome probabilities given the value of a single hydroclimate predictor. Then the predictions of climate change models on lake hydroclimate were input to the lake water quality model to project future water quality conditions.

Climate Change Projections

Two carbon dioxide increase scenarios and three associated climate change projections were used to generate projected hydroclimate changes in the Lake Cachuma region. The first scenario was a standardized one percent per year increase in carbon dioxide concentration starting at present concentration (i.e., transient scenario) using the HadCM2 and PCM model projections for future periods 2010–2039, 2050–2079, and 2080–2099, with reference mid-points of 2025, 2065, and 2090 (Miller et al. 2003). The second scenario was a carbon dioxide doubling relative to present carbon dioxide concentration (i.e., step-change condition) using the CCM3 projection (Snyder et al. 2002). Table 5-2 provides a summary of the climate model projections.

Both the HadCM2 and PCM projections simulated development as carbon dioxide increases until a point of carbon dioxide doubling, around simulation year 70. After that, the simulation continues until the climate equilibrates with the doubled carbon dioxide concentration. CCM3 makes no reference as to when carbon dioxide doubling occurs. The relative differences among the three climate change projections are as follows:

- Southern California warming is projected to be similar in the PCM and CCM3 projections but is projected to occur at a faster rate in the HadCM2 projection.
- HadCM2 suggests that there will be vastly increased precipitation over Southern California.
- PCM suggests that precipitation will decrease over Southern California under the same scenario assumption.
- CCM3 projects no change in precipitation over Southern California.

Table 5-2. Summary of climate model projections

Climate Model	Forecast Period Mid-Point 2025	2065	2090	No Date
		Precipitation Ratios*		
HadCM2	1.4	1.5	1.9	—
PCM	0.9	0.8	0.7	—
CCM3	—	—	—	1.0
		Air Temperature Shifts (°C)†		
HadCM2	1.2	2.2	3.2	—
PCM	0	1.2	2.0	—
CCM3	—	—	—	2.0

*Ratio of mean annual precipitations for future period to baseline period of 1963–1992.

†Mean annual air temperature shift (future period minus baseline period of 1963–1992).

Projected Climate Change Impacts on Water Quality

Table 5-3 summarizes the climate change impacts on lake water quality.

These impacts include

- In the main seasons of precipitation occurrence (i.e., wet season, winter, and spring), the dominant water quality impacts are associated with turbidity and apparent color. These parameters are related to rainfall–runoff events.
- In the main seasons of minimal precipitation occurrence (i.e., dry season and summer), the dominant water quality impacts are related to taste and odor (i.e., TON, water temperature, pH, DO, and DO saturation).
- No correlations during autumn when the lake turnover occurs.

SHORT-TERM IMPACTS OF CLIMATE VARIABILITY ON LAKE WATER

ENSO indicators were assessed for their ability to predict short-term future seasonal to annual water quality changes in Lake Cachuma. Figure 5-3 illustrates the conceptual lead–response ENSO teleconnection model for relating distant climate variability in a previous season to future local hydroclimate.

Two-stage discriminant analysis models of ESNO indices–hydroclimate-lake water quality were developed. Forecasting opportunities were assessed through analysis of historical lead–response teleconnections involving ENSO indices and lake hydroclimate. The ENSO index becomes the predictor of lake hydroclimate variable (stage 1 forecast model), which becomes the predictor of a lake water quality parameter (stage 2 forecast model).

Many teleconnection scenarios were developed for combinations of ENSO indicators, ENSO periods, hydroclimate indicators, and hydroclimate periods. Trends from this modeling include

- The most common ENSO indicators that passed teleconnection scenario criteria were standardized SOI (SOI_{stan}) and Nino4.
- There was greater success in identifying scenarios where the hydroclimate period was the wet season or winter rather than the dry season, spring, summer, or autumn.

Table 5-3. Water quality response to climate change predictions

Water Quality Parameter	Water Quality Season(s)	Model Prediction (increased likelihood)	
		HadCM2 (warm/wet)	PCM (warm/dry)
	Hydroclimate Predictor Related to Air Temperature		
Water temperature	Dry, spring, and summer	Increased likelihood of high water temperature	Increased likelihood of high water temperature
Odor (TON)	Dry and spring	Increased likelihood of high TON	Increased likelihood of high TON
DO and % DO saturation	Summer	Increased likelihood of high DO and % DO saturation	Increased likelihood of high DO and % DO saturation
pH	Dry, spring, and summer	Increased likelihood of high pH	Increased likelihood of high pH
	Hydroclimate Predictor Related to Precipitation		
Turbidity	Wet, winter, and spring	Increased likelihood of high turbidity	Increased likelihood of low turbidity
Apparent color	Winter and spring	Increased likelihood of high color	Increased likelihood of low color

- There was greater success in identifying scenarios where the hydroclimate variable was related to precipitation or runoff rather than air temperature.

Table 5-4 provides a summary of the four two-stage models that were found to have consistent ENSO indicators and hydroclimate variables that forecast water quality parameters with greater than 50 percent reliability of being categorically correct (i.e., above median or below the median parameter value). Each model involved the ENSO indicator SOI_{stan} and the hydroclimate variable DP10 (monthly days with precipitation > 1.0 in.). Water quality parameters that could be forecast included wet season color and manganese, winter turbidity, and spring color.

ASSESSMENT TOOLS

The following four spreadsheet assessment tools were developed in this study:

- ASSESSMENT_TOOL_CLIMATE_CHANGE presents the results of the climate change scenario evaluations.
- ASSESSMENT_TOOL_WATERBUDGET presents the Lake Cachuma water budget.
- ASSESSMENT_TOOL_HC presents selected models for climate change impact assessments where the hydroclimate period leads the water quality period, thus allowing for the possibility of seasonal water quality forecasting.
- ASSESSMENT_TOOL_ENSO presents selected two-stage water quality forecasting models linked to ENSO teleconnections.The assessments tool can assist Goleta Water District and other water districts dependent on Lake Cachuma water supply in understanding the relationship among climate change, climate variability, and source water quality. The tools can be used by the district on an ongoing basis for evaluating source-water quality improvement options, future water treatment alternatives, operational planning, and financial planning. The availability of the assessment tools will

also allow the district to incorporate potential climate change and anticipated climate variability impacts into its facility and operations planning decisions.

CONCLUSIONS

The climate change assessment indicates that warming would increase the probability for increased water temperature, TON, DO, and pH during the dry season. These water quality forecasts suggest that agencies, such as the Goleta Water District, that use Lake Cachuma for water supply will incur increased taste and odor events in the future. Because the climate change models HadCM2 and PCM predict different outcomes for precipitation, changes in turbidity and color are less certain. However, during the main seasons of precipitation (i.e., wet season, winter, and spring), the dominant water quality impacts are associated with turbidity and apparent color, which are related to rainfall–runoff mechanisms.

The assessment of ESNO indicators for predicting short-term climate changes indicate that ENSO is a strong predictor of regional precipitation during the wet season, with opportunities to forecast turbidity, color, and manganese levels in the lake during the wet season.

ACKNOWLEDGMENTS

The paper is based on work performed according to USEPA Cooperative Agreement R-8298601-2 between the USEPA Global Change Research Program and the Goleta Water District, with Kennedy/Jenks Consultants as a subcontractor. The views expressed in this paper are those of the authors and do not necessarily reflect those of USEPA. Catriona Rogers is the USEPA project officer for this project. Levi Brekke was an employee of Kennedy/Jenks Consultants when the analyses were performed.

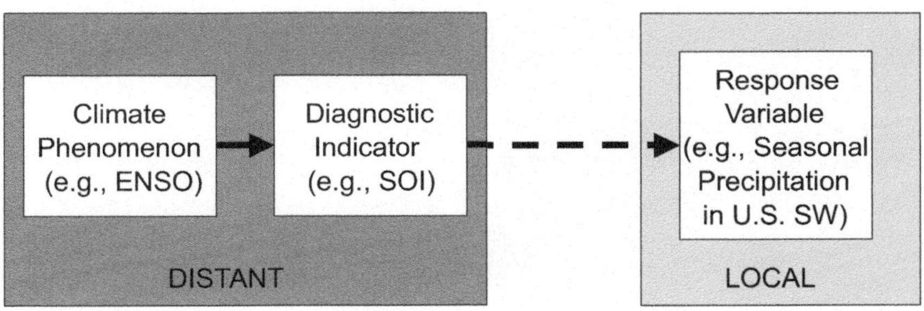

Figure 5-3. ENSO teleconnection model

Table 5-4. Water quality response to enso teleconnection

ENSO Indicator	Hydroclimate Indicator	Water Quality Response
Previous dry season SOI	Annual DP10	Wet season color Wet season manganese Winter turbidity Spring color

REFERENCES

Kiehl, J.T., J.J. Hack, G.B. Bonan, B.A. Boville, D.L. Williamson, and P.J. Rasch. 1998. "The National Center for Atmospheric Research Community Climate Model: CCM3." *Journal of Climate,* 11(6), 1131–1149. Victoria, B.C.

Miller, N.L., K. Bashford, and E. Strem. 2003. "Potential Climate Change Impacts on California Hydrology." Submitted to the *Journal of the American Water Resources Association.* Middleburg, Va.

Piechota, T.C. 1997. "Long-lead Seasonal Streamflow Prediction and the El Nino /Southern Forecasting." Ph.D. Dissertation, University of California–Los Angeles.

Redmond, K.T., and R.W. Koch. 1991. "Surface Climate and Streamflow Variability in the Western United States and Their Relationship to Large-Scale Circulation Indexes." *Water Resources Research,* 27(9), 2381–2399. Washington, D.C: American Geophysical Union.

Snyder, M.A., J.L. Bell, and L.C. Sloan. 2002. "Climate Responses to a Doubling of Atmospheric Carbon Dioxide for a Climatically Vulnerable Region." *Geophysical Research Letters,* 29(11), 9–1:4. Washington, D.C.: American Geophysical Union.

Originally a 2004 ACE proceeding.

– 6 –

The Climate Footprint and the Practical Application at Water Companies in the Netherlands

Peter Wiers, Henk de Boer, Theo Janse, Leo Joosten, Jan Peter van der Hoek, and Dirk van der Woerdt

INTRODUCTION

This article provides a brief introduction to the issue of climate change and the steps three Dutch water companies are taking to calculate and reduce their climate Climate Footprint.

The Netherlands will be affected by the effects of global warming and sea level rise. The Netherlands literally means "The Low Countries." Roughly 60% of the people live below sea level, and two-thirds of the gross national product is earned there!

Climate change will affect the Dutch water sector in particular, through floods, droughts and changing water quality of the water sources. The Dutch water sector is actively involved in adaptation measures. But ironically, these adaptation measures may lead to more demand for energy and building materials, causing further greenhouse gas emissions.

The water sector is an emitter of greenhouse gasses[1]. In addition to adaptation, it can also play a role in mitigation, and help to prevent dangerous climate change.

Assisted by external experts from Climate Partners and Climate Chance, three Dutch water companies have calculated their Climate Footprints. As can be seen in Figure 6-1, these are:

- Waterbedrijf Groningen in the north of the Netherlands,
- Waternet in the city of Amsterdam and its surroundings,
- Brabant Water in the south.

CLIMATE FOOTPRINT METHODOLOGY

The Climate Footprint is a practical tool, developed by Climate Partners/PW Advies, upon initiative from the Cooperative Programme on Water and Climate. The Climate Footprint largely coincides with the energy part of the Ecological Footprint which is used in the WWF Living Planet Report[2] and was developed in co-operation between large corporations, NGOs, and leading scientists in the USA[3]. Figure 6-2 shows how it is our increasing fossil energy

1 Prof. Dr. H. Gijzen, UNESCO-IHE, personal communications: "between 2 and 5% of world-wide GHG emissions."

2 WWF (2004, 2006, 2008).

3 Wackernagel and Rees (1995).

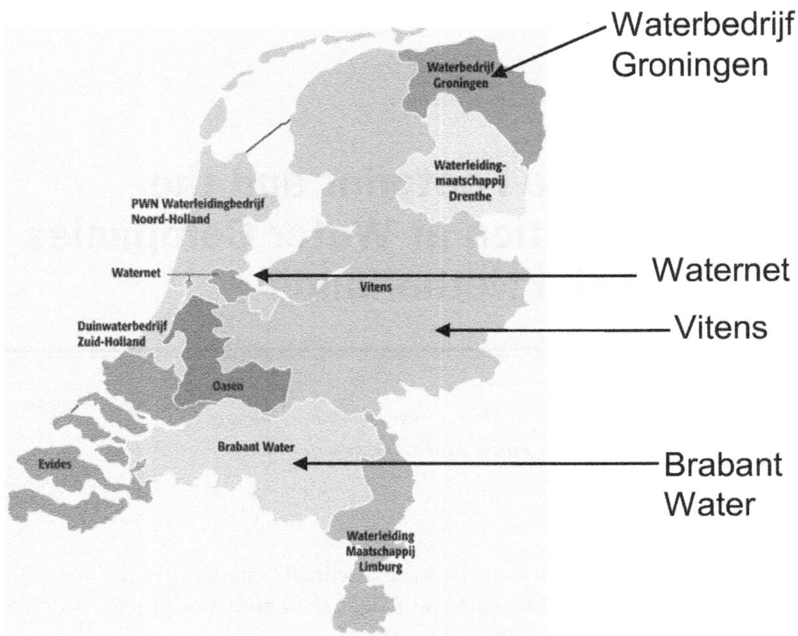

Figure 6-1. The three Dutch water companies mentioned in this article

consumption (the "carbon" footprint) that causes humanity to exceed the bio-capacity of the earth.

The Climate Footprint is supported by an increasing number of organisations and offers a specific, measurable, and transparent frame of reference for reducing greenhouse gas emissions and helping companies to become climate neutral.

ORGANISATIONAL SCOPE

The first step in measuring any organisation's Climate Footprint is to decide on the organisational scope, i.e., which organisational entities to include and which not to include?

Greenhouse gas emissions emanate from all parts of the water cycle (Figure 6-3). Most emissions are associated with heating the water for usage, e.g., for taking a shower or washing. This may well be a factor of four higher[4] than all the other factors combined. Water companies can help to raise awareness and stimulate energy-efficient consumer behaviour, but they do not have a direct influence on it. For this reason, energy use by consumers has not been taken into account.

The following steps are included in calculating the companies' Climate Footprints:

- Drinking water production and distribution,
- Sewerage treatment,
- Wastewater treatment,
- Surface and groundwater management (water system).

4 Nicole Zantkuijl, VEWIN, personal communications; Milieucentraal.

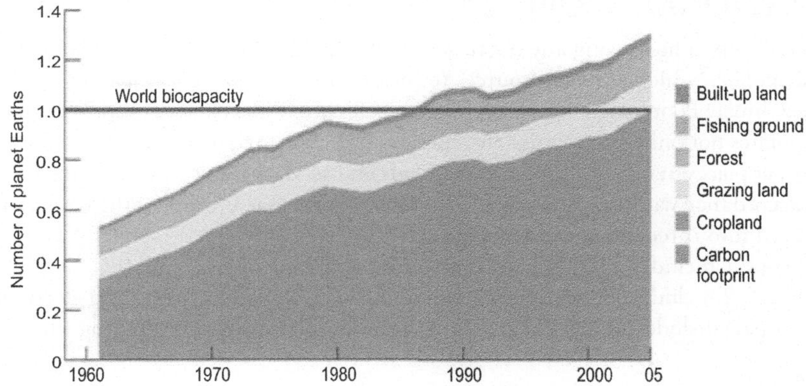

Figure 6-2. Increased fossil energy use causes humanity to exceed the bio-capacity of the Earth

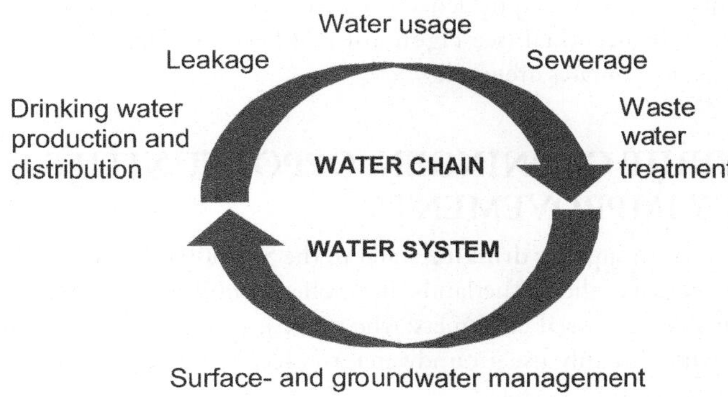

Figure 6-3. The water cycle

	Drinking Water Production	Consumer	Sewerage	Sewage Treatment	Water System
Waterbedrijf Groningen	■				
Brabant Water	■				
Waternet	■		■	■	■

Figure 6-4. Companies studied and the part of the water cycle in which they operate

Water companies in the Netherlands only operate within a small part of the water cycle. The Dutch water sector is highly fragmented, between drinking water companies, municipal sewerage departments, and water boards. As can be seen in Figure 6-4, Waterbedrijf Groningen and Brabant Water are purely drinking water companies. Waternet is an exception, being an integrated water company that spans the whole municipal water cycle. Hence its name.

OPERATIONAL SCOPE

The second step in measuring any organisation's Climate Footprint is to decide on the operational scope, i.e., which emission sources to include and which not to include?

The Climate Footprint has certain characteristics of a life-cycle and a supply chain analysis. It incorporates not only direct emissions, but also indirect emissions emitted at other moments and in other places in order to produce the goods and services needed by the company studied.

In accordance with the WRI/WBCSD Greenhouse Gas Protocol,[5] the emission sources are grouped into three categories (see Figure 6-5):

- Scope 1 includes all direct emissions from company owned assets,
- Scope 2 includes indirect emissions from generating and distributing electricity,
- Scope 3 includes all indirect emissions, from business travel, commuting, chemicals, etc.

CONVERTING DATA TO CO_2 EQUIVALENTS

The third and final step is to convert all the emissions of greenhouse gasses (e.g. CO_2, N_2O, and CH_4) into CO_2 equivalents and add them all up.

Data are converted to CO_2 equivalents using conversion data from public sources likethe WRI, IPCC, and Milieucentraal (see Figure 6-6). Where no public data are available, data from other sources or estimates are used.[6]

WATERBEDRIJF GRONINGEN: OPPORTUNITIES FOR EFFICIENCY IMPROVEMENTS

Waterbedrijf Groningen supplies drinking water to the 575,000 inhabitants of the province of Groningen, in the north of the Netherlands. It supplies 45 million m^3/year, to 265,000 household connections and industrial customers. The water is produced at six drinking water purification plants which mainly use groundwater as a feedstock. About 15% of the feedstock is surface water.

In its mission statement, Waterbedrijf Groningen mentions its commitment to manage the water interests in the region in a sustainable way. One of the aspects of sustainability is to reduce the effect of the activities of the company on climate change, i.e., to reduce the Climate Footprint. Before being able to do so, it is necessary to know what this footprint amounts to.

In 2007, the Climate Footprint of Waterbedrijf Groningen amounted to 9,300 tonnes of CO_2 equivalent. The emissions originate from the following sources:

- 49% are the result of direct emissions of CO_2 and methane, which are extracted during groundwater purification,
- 30% are indirect emissions due to the use of raw materials and infrastructure maintenance,
- 9% are indirect emissions due to transport, including commuting from home to work,
- 12% are from other sources, such as natural gas, heating oil, and indirect energy needed to build the renewable energy infrastructure.

5 See: www.ghgprotocol.org.

6 Such as energy consultants Ecofys and Life Cycle programme's like Simapro en EcoInvent.

Emission Source	Examples
Scope 1— Direct emissions	• Combustion of natural gas, heating oil • Emissions from company cars • Direct emissions of CH_4, N_2O in production process
Scope 2— Indirect emissions from electricity	• Non-renewable • Renewable
Scope 3— Other indirect emissions	• Emissions from third-party trucks • Business travel • Commuting • Chemicals used in production process • Waste and replacement of infrastructure

Figure 6-5. Greenhouse gas protocol operational scopes

Emission Scope	Description	Conversion Factor
Scope 1— Direct emissions	• Combusting of natural gas, heating oil • Emissions from company cars (diesel) • Direct emissions of CH_4 • Direct emissions of N_2O	• 1.8 kg CO_2/m^3 gas • 2.68 kg CO_2 • 23 kg CO_2/kg CH_4 • 300 kg CO_2/kg N_2O
Scope 2— Electricity	• Electricity purchased (non-renewable)	• 600 kg CO_2/MWh
Scope 3— Indirect emissions	• Chemicals used in production process • Replacement of infrastructure	• Variable, but 1.15 kg CO_2/kg base chemicals • Variable, but 1 kg CO_2/kg product as a first estimate

Figure 6-6. Converting data to CO_2 equivalents

When normalised by the amount of drinking water delivered, the Climate Footprint amounts to 200 g of CO_2 per m^3 of water. This amounts to 22 kg of CO_2 per household connection per year. For an average inhabitant, these emissions associated with the consumption of drinking water are ca. 0.1% of the overall per capita emissions of greenhouse gasses.

Part of the reason that the Climate Footprint of Waterbedrijf Groningen is so small lies in the fact that it uses renewable electricity.[7] Switching to renewable electricity has more than halved its Climate Footprint. Another reason is the company's relentless focus on efficiency improvements.

7 With "fossil-based" electricity, it would amount to 490 g of CO_2 eq/m^3.

OPPORTUNITIES WITHIN THE COMPANY'S PROCESSES

Within its own processes, Waterbedrijf Groningen optimises the energy use in its purification plants and its use of chemicals.

Within its purification plants, the company has switched to, and continues to look for, the most efficient rotating equipment that is available. The largest energy consumers are the pumps that pump the drinking water into the distribution system to transport it to the customers. Optimization of this efficiency is an ongoing process. Also, the amount of space heating is reduced to a minimum, as the company operates its purification plants fully automatically. The temperature of the groundwater flowing through the pipes inside the plant provides a constant temperature of 12°C, summer and winter. Only the rooms where personnel are staying are heated at office temperatures.

Waterbedrijf Groningen has reduced and continues to reduce its use of chemicals, as well as the Climate Footprint of the chemicals it uses, such as NaOH and milk of lime. Milk of lime, for instance, was previously pre-mixed in Germany, but this is now done locally. By purchasing dry CaO and mixing it locally, the transborder transport of many truck-loads of essentially water is now avoided.

OPPORTUNITIES OUTSIDE THE COMPANY'S PROCESSES

In addition to reducing its own Climate Footprint, Waterbedrijf Groningen is also working on opportunities outside its own processes.

It is involved in a pilot where algae are used to purify water. The nitrates and phosphates in the water, as well as the CO_2 from a local power plant, are used as nutrients. And the waste heat from the power plant is used to provide a pleasant temperature whereby the algae flourish. If everything works, the algae turn all these waste products into clean water, fresh oxygen, valuable minerals, and renewable energy.

Waterbedrijf Groningen is also involved in a heat and cold storage project, whereby energy from groundwater is used to heat homes in winter and to cool these homes during hot summer days (Figure 6-7).

BRABANT WATER: CLIMATE NEUTRAL IN 2013?

Brabant water supplies drinking water and industrial water to over 2.4 million individuals and companies, mainly in the province of North Brabant in the southern part of the Netherlands.

Brabant water collects, purifies, and distributes water of excellent quality at the lowest possible cost and with high delivery performance levels. In these activities, Brabant Water tries to take into account various ecological and environmental aspects as much as possible.

In its mission statement, Brabant Water mentions sustainability as an important issue. Until recently, however, Brabant Water did not have a big ambition toward climate neutrality. But the company is standing at a turning point. In 2008, Brabant water calculated its Climate Footprint for the first time, based on 2007 data.

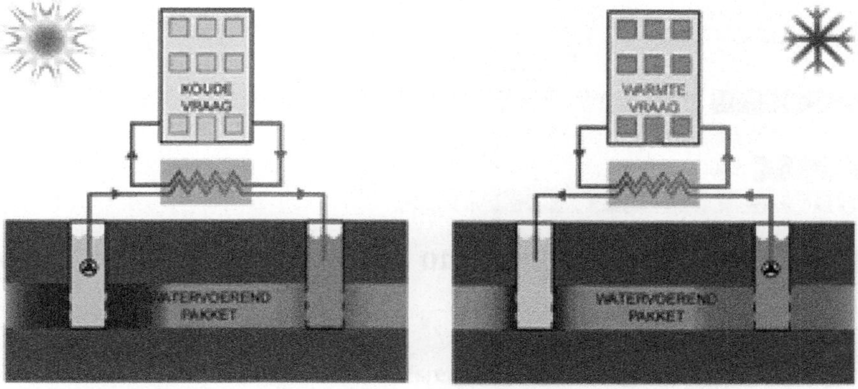

Figure 6-7. Using groundwater for cooling in summer and heating in winter

Current Climate Footprint

In 2007, the Climate Footprint of Brabant Water amounted to ca. 78,000 tonnes of CO_2 equivalent. This roughly equals the emissions of all the households in a village of 18,000 people.[8] It amounted to 450 grams of CO_2 per m^3 of drinking water.[9]

As can be seen in Figure 6-8, the bulk majority of emissions comes from the use of electricity. More than 60% of the Climate Footprint is electricity-related, as Brabant Water still used nonrenewable electricity up to then. Another big factor are the process-related emissions, mainly methane. At Brabant Water, these emissions account for 9,000 tonnes of CO_2 equivalent.

Project Objective

As part of the Climate Footprint study, Brabant Water has also looked into the possibilities to:

- reduce the company's Climate Footprint;
- contribute to reducing the Climate Footprint of the province of North Brabant;
- become climate neutral in 2013.

After determining its Climate Footprint, Brabant Water has stated its ambition to become climate neutral in five years, if it is financially justifiable and technically possible. A very important proviso is that customer comfort may not suffer from the measures taken.

Brabant Water does not only want to reduce and offset its own greenhouse gas emissions, it also wants to make a positive contribution to reducing the Climate Footprint of the province of North Brabant. For instance, by stimulating hot and cold storage projects for heating and cooling buildings, or by producing bio-fuels in cooperation with other stakeholders. A very important issue for Brabant Water is raising awareness amongst its employees so that they get involved and make their own contributions too; and not only at the office, but also at home.

8 Assuming 2.2 people and 16,000 ca km, 3,300 kWh and 1,800 m^3 gas per household.

9 As from January 2009, ca. 140 gram per m^3, now that Brabant Water has switched to renewable electricity.

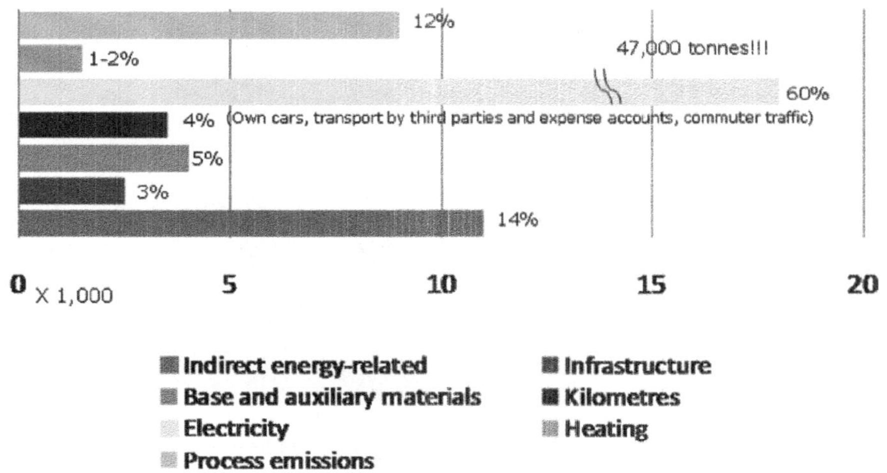

Figure 6-8. Brabant Water 2007 Climate Footprint, by source of emission

Planned Approach

To reach the objectives, two tracks are planned:

- immediate (technical) actions;
- involving the employees.

Immediate (Technical) Actions

Immediate actions consist of:

- purchasing renewable electricity,
- screening windows,
- optimising the pressure, energy savings within the pumping system,
- methane collection.

As of 1 January 2009, Brabant Water has "greened" its electricity supply, by buying renewable electricity certificates. Later on it will decide whether and how much renewable electricity to produce itself.

In all new construction and reconstruction projects, solar and wind energy will be taken into consideration. In recent plans for the reconstruction of the pumping station in Veghel, it was decided to use solar collectors to generate the energy needed for the air-dryers.

Another thing that's being examined is screening the windows on the sun-facing side of the building. It will reduce cooling energy by approximately 14%.

Pressure optimisation is also under study. It won't be pressure minimisation, because customer satisfaction must stay at a high level. The idea is to improve energy efficiency by reducing the pressure at the pumping station and raising the water pressure locally. A pilot study will be done in the first half of 2009.

Recently, Brabant Water started a study into energy saving opportunities within the pumping system and a study into the collection of methane.

Involving the Employees

The second track concerns involving the employees. They possess the power to change things. They are the experts! Brabant Water is involving them to conjure up and implement the ideas they care about.

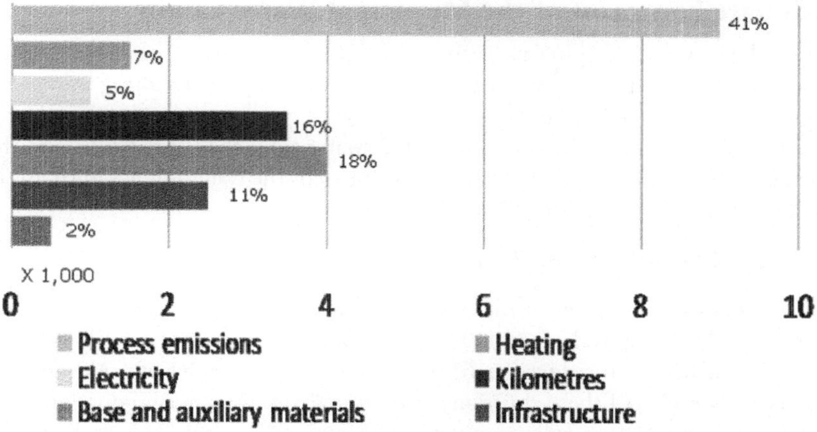

Figure 6-9. 2007 Climate Footprint (anticipating on switch to renewable electricity)

The practical solutions brought forward by employees include the subject of mobility, video-conferencing, etc., and some of the ideas are fun, too, such as the introduction of a woolly jumper day in the office.

The important thing is that these ideas are ideas suggested by the employees themselves, through a bottom-up process. By empowering its employees, Brabant Water is mobilising the knowledge and energy of all the employees and not only a few members of staff. That way, climate neutrality is internalised rather than imposed by rigid rules.

To get the employees enthusiastic, they must be involved, informed, and receive feedback, so that they have the feeling they can make a difference. Brabant Water is communicating its 2007 Climate Footprint as if it were already using renewable electricity. That way, mobility counts for 16% instead of 4% of the Climate Footprint, and reducing the number of kilometres driven starts to make sense (see Figure 6-9).

WATERNET: THE CLIMATE FOOTPRINT OF THE TOTAL WATER CYCLE

Waternet, the Amsterdam water company, is an operational organisation governed by two public authorities: the city of Amsterdam and the Waterboard Amstel, Gooi and Vecht.

For the city of Amsterdam, Waternet is responsible for:

- Sewerage,
- Groundwater,
- Drinking water.

For the Waterboard Amstel, Gooi en Vecht, Waternet is responsible for:

- Surface water management,
- Wastewater treatment.

Waternet was the first water company in the Netherlands to calculate a Climate Footprint, for the Amsterdam municipal water chain. Using data from 2007, Waternet has recently recalculated this Climate Footprint, and also calculated the Climate Footprint of the water system within its area. Together, these two Climate Footprints make up the Climate Footprint of the total water cycle.

The Climate Footprint of the Municipal Water Chain

The Amsterdam municipal water chain is an example of a western urbanised water chain, delivering high quality drinking water to the home, collecting the resulting wastewater, and turning it into an acceptable effluent.

As a raw water source, Waternet uses surface water from the River Rhine and from the Bethune polder. Figure 6-10 shows the main process steps the water follows until it is eventually discharged as a clean effluent.

The Climate Footprint Over Time

In 1990, the Climate Footprint of the Amsterdam municipal water chain amounted to ca. 90,000 tonnes of CO_2 equivalent, or 110 kg per inhabitant.

In 2004 this had been reduced to 65,000 tonnes of CO_2 equivalent. As can be seen in Figure 6-11, about 40% of these emissions were from drinking water production and distribution, 15% from wastewater collection (sewerage), and 45% from wastewater treatment.[10]

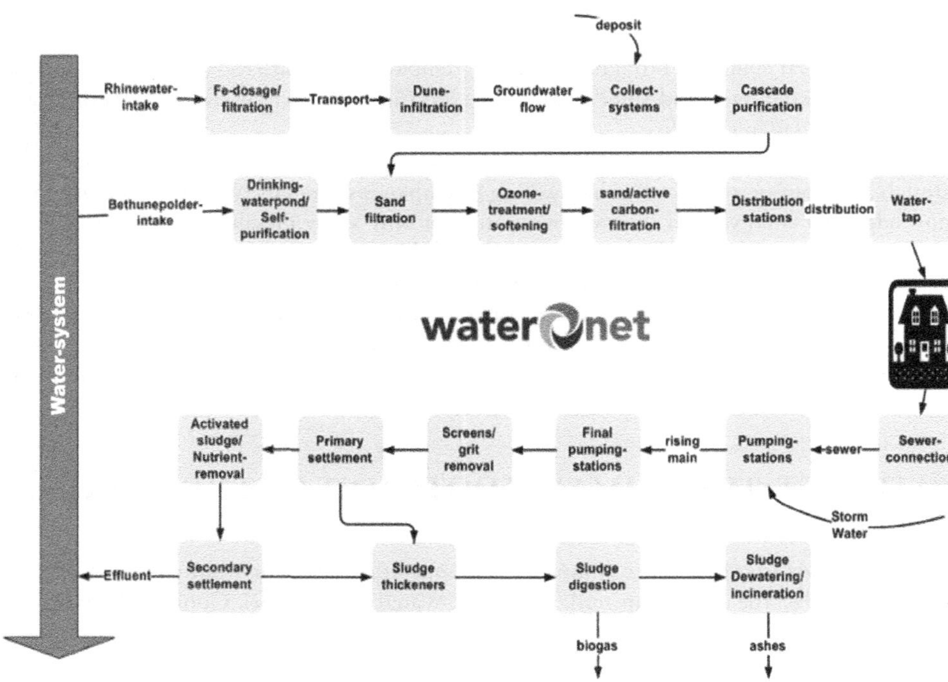

Figure 6-10. The Amsterdam water chain and its main process steps

10 Wastewater excluding short-cyclic CO_2 emissions, because these are renewable and not of fossil origin.

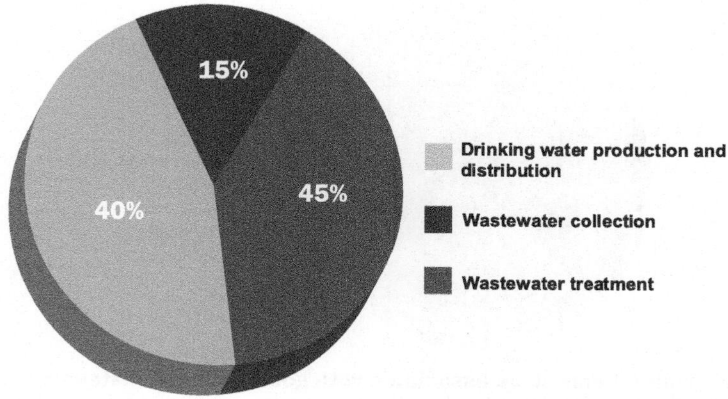

Figure 6-11. Distribution of the Climate Footprint over various parts of the water chain

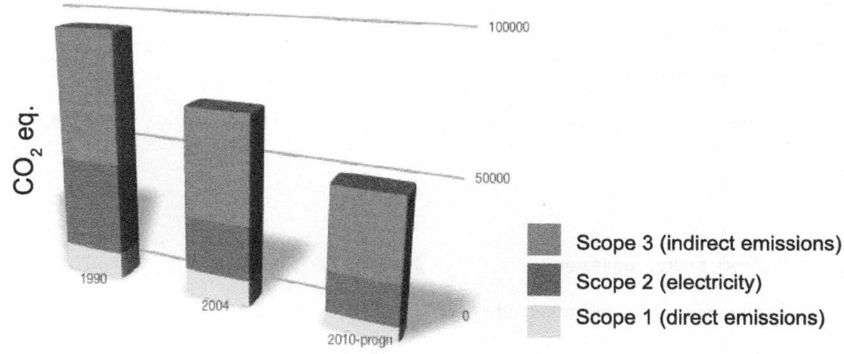

Figure 6-12. Waternet: 2004 results and 2010 targets

At a time when politicians still talked about emission reductions of 6% in 2012 (NL), 20% by 2020 (EU), and 40% in 2025 (city of Amsterdam), Waternet already set itself the target to reduce the Climate Footprint of its water chain even further and aim for a 50% reduction compared to 1990; and to achieve this target no later than in 2010 (see Figure 6-12).

A surprising finding of the Climate Footprint study was the large amount of indirect emissions from goods and raw materials purchased. Included in these are chemicals used in the production process and materials used for construction. As can be seen in Figure 6-13, chemicals used during drinking water treatment, such as NaOH, FeCl, and HCl, constitute the largest part. The remainder is from construction materials, such as iron, steel, copper, PVC, and concrete. For these, estimates had to be made regarding average annual volumes, often based on the lifetime of the particular infrastructure.

To translate the use of materials into tonnes of CO_2, the expertise of Climate Partners has been invoked. Wherever possible, standard conversion factors from public sources and LCA programmes[11] are used. Where not available, one has to resort to data and estimate from other sources.

11 e.g., Simapro Eco-Indicator 99.

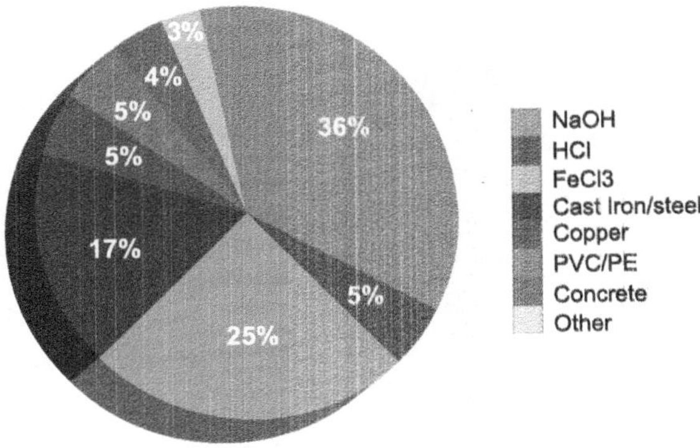

Figure 6-13. Indirect emissions associated with goods and raw materials purchased

	Tonne CO_2 equivalents		
	Total 1990	Total 2004	Total 2007
Scope 1			
Gas, heating	5.296	5.833	1.850
Transportation	631	707	679
Process-related emissions	4.031	5.920	16.249
Scope 2			
Electricity purchased	41.964	16.673	2.386
Scope 3			
Indirect energy-emission	10.040	4.331	970
Commuting and public transport	1.263	1.534	1.465
Goods and raw materials	26.060	29.911	29.438
TOTAL	89.285	64.909	53.038

Figure 6-14. The Climate Footprint of the Amsterdam Municipal Water Chain

Results in 2007

In 2007, the Climate Footprint of the Amsterdam municipal water chain was further reduced to 53,000 tonnes of CO_2.

As can be seen in Figure 6-14, emissions from gas and heating have decreased to 1,850 tonnes. Since the opening of the new wastewater treatment plant adjacent to the Amsterdam waste company, the waste heat from biogas and sludge combustion has been efficiently used to heat buildings and to speed up the sludge digestion process, in which hitherto precious natural gas was used.

Direct process related emissions of N_2O and CH_4 have not physically increased, but have been recalculated according to a new official formula.[12] There are some doubts about this

12 VROM-protocol 7138 wastewater 6B, "CH_4 and N_2O from wastewater" (15-02-2007).

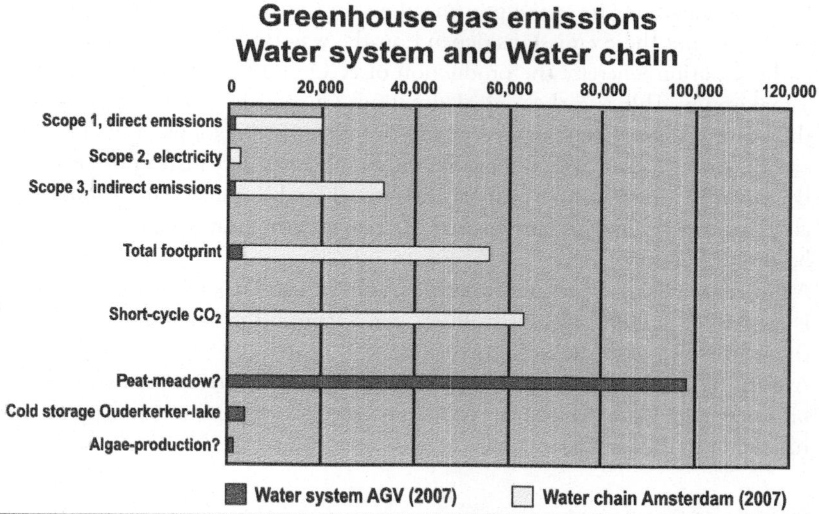

Figure 6-15. Waternet: Greenhouse gas emissions from the total water cycle

formula, however, and the newly-calculated 16,000 tonnes of CO_2 equivalent are probably on the high side.

Since the move to 100% renewable electricity, emissions from electricity have further decreased to ca. 2,400 tonnes of CO_2.

Waternet aims to reduce the emissions from goods and raw materials purchased, too. This will be achieved in two different ways:

- improved production processes (using fewer chemicals),
- sustainable procurement (buying from more environmentally-friendly suppliers).

The Climate Footprint of the Total Water Cycle

Based on 2007 data, Waternet has also calculated the Climate Footprint of the total water cycle. This amounted to 56,000 tonnes of CO_2 equivalent.

Of the two components that make up the water cycle, the water chain is by far the largest. The water system is relatively small, as can be seen in Figure 6-15:

- Direct emissions are low and mainly the result of transportation,
- Electricity use is much lower, and occurs mainly at the pumping stations.[13]
- There are also few indirect emissions. The largest item so far is the use of chemicals at the dephosphorisation plant.

Figure 6-15 also puts into perspective some factors in which Waternet is involved, but that are not included in the Climate Footprint:

- Short-cyclic CO_2 emissions from the wastewater treatment process. These emissions are not of fossil origin, and do not add extra CO_2 to the atmosphere. They are therefore excluded;

13 The amount of electricity used in the water system is relatively small and since renewable energy is used, its emissions are almost negligible.

- Potential emissions from the peat meadows in the area where Waternet manages the water system. These emissions depend on the way the system is managed. Indicated is the situation whereby the production of N_2O and CH_4 from these meadows is at a maximum. This is a theoretical situation. Current emissions are certainly less. If the water system were managed purely to minimise emissions, it could even act as a carbon-sink (in that case, the bar should point to the left instead of far to the right). But turning the area into a peat-forming marshland is unlikely to be looked at favourably by property owners and farmers. This makes the peat meadows a factor which is hard to influence;
- A project in which cold water from the Ouderkerkerplas, rather than air-conditioners, is used to cool buildings. This project saves energy (the bar should actually point to the left instead of to the right);
- A potential pilot to grow algae by using effluent as a nutrient, which would produce bio-fuels and therefore reduce emissions (the bar should actually point to the left instead of to the right).

CONCLUSIONS AND RECOMMENDATIONS

Conclusions

The Climate Footprint provides a good overview of all the greenhouse gas emissions within a company's sphere of influence. And it is a useful tool for benchmarking companies against themselves, over time.

In 2007, the combined Climate Footprints of the three companies studied amounted to 143,000 tonnes of CO_2 equivalent. This is 20% less than in 1990.

As the examples of these companies show, even more impressive improvements are possible. By 2010, these companies will have reduced their Climate Footprints by more than 50% compared to 1990. The EU and others have urged to adopt a 50% emission reduction target by 2050, in order to stabilise greenhouse gas levels at 400 ppm and keep average temperature rises limited to 2°C.[14] The achievements of these three companies suggest that, at least for Dutch water companies, this emission reduction target is quite achievable.

The Climate Footprint also makes clear what the absolute and relative contributions of individual greenhouse gas emissions are. This shows that each company is very different, even within the drinking water sector of a small country like the Netherlands. Waterbedrijf Groningen and Brabant Water, for instance, use mainly groundwater as raw water. Many of their sources contain methane and CO_2, which tends to get into the atmosphere during drinking water treatment. Waternet, on the other hand, uses mainly surface water derived from the river Rhine, and needs to use more chemicals and energy during treatment.

Recommendations

Several recommendations can be made, using the following questions as a guideline:

- How to use (and not use) the Climate Footprint?
- How to organize the internal process within the company?
- How to organize the relations with the external world?

14 M.G.J. Den Elzen and M. Meinshausen (2005).

How to Use the Climate Footprint?

Risk management

By virtue of the scope of the Greenhouse Gas Protocol, the Climate Footprint can be used to identify those emissions that carry (potential) legal responsibilities and risks, and may therefore need extra attention.

Efficiency improvements

Perhaps most of all, the Climate Footprint is a valuable tool for prioritising research and targeting efficiency improvement measures. Knowing where emissions are means knowing where to save on energy and chemicals, leading to further cost reductions.

Benchmarking

Can the Climate Footprint also be used as an instrument for benchmarking?

As we have seen, the Climate Footprint is a very useful tool for benchmarking a company against itself, over time. It can be used to communicate where the company stands now, how much progress has already been made, and where the targets are that need to be achieved.

At this stage, we recommend to use the Climate Footprint for benchmarking companies against themselves over time, but not against each other. A company with a small Climate Footprint, even if normalized, is not necessarily performing better than a company with a large Climate Footprint; for instance, if the other company has an inherently higher methane content in its groundwater, or if the other company has to rely on the treatment of surface water. To make a more relevant comparison, one should compare individual items such as emissions from electricity. And although technically possible (see Figure 6-16), we think this approach is fraught with difficulties.

Becoming climate neutral

The Climate Footprint is an essential instrument for companies who want to become climate neutral. We recommend companies to follow the so-called "Trias energetica" (see Figure 6-17). First, try to save as much energy as possible. This actually saves money, which can subsequently be reinvested in further improvements. Secondly, use renewable energy and climate-friendly goods and services. Thirdly, use clean fossil energy[15] and compensate remaining emissions. Compensation can be very motivating if it is done through projects that fit the company's overall strategy and objectives, and by involving the employees. Many Dutch water companies already use their expertise to support water and sanitation projects in developing countries. If they could help their international partners to save energy too, that might perhaps be an attractive compensation option.

How to Organize the Internal Process Within the Company?

A Climate Footprint can be used to show where a company comes from, where it stands, and where it wants to go in terms of greenhouse gas emissions.

But the key question is: which steps are we going to take to get there?

15 Assuming there will some day be clean fossil energy, e.g., from projects where the carbon is stored.

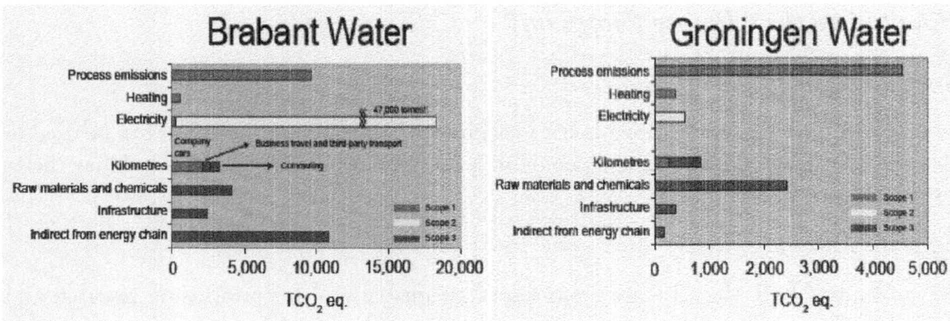

Figure 6-16. Theoretical possibility to benchmark Climate Footprints between companies

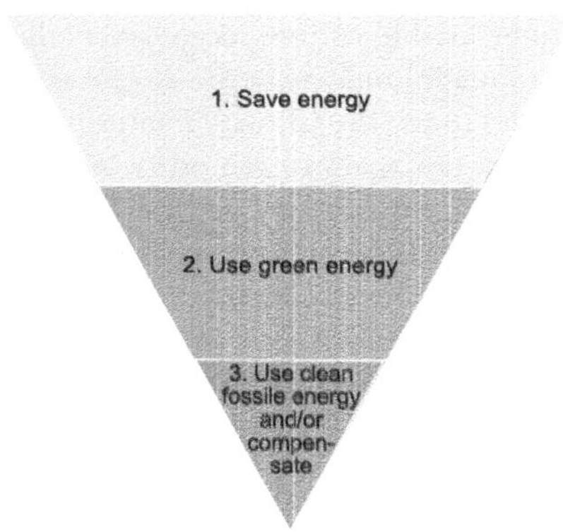

Figure 6-17. Trias Energetica

In the water-sector, we often think in technical terms: how can we solve the problem as quickly as possible against the lowest possible costs? But embarking on an ongoing mitigation programme is a continuous learning process. In tackling climate change, it's not only about technical measures, rules and procedures. It's also about people, culture, and behaviour (see Figure 6-18).

For real changes, it is essential to involve the employees: are they concerned about climate change? What does it mean to them? How would they want to translate that into action, both in their job and in their personal life? A company which takes climate change seriously also invests in a dialogue with its people before embarking on a course of emission reduction or a policy of climate neutrality. Both Waternet and Brabant Water have shown they want to invest in involving their employees. Being able to make a difference in reducing greenhouse gas emissions at work proves very stimulating for employees. It can be an important weapon in the war for talent.

How to Organize the Relations with the External World?

When dealing with climate change, one should not forget the external world (see Figure 6-19).

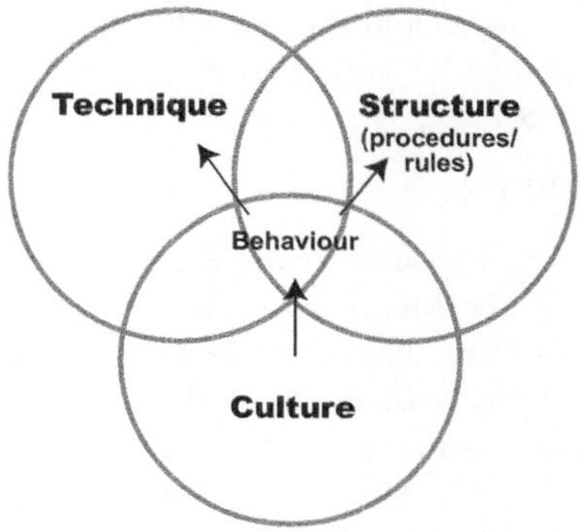

Figure 6-18. Real change is also about people, culture, and behaviour[16]

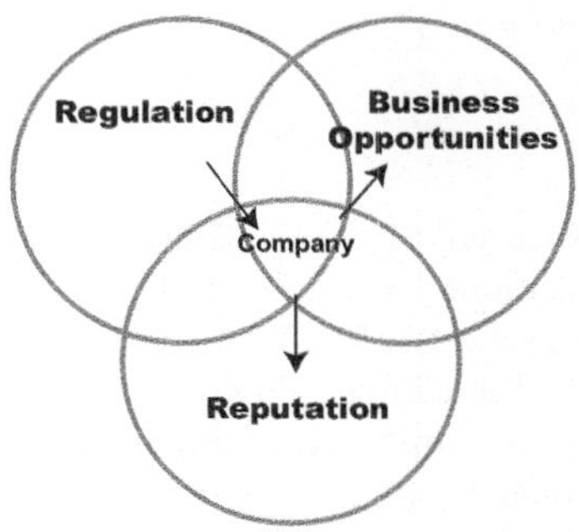

Figure 6-19. Relations with the outside world

Government regulations

There are governments out there who seek to regulate. Better be prepared, so that if regulation comes it is as much as possible on your conditions, and not only theirs.

Reputation

Measuring your Climate Footprint and taking action to reduce it helps to improve your public image. Climate change is a "hot" topic in public opinion, and mitigation efforts can strengthen a company's reputation in the outside world and with your customers. Ongoing mitigation efforts and actions to become climate neutral can become an important communication theme.

Business opportunities

Last but not least, climate change brings in new business opportunities. For water companies, this is especially so in the field of interaction between water and energy, as all three companies are keen to explore.

Not surprisingly, communication is a key instrument in dealing with both the internal process and relations with the external world. In 20 years time, the world will look a lot

16 Source: Global Action Plan.

different from now in terms of energy and water. Keen companies don't wait, but proactively manage the issue of climate change.

ACKNOWLEDGMENTS

As the lead author to this article, I wish to thank Mr. Henk van Schaik of the Co-operative Programme on Climate and Water, Ms. Ase Johannessen of IWA, and Ms. Carol Howe of UNESCO-IHE/ Switch. The CPWC has taken the initiative to develop and apply the Climate Footprint methodology in the Netherlands and has actively stimulated its dissemination at international conferences, such as the Aquatech in Amsterdam, the AWWA Conference in Atlanta, and the World Water Week in Stockholm.[17] I am most grateful for their inspirational comments and efforts to put sustainable development on the water community's agenda.

Furthermore, I wish to thank my co-authors from Waterbedrijf Groningen, Brabant Water, Waternet, and Climate Chance, for their willingness to share their practical experiences. I hope you find their stories useful and stimulating, and that they will inspire you to measure and reduce your Climate Footprint, too.

REFERENCES

Den Elzen, M.G.J., and Meinshausen, M. 2005. Meeting the EU 2°C climate target: global and regional emission implications Janse. Th., and Wiers, P., 2006, Broeikasgasemissie vanuit de Amsterdamse Waterketen, in: H_2O #18, pp. 87–90.

Janse, Th., and Wiers, P. 2006. The Climate Footprint: A practical tool to address climate change, in Aquatech Conference. Amsterdam September 2006.

Wiers, P. 2008. Mitigating Climate Change in the Netherlands. Reducing the Climate Footprint of Water. AWWA/IWA Conference. Atlanta, Ga. June 10, 2008.

Wiers, P. 2008. The Climate Footprint: An effective tool to promote energy efficiency and development. IWA/UN Habitat, World Water Week. Stockholm. 21 August 2008.

Wackernagel, M., and Rees, W. 1995. "Our Ecological Footprint: Reducing Human Impact on the Earth."

WWF. 2008. Living Planet report.

17 For more details, see the paragraph on 'References'

— 7 —

Climate Footprint and Mitigation Measures in the Dutch Water Sector

Jos Frijns, Mirabella Mulder, and Jelle Roorda

INTRODUCTION

The water sector faces a challenge to cope with the consequences of climate change. Climate is on top of the agenda and the water sector would like to have an idea of its contribution to the greenhouse effect. The water sector can set an example by adopting climate mitigation measures.

For the production of drinking water, and treatment of sewerage and wastewater, energy is required and greenhouse gases are released. In this paper, the climate footprint of the Dutch water sector is presented. Measures are proposed to adjust the water cycle in order to minimise the negative impact on the greenhouse effect.

The results presented have been brought about by a KWR–Grontmij project in collaboration with the drinking water companies Brabant Water and Evides, the water boards Brabantse Delta and Delfland, and the municipality of Delft (see Frijns et al. 2008). The project was commissioned by the Dutch Ministry of Environment with support from STOWA.

In assessing the climate footprint, it is important to define the systems boundary. This defines the scope of the water sector: drinking water, sewerage and wastewater treatment, including stormwater and sludge treatment. The assessment was limited to the domestic water cycle. The industrial water cycle with its own water sources and treatment facilities was not included.

The contribution to the greenhouse effect should also be clearly defined. Firstly, CO_2 emissions from energy consumption contribute to the carbon footprint. The production of energy from biogas production in wastewater treatment systems needs to be accounted for. Secondly, the direct emissions of greenhouse gases should be incorporated. Methane gases (CH_4) are released in degassing groundwater and in wastewater and sludge treatment. Furthermore, the strong greenhouse gas nitrous oxide (N_2O) is released in wastewater treatment and effluent discharges. Carbon dioxide (CO_2) process emissions, however, are excluded as these are short-cycle carbons. Finally, indirect emissions have to be accounted for. Indirect emissions are a result of the embedded energy from chemicals used and the energy use related to the organisation (i.e., office and transport) of the water sector. The contribution of material use to the carbon footprint was disregarded, as over the total lifespan of the infrastructure the amount is very limited (see also Strutt et al. 2008).

The climate or carbon footprint is expressed in CO_2-equivalents or the global warming potential (GWP). In accordance with the Kyoto-protocol (IPCC 1996, MNP, 2007). a GWP of 21 is used for methane (CH_4) and 310 for nitrous oxide (N_2O). The CO_2-equivalents used for energy are 0.59 kg CO_2/kWh electricity and 1.8 kg CO_2/m^3 natural gas.

CLIMATE FOOTPRINT

In the Netherlands, 1,210 million m^3/year drinking water is produced from ground- and surface water, which after domestic usage is discharged to the sewer system. Together with storm water, 1,854 million m^3/year is transported to the municipal wastewater treatment plants (wwtp). Here, 26.8 million pollution equivalents of wastewater is treated and released to the surface water (CBS 2006).

The climate footprint of the Dutch water sector was projected to be 1.67 million tonnes CO_2-eq. per year (see Table 7-1). Figure 7-1 shows that the largest contribution to the GWP is from wastewater treatment.

Figure 7-1 also shows the distribution of energy use, direct, and indirect emissions to the GWP. This is further discussed below.

ENERGY CONSUMPTION

Obviously, the required energy use has a large share in the greenhouse effect of the water cycle: 56% which stands for 934,840 tonnes CO_2-eq. per year. The energy consumption figures were largely based on the amount reported by water companies in benchmark reports, environmental reports and statistic databases (CBS 2006, UvW 2007, Vewin 2007). The average energy use for the production and distribution of drinking water is 0.47 kWh/m^3 and for wastewater treatment, 26.6 kWh/population equivalent removed. The energy production (biogas from sludge digestion) is taken into account. In Table 7-2, the specific energy requirements of technologies applied are listed.

Table 7-1. Total yearly global warming potential of the Dutch water sector

Drinking water	436,875 tonnes CO_2-eq.
Sewerage	123,620 tonnes CO_2-eq.
Wastewater	1,114,310 tonnes CO_2-eq.
Total	1,674,805 tonnes CO_2-eq.

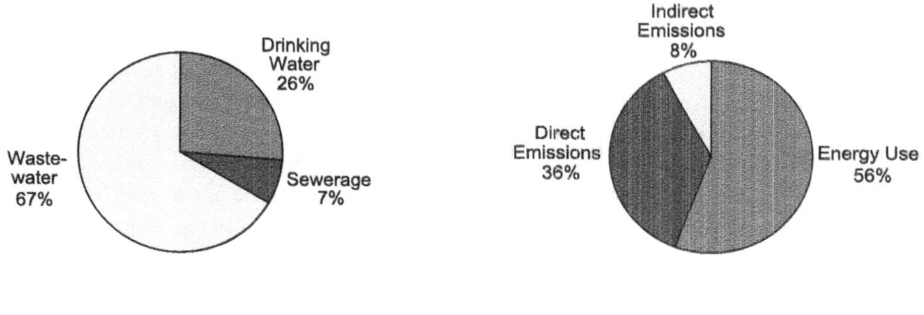

Figure 7-1. GWP distribution

Table 7-2. Specific energy requirement of key process technologies

Process	Specific Energy Use (on Average)
Drinking Water	
Production and distribution total	0.47 kWh/m^3
Intake	0.055 kWh/m^3
Tower aeration	0.08 kWh/m^3
Softening	0.020 kWh/m^3
UV disinfection	0.062 kWh/m^3
Membrane filtration	0.5–1.5 kWh/m^3
Distribution	0.11 kWh/m^3
Sewerage	
Transport	0.11 kWh/m^3 *
Wastewater Treatment	
Treatment total	26.6 kWh/i.e.
Aeration	14.8 kWh/i.e.
• Bubble aeration	4.0–5.0 kg O_2/kWh
• Point aeration	1.5–2.5 kg O_2/kWh
Sludge dewatering	0.12 kWh/kg dry solids
• Centrifuge	0.04–0.065 kWh/kg dry solids
• Belt press	0.02–0.035 kWh/kg dry solids

Sources: Loeffen et al. (2005); de Moel et al. (2006).
* Estimate.

DIRECT EMISSIONS

CO_2 is released from biological treatment processes, but these do not contribute to the greenhouse effect as these are short-cycle carbon emissions. However, emissions of methane and nitrous oxide add to the climate footprint. Together, these direct emissions have a surprisingly large share (36%) of the total GWP of the water sector: 610,345 tonnes CO_2-eq. per year (see Table 7-3). This is mainly caused by methane emissions from degassing groundwater and process emissions of methane from wastewater and sludge treatment, and particularly from nitrous oxide emissions from wastewater treatment and effluent discharges.

During wastewater treatment, methane and nitrous oxide emissions occur in biological reactors, sludge digestion, sludge storage, and end-processing of sludge like incineration, drying, and composting. Scientific research on nitrous oxide emissions during wastewater treatment focuses mainly on the microbiological mechanisms which trigger these emissions. Process conditions of advanced treatment methods for nitrogen removal appear to have a large impact on nitrous oxide emission. The following factors seem to increase nitrous oxide emissions from nitrification and denitrification processes in wastewater treatment plants (see also Kampschreur et al. 2007):

Nitrification:

- low oxygen concentrations caused by, for example, insufficient aeration and/or gradients in sludge flocks
- high nitrite (NO_2) levels caused by suboptimal process conditions such as insufficient aeration, low sludge age, toxic compounds, low temperature, high ammonium concentrations (NH_4), and so on.

Denitrification:

- high oxygen levels caused by, for example, recycling aerated streams from nitrification compartments
- high nitrite (NO_2) levels caused by, for example, a shortage of biological degradable organic matter
- low ratio of biochemical oxygen demand (BOD) and nitrogen.

In the IPCC-directives, some guidelines are given to calculate the emissions (IPCC 2006):

CH_4 (wwtp) = CH_4 (waterline) + CH_4 (sludge digestion) = 0.0085 * CZV (influent)

N_2O (wwtp) = 3.2 g N/population equivalent/year

N_2O (effluent wwtp) = 0.005 * Ntot (effluent) * 44/28

There is still debate on the sewerage and wastewater treatment amounts of CH_4 emissions and the amounts and mechanism of N_2O emissions, and there is also still ambiguity in the standard to be applied. Only few data of measured methane and nitrous oxide emissions from full-scale wastewater treatment plants are available. In 2009, a measuring campaign

Table 7-3. Overview of yearly direct emissions

Emissions	Amount per Year (kg)	GWP (tonnes CO_2-eq.)
Drinking Water		
CO_2 (degassing, long cycle)	10,000,000	10,000
CH_4 (degassing)	1,770,000	37,170
N_2O (ozonation)	2,400	745
Sludge treatment	n.a. (reuse)	0
Sewerage		
CH_4: incorporated in CH_4 water		
N_2O sewer overflow	10,700	3,320
Wastewater Treatment		
CH_4 water section	6,565,020	137,865
CH_4 sludge section	1,406,790	29,545
N_2O water section	429,210	133,055
N_2O from effluent	149,730	46,415
N_2O non-flared biogas	201,745	62,740
CH_4 flared biogas	990,950	20,810
N_2O flared biogas	19,820	6,145
CH_4 sludge treatment	428,980	9,010
N_2O sludge treatment	366,210	113,525
Subtotal		***610,345***

is to be started by STOWA on methane and nitrous oxide emissions. While awaiting these results, in this project the methane emission of wastewater treatment was based on the IPCC guidelines.

For nitrous oxide, however, the IPCC guidelines were not applied. The average value of N_2O release from wastewater treatment plants found in literature, in which measurements were carried out on full-scale wastewater treatment plants, is 0.5% of the Nkj-load of the influent (UKWIR 2008). This value is much higher than the IPCC-guideline of 3.2 g N/p.e./year which for the Dutch situation equals 0.01%. Thus, for this project the following calculation method was used:

$$N_2O \text{ (wwtp)} = 0.005 * \text{Nkj (influent)}$$

As a result, the nitrous oxide emissions from the water section of wastewater treatment plants is 133,055 tonnes CO_2-eq. per year.

INDIRECT EMISSIONS

The energy required to produce the chemicals that are used in the water sector also adds to the GWP of water. On average, this contribution is only 3% of the total GWP. The chemical use varies substantially between water companies; certain surface water abstraction sites use a relatively large amount of chemicals, e.g., sodium hydroxide and ferrous sulfate. The GWP-contribution of material use is negligible. Finally, a limited contribution results from the energy use related with the office and transport of the water sector organisations. In total, indirect emissions contribute to 129,620 tonnes CO_2-eq. per year (8% of the total GWP).

OVERALL

Compared to the total yearly 212 million tonnes GWP of the Netherlands (MNP 2007). the 1.67 million tonnes GWP of the water sector is limited (0.8%). A small number of sectors such as the energy sector, heavy industry, and transport account for the largest share of the total GWP, and about 20 other sectors each account for a small percentage. The water sector is one of these and they have an interest in setting an example, especially if an appeal is made to households to reduce their warm water consumption.

The GWP of water for domestic use is 1.5 kg CO_2-eq./m^3 water (see Table 7-4). This equates to 160 kg CO_2-eq. per year per household (3.3% of the GWP from energy use in households). The GWP for heating tap water, however, is four times as large (693 kg CO_2-eq. per year per household). The amount of hot water used for showering increased last year in the Netherlands, and here an important action point is anticipated to reduce water use and simultaneously conserve energy.

MITIGATION MEASURES

Energy Efficiency Measures

To reduce the climate footprint several mitigation measures can be taken, some of which are already being implemented. First and foremost, an efficient and optimal functioning process for drinking water, and sewerage and wastewater treatment is required. The climate footprint is related to the applied treatment steps, and as it is expected that more advanced treatment will be needed to deal with substances such as hormones and medicine residues, the climate

Table 7-4. The GWP of water for households

Drinking water	0.36 kg CO_2-eq./m^3 water
Sewerage	0.07 kg CO_2-eq./m^3 water
Wastewater	1.07 kg CO_2-eq./m^3 water
Total	1.50 kg CO_2-eq./m^3 water

footprint is likely to increase. The challenge, thus, is to improve the energy efficiency of water production, transport, and treatment.

Examples of efficiency and optimisation measures per activity are: low pressure UV, minimising scaling of membranes, drinking water demand modelling, central softening, avoiding groundwater intrusion in sewer systems, aeration control systems, bubble aeration, sludge digestion, and biogas use and/or efficient biogas conversion (for an extensive overview see Frijns et al. 2008). Reduction of direct emissions is less easy to implement. Possibly, methane collection at groundwater abstraction sites will become feasible. Little is known about the possibilities to reduce nitrous oxide gas emissions.

Mitigation Measures in the Water Cycle

A more substantial gain in GWP reduction is expected from a number of climate mitigation measures that affect the whole water cycle. These so-called energy-effectiveness measures relate to new concepts, design and optimisation of the whole water cycle (Table 7-5).

Tuning of Sewerage and Treatment

In recent years, several optimisation studies of the wastewater collection and treatment system have been undertaken in the Netherlands. By disconnecting part of the stormwater and real-time control of the sewerage pumping system, less stormwater is transported to the municipal wastewater treatment plants. This reduces the energy required for transport as well as treatment.

Conservation of (Warm) Water at Households

Water conservation at households implies less water production, use, transport, and treatment and thus reduced greenhouse gas emissions from households and from the water sector. For example, by applying rainwater for toilet flushing, about 80 L of drinking water per day can be saved. If the rainwater flows under gravity conditions only from the roof to the toilet, the drinking water saving results is a reduction of 10 kg CO_2-eq. per household per year (or 6.5% of the GWP of domestic water use).

The use of water-saving devices for showers not only reduces the water use by 10 m^3 per year, but also saves 45 m^3 natural gas. Together, this substantially reduces the GWP by 95 kg CO_2-eq. per year per household. This underlines the potential savings to be achieved by households, and thus the interest for the water sector to set an example if they successfully want to make an appeal to households to reduce water and energy use from showering.

Reuse of Warmth Content

There are several options to reuse the energy/warmth that is embedded in water, both at households and at treatment plants (Koetse et al. 2008). One can think of recovery of warmth

Table 7-5. Mitigation measures in the water cycle

1. Tuning of sewerage and treatment:
   - optimisation of wastewater system
2. Conservation of (warm)water at households:
   - toilet flushing with rainwater
   - warm tap water conservation
3. Reuse of warmth content:
   - recovery of warm water from showers
   - application of heat pumps on wastewater in houses
   - recovery of warmth from influent (or effluent) at wwtp
4. Modern sanitation:
   - urine separation and/or black-water collection systems
   - reuse of nutrients from wastewater

from showers (30% gas saving for showering) and from domestic wastewater used for central heating (50% gas saving on heating).

Recovery of warmth from the influent (or effluent) of wastewater treatment plants for heating of nearby neighbourhoods or office buildings (as, for example, applied in Apeldoorn) can potentially result in 70 m^3 methane gas saving per year or 125 kg CO_2-eq. per household per year.

Modern Sanitation

A promising development is the introduction of new sanitation concepts based on the separate collection of urine and/or black water. These systems show a positive energy balance (Roorda et al. 2008), as less water is used for flushing and transport, and less organics and nutrients need to be oxidised at wastewater treatment plants. Moreover, nutrients can be recovered. The GWP reduction of source-separated systems is about 6 kg CO_2-eq. per household per year compared to conventional toilet systems.

If black water (in nonvacuous systems) is digested, about 500 MJ net energy per household per year can be produced. Combined with the reduced water use, the GWP gain of black-water systems can be 38 kg CO_2-eq. per household per year, or 24% reduction of the GWP of domestic water use. Co-digestion with organic solid waste will further increase the energy gain.

CONCLUSIONS AND RECOMMENDATIONS

The assessment of the climate footprint provides an insight into the energy use and greenhouse gas emissions of the Dutch drinking water, sewerage, and wastewater treatment sector. Although there is need for further knowledge on nitrous oxide emissions, the contribution of direct emissions of greenhouse gases was shown to be relatively high. For estimating nitrous oxide emissions, but also in general, a standardised approach for the climate footprint methodology would be welcome, especially for benchmarking purposes (see Frijns and Mulder 2009).

It can also be concluded that there are numerous options to reduce the climate footprint of the water sector. That is particularly the case for energy conservation. Still little is known about the possibilities to reduce direct greenhouse gas emissions and additional research is required. Priority can be given to the implementation of energy-efficiency measures, especially because such measures match with efficiency improvements and process optimisation.

To achieve a substantial reduction of the climate footprint, mitigation measures at the level of the whole water cycle will be required. It is recommended to approach the water cycle from a climate perspective to gain an insight into ways to improve both the climate footprint and the design of the water cycle. One can also endorse combinations with other cycles, such as the energy and solid waste cycle. Involve the users in the cycle analysis, and motivate households to conserve warm water use. The water sector should set an example by implementing mitigation measures. Such climate neutral optimisations in the water cycle require collaboration between the water organisations.

REFERENCES

CBS (2006). StatLine database Netherlands.

Frijns, J., and M. Mulder (2009). Benchmarking sustainability of the water sector: carbon footprint assessment, IWA Benckmarking Water Services conference, Amsterdam, 12–13 March 2009.

Frijns, J., M. Mulder, and J. Roorda (2008). Towards a Climate Neutral Water Cycle, Stowa nr 2008–17, Stowa, Utrecht, Netherlands (in Dutch).

IPCC (1996). The Second Assessment Report of the Intergovernmental Panel on Climate Change (IPCC). Cambridge University Press, Cambridge, UK.

IPCC (2006). Guidelines for National Greenhouse Gas Inventories, Vol, 5, Waste; Chapter 6, Waste water, IGES, Hayama, Japan.

Kampschreur, M.J., C. Picioreanu, N. Tan, R. Kleerebezem, M. Jetten, and M. van Loosdrecht (2007). Unraveling the source of nitric oxide emission during nitrification, *Water Environment Research*, 79(13). 2499–2509.

Koetse, E., A. Mels, I. Bisschops, and J. Kunst (2008). Wastewater Treatment System IJburg 2nd Phase; Options for Modern Sanitation, Grontmij/LeAF report (in Dutch).

Loeffen, P.G.M., W. van Rij, and J. Frijns (2005). Energy Conservation Measures at WWTPs, Grontmij I&M-99048741 (in Dutch).

MNP (2007). Greenhouse Gas Emissions in the Netherlands 1990–2005, National Inventory Report 2007, MNP report 500080 006, MNP, Bilthoven, Netherlands.

Moel, P.J. de, J. Verberk, and J. van Dijk (2006). *Drinking Water—Principles and Practices*, World Scientific Publishing Company, Singapore.

Roorda, J.H., E. Koetse, J. Kunst, A. Buunen-van Bergen, A. Mels, I. Bisschops, P. Piekema, and R. Neef (2008). Modern sanitation for the developed world: an example for new areas, IWA Sanitation Challenge conference, Wageningen, 19–21 May 2009.

Strutt, J., S. Wilson, H. Shorney-Darby, A. Shaw, and A. Byers (2008). Assessing the carbon footprint of water production, *Journal AWWA* 100, no. 6, pp. 80–91.

UKWIR (2008). Carbon Accounting in the UK Water Industry: Methodology for Estimating Operational Emissions. Report UKWIR/08/CL/01/5, UKWIR, London, UK.

UvW (2007). Clean Wastewater 2006, Report UvW, Unie van Waterschappen, Den Haag, Netherlands (in Dutch).

Vewin (2007). Water in Sight 2006, Report Vewin 2007/79/6292, Vewin, Rijswijk, Netherlands (in Dutch).

– 8 –

The Water Footprint of Bio-Energy

A.Y. Hoekstra, P.W. Gerbens-Leenes, and Th.H. van der Meer

INTRODUCTION

The water sector is becoming more energy-intensive. Think, for example, of the energy needed for desalination of salt or brackish water, for pumping groundwater from deeper and deeper and for large interbasin water transfer schemes. At the same time, the energy sector is becoming more water-intensive—especially because of the increasing focus on biomass as a source of energy. All energy scenarios for the coming decades show a shift toward an increased percentage of bio-energy. This chapter focuses on the question of how much water is involved in the production of bio-energy.

The source of bio-energy can be crops specifically grown for that purpose, natural vegetation, or organic wastes (Minnesma and Hisschemöller 2003). Many of the crops used for bio-energy can also—alternatively, not at the same time—be used as food or feed. Biomass can be burnt to produce heat and electricity, but it can also be used for the production of bio-ethanol or biodiesel, biofuels that can displace fossil energy carriers in motor vehicles (Hughes et al. 2007).

At present, the production of biomass for food and fibre in agriculture requires about 86% of the worldwide freshwater use (Hoekstra and Chapagain 2007, 2008). In many parts of the world, the use of water for agriculture competes with other uses such as urban supply and industrial activities, while the aquatic environment shows signs of degradation and decline (Postel et al. 1996). An increase of demand for food in combination with a shift from fossil energy towards bio-energy puts additional pressure on freshwater resources. For the future, hardly any new land is available, so all production must come from the natural resource base currently available (FAO 2003), requiring a process of sustainable intensification by increasing the efficiency of land and water use (Fresco 2006).

Globally, many countries explore options to replace gasoline by biofuels (Hughes et al. 2007). The EU and the US have set targets for the replacement. When agriculture grows crops for bio-energy, however, it needs additional water that cannot be used for food anymore. Large-scale cultivation of biomass for fossil fuels substitution influences future water demand (Berndes 2002). An important question is whether we should apply our freshwater resources for the production of bio-energy or food crops. The FAO estimates that in 2007 alone, before the food price crisis hit, 75 million more people have been pushed into undernourishment as a result of higher prices, bringing the total number of hungry people to 923 million (FAO 2008a). Moreover, the FAO reports biofuels increasing food insecurity (FAO 2008b). The World Bank recognizes biofuels production as a major factor driving food prices. It estimates that 75% of the increase of food prices in the period 2002–2008 was due to biofuels (Mitchel 2008).

The replacement of fossil energy by bio-energy generates the need for detailed information on water requirements of this new energy source. A tool for the calculation of water needs for consumer products is the concept of the water footprint (WF) (Hoekstra and Hung 2002; Hoekstra and Chapagain 2007, 2008), defined as the total annual volume of freshwater used to produce goods and services for consumption.

The objective of this chapter is to give a global overview of the WF per unit of bio-energy (m^3/GJ), including heat, electricity, bio-ethanol, and biodiesel. It covers the 12 main crops that together contribute to 80% of global crop production. In addition, the study includes jatropha, a plant species often mentioned in the context of bio-energy. Research questions are: (1) what are the green and blue WFs (m^3/GJ) for heat and electricity derived from the combustion of biomass per crop per country, and (2) what are the WFs (m^3/GJ) for transport fuels (bio-ethanol and biodiesel) per crop per country? The study excludes organic wastes, such as manure or crop residues, biogas, and energy from algae.

The study builds on two earlier studies: one that estimated the WFs of a large variety of food and fibre products (Hoekstra and Chapagain 2007, 2008) and one that estimated the WF of heat from biomass (Gerbens-Leenes et al. 2009). It refines the study of Hoekstra and Chapagain by taking precise production locations into account for the calculation of crop water requirements and by using local estimates for the start of the growing season based on an analysis of when weather conditions at specific locations are most favourable. An additional refinement is that the study differentiates between blue and green water. Next, it extends the study of Gerbens-Leenes et al.—which focussed on the water footprint of heat from biomass—to the water footprint of bio-electricity and biofuels.

BIO-ENERGY

Energy derived from biomass is termed bio-energy. The FAO defines biomass as material of organic origin, in nonfossilized form, such as agricultural crops and forestry products, agricultural and forestry wastes and by-products, manure, microbial biomass, and industrial and household organic waste. Biomass is applied for food or feed (e.g., wheat, maize, sugar), materials (e.g., cotton, wood, paper), or for bio-energy (e.g., maize, sugar, jatropha). Figure 8-1 shows that biomass can provide different forms of bio-energy: heat, electricity, and biofuels like ethanol and biodiesel. First-generation biofuels concern presently available biofuels produced using conventional technology, i.e. fermentation of carbohydrates into ethanol, and extracting and processing oil from oil crops into biodiesel. Biomass not only contains starch, sugar, and oil that can be processed into biofuels, it also contains large amounts of cellulosic biomass. So far, the cellulosic fraction could be used for energy by burning it to provide heat and produce electricity. It is expected that these cellulosic fractions will form an attractive source for the production of next-generation biofuels. Next-generation biofuels concern future available biofuels produced using new technology under development that aims to also convert cellulosic fractions from crops into biofuels, e.g., ethanol (Worldwatch Institute 2007). In this way, the production of biofuel per unit of crop can increase substantially.

WATER FOOTPRINT

The WF of a product is defined as the volume of freshwater used for production at the place where it was actually produced (Hoekstra and Chapagain 2008). In general, actual water contents of products are negligible compared to their WF, and water use in product life cycles are dominated by the agricultural production stage. The WF consists of three components: the green, blue, and gray WF. The green WF refers to rainwater that evaporated during

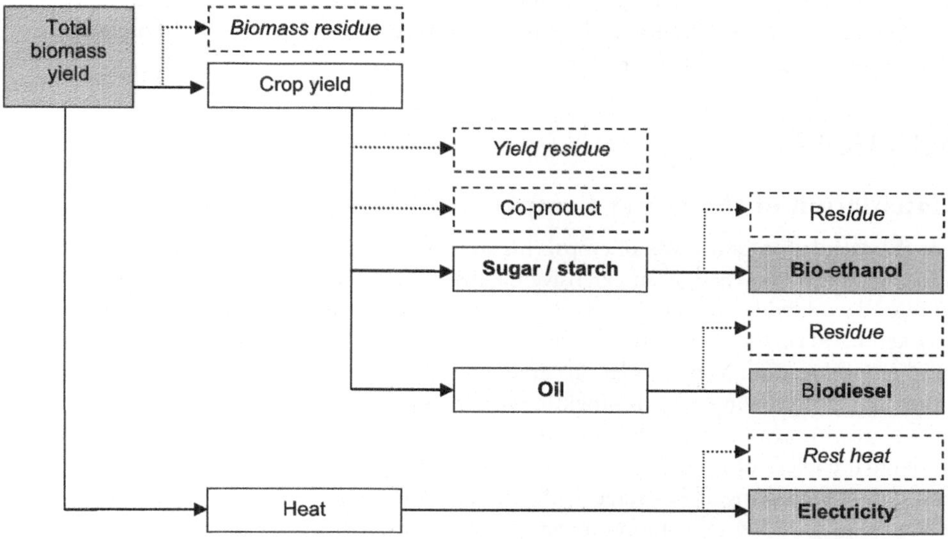

Figure 8-1. Total biomass yield can be converted into heat and subsequently into electricity. Alternatively, the crop yield, which is part of the total biomass, can be converted into bio-ethanol (in the case of starch and sugar crops) or biodiesel (in the case of oil crops). In every step in the production chain, residues or rest heat are generated.

production, mainly during crop growth. The blue WF refers to surface- and groundwater for irrigation evaporated during crop growth. The grey WF is the volume of water that becomes polluted during production, defined as the amount of water needed to dilute pollutants emitted to the natural water system to the extent that the quality of the ambient water remains beyond agreed water quality standards.

CROP COVERAGE

Globally, a limited number of crops determines total production. Theoretically, all crops can be applied for bio-energy. In practice, only some crops dominate production: sugar cane, sugar beet, maize, rapeseed, and soybean (Worldwatch Institute 2007). Since this study aims to provide a global overview of the WFs of the main crops that can be used for bio-energy, it included the 12 crops that contribute to 80% of total global crop production: sugar cane, maize, wheat, rice, potato, sugar beet, rye, cassava, soybean, barley, sorghum, and rapeseed. Additionally, the study included jatropha curcas, a tree species that provides oil from its seeds (Banerji et al. 1985).

The composition of biomass determines the availability of energy from a specific biomass type, resulting in differences in combustion energy and options for biofuel production. This study includes four categories of biomass: starch crops (cereals: barley, maize, rice, rye, sorghum, wheat; and tubers: cassava, potato,); sugar crops (sugar beet, sugar cane); oil crops (rapeseed, soybean); and trees (jatropha). For the assessment of the WF of bio-energy, it follows the method of Hoekstra and Chapagain (2008) to arrive at estimates of the WF of crops and the method of Gerbens-Leenes et al. (2009) to translate the WF of crops (m^3/ton) into a WF of bio-energy (m^3/GJ). Data were obtained from agricultural studies. Based on sugar or starch contents, we calculated the amount of energy that the crop could provide in the form of ethanol. For each oil crop, based on the oil content, we calculated the amount of energy that the crop could provide in the form of biodiesel.

METHOD

Calculation of the Water Footprint of Crops

For the calculation of the WF of crops, this study used the methodology from Hoekstra and Chapagain (2008). There is an extensive database that includes the WF of almost all crops produced worldwide (m^3/ton) based on average, national meteorological data (Chapagain and Hoekstra 2004). The current study, however, assessed the WFs of crops more specifically per production location. WF calculations are made by summing daily crop evapotranspiration (mm/day) over growing periods providing information on crop water requirements. The start of the growing season depends on climatic conditions, on the production location, and on individual choices of farmers. For the start of the growing season, the study considered the first option for sowing after winter or after a dry season, assuming that growing seasons start when mean monthly maximum temperatures are above 10°C and when sufficient rain and global radiation is available.

For the main producing countries, this study calculated crop water requirements for the 12 crops and for jatropha, distinguishing between the green and blue WF, but excluding the gray WF. Next, it selected the main producing countries. For jatropha, it considered production in Brazil, Guatemala, Indonesia, and Nicaragua, countries for which data were available (Daey Ouwens 2000). Next, it selected agricultural production locations. Information was derived from the Madison Center for Sustainability and the Global Environment at the University of Wisconsin (2008). For these areas, it selected weather stations providing climatic data that were used as input for the calculations. Data were derived from Müller and Hennings (2000).

The calculation of crop water requirements (mm/day) has been done per major production region, using the calculation model CROPWAT 4.3 (FAO 2007) based on the FAO Penman–Monteith method to estimate reference crop evapotranspiration (Allen et al. 1998), and a crop coefficient that corrects for the difference between actual and reference crops.

Calculations for green and blue WFs (m^3/ton) have been done using the method of Hoekstra and Chapagain (2008). Green water use (m^3/ha) over the length of the growing period is calculated as the sum of daily volumes of rainwater evapotranspiration. The latter is equal to the crop water requirement except if effective precipitation is less than the requirement. In that case, rainwater evapotranspiration is equal to effective precipitation. Blue water use (m^3/ha) over the length of the growing period is calculated as the sum of daily volumes of irrigation–water evapotranspiration. The latter is equal to the irrigation requirement if this requirement is actually met, and otherwise it is equal to actual effective irrigation. The irrigation requirement is defined as the crop water requirement minus effective precipitation. In the calculation, it has been assumed that irrigation requirements are actually met. The green WF of a crop (m^3/ton) is the total green water use over the length of the growing period (m^3/ha) divided by the crop yield (ton/ha). The blue WF (m^3/ton) is the total blue water use over the length of the growing period (m^3/ha) divided by the crop yield (ton/ha). In general, yields show variation among years. The study therefore calculated average yields over five production years (1997–2001) using data from the FAO.

Calculation of the WF of Heat and Electricity from Biomass

The energy content of biomass is expressed in terms of combustion values. Energy analysis defines the energy content of a substance as the amount of heat produced during combustion at 25°C at 1 bar. It distinguishes between the higher heating value (HHV) and the lower heating value (LHV) (Blok 2006). For the HHV, energy analysis measures the heat content of water that is the product of the combustion process in the liquid form; in the case of LHV it measures the heat content in the gaseous form. For the calculation of the WF of heat from biomass, the study has followed the method of Gerbens-Leenes et al. (2009) that calculates the energy yield of a crop (GJ/ton) by combining data on the heat of combustion of plant components with information on composition, harvest index, and dry mass fraction of a crop as shown in Table 8-1.

$$E_{\text{heat}}(c) = HI(c) \times DMF_y(c) \times \sum_{i=1}^{5} \left(f_{y,i} \times HHV_i\right) + \left(1 - HI(c)\right) \times DMF_r(c) \times \sum_{i=1}^{5} \left(f_{r,i} \times HHV_i\right)$$

Eq. 8-1

$E_{\text{heat}}(c)$ is the energy yield of crop c in the form of heat (GJ/ton), $HI(c)$ is the harvest index of crop c (gram/gram), $DMF_y(c)$ is the dry mass fraction of the crop yield (gram/gram), $DMF_r(c)$ is the dry mass fraction in the rest fraction (i.e., in the residue biomass), $f_{y,i}$ is the fraction of component i in the dry mass of the crop yield (gram/gram), $f_{r,i}$ is the fraction of component i in the dry mass of the rest fraction (gram/gram), and HHV_i is the higher heating value of component i (kJ/gram).

For the generation of electricity from biomass, industry can apply the heat that becomes available from the combustion of total biomass. The energy in the form of electricity from crop c (GJ/ton) depends on the efficiency with which energy in the form of biomass heat can be transformed into electricity:

$$E_{\text{electr}}(c) = \eta \times E_{\text{heat}}(c) \qquad \text{Eq. 8-2}$$

For the value of the efficiency η, the study applied a value of 59%, based on the maximum efficiency based on Carnot and the technology of Biomass-fired Integrated Gasifier Combined Cycle (BIG/CC) operated at a temperature of 720 K (Blok 2006, Faay 1997).

The WF of heat from a crop c (m^3/GJ) is calculated by dividing the WF of the crop (m^3/ton) by the heat content of the crop (GJ/ton). The WF of biomass electricity from a crop c (m^3/GJ) is calculated by dividing the WF of the crop (m^3/ton) by the electricity output per crop unit (GJ/ton):

$$WF_{\text{heat}}(c) = \frac{WF(c)}{E_{\text{heat}}(c)}; \quad WF_{\text{electr}}(c) = \frac{WF(c)}{E_{\text{electr}}(c)} \qquad \text{Eq. 8-3}$$

Calculation of the Wf of First-Generation Biofuels

At present, bio-ethanol is produced from sugars from sugar cane or sugar beet, or from starch hydrolysed into sugars derived from maize, wheat, or cassava (Worldwatch Institute 2007). Under anaerobic conditions, sugar naturally ferments into acids and alcohols (mainly ethanol). For thousands of years, people have applied yeasts to enhance fermentation. The main metabolic pathway involved in ethanol fermentation is glycolysis through which one molecule of glucose is metabolized and two molecules of pyruvate are produced (Verkerk et al. 1986; Bai et al. 2008). Under anaerobic conditions, pyruvate is further reduced to ethanol

Table 8-1. Main characteristics of twelve crops

	Cassava	Barley	Maize	Paddy Rice	Potato	Rapeseed	Rye	Sorghum	Soybean	Sugar Cane	Sugar Beet	Wheat
Harvest Index	0.70[a]	0.42[a]	0.45[a]	0.42	0.70[a]	0.32[a]	0.42	0.42	0.40[a]	0.60[a]	0.66[a]	0.42[a]
Economic yield	Tuber[b]	Ear + Grain[b]	Whole Tops[b]	Inflor + Grain	Tuber[b]	Inflor + Seed[d]	Ear + Grain[b]	Ear + Grain[b]	Beans[a]	Whole Tops[a]	Beet[a]	Ear + Grain[b]
Dry mass[b]	0.38	0.85	0.85	0.85	0.25	0.74	0.85	0.85	0.92	0.27	0.21	0.85
Composition dry mass (g /100 g)[c]												
Carbohydrates	87	76	75	76	78	7	76	76	29	57	82	76
Proteins	3	12	8	8	9	22	12	12	37	7	5	12
Fats	1	2	4	2	0	42	2	2	18	2	0	2
Lignin	3	6	11	12	3	2	6	6	6	22	5	6
Organic acids	3	2	1	1	5	1	2	2	5	6	4	2
Minerals (K, Ca, P, S)	3	2	1	1	5	26	2	2	5	6	4	2
Rest fraction	Leaves	Shells	Stems	Stems	Leaves	Leaves	Stems	Stems	Leaves	Stems	Leaves	Stems
Dry mass[b]	0.38	0.85	0.85	0.85	0.13	0.13	0.85	0.85	0.15	0.27	0.21	0.85
Composition dry mass (g /100 g)[c]												
Carbohydrates	52	62	62	62	52	52	62	62	52	62	52	62
Proteins	25	10	10	10	25	25	10	10	25	10	25	10
Fats	5	2	2	2	5	5	2	2	5	2	5	2
Lignin	5	20	20	20	5	5	20	20	5	20	5	20
Organic acids	5	2	2	2	5	5	2	2	5	2	5	2
Minerals (K, Ca, P, S)	8	4	4	4	8	8	4	4	8	4	8	4

Data on composition, harvest index and dry mass are averages of existing crops. Data were derived from agricultural studies.

a. Source: Goudriaan et al. (2001)
b. Source: Penning de Vries et al. (1989)
c. Source: Habekotté (1997)
d. Source: Akthar (2004)

with the release of CO_2. The overall reaction is $C_6H_{12}O_6$ ◊ $2C_2H_5OH + 2CO_2$. Theoretically, the maximum yield of ethanol is 511 g of ethanol and 489 g of carbon dioxide per kg of glucose metabolized (or 530 g of ethanol per kg of starch). Often, various by-products are also produced, for example, glycerol (Bai et al. 2008). During ethanol fermentation, yeast cells suffer from stresses such as ethanol accumulation inhibiting yeast cell growth and ethanol production. The final ethanol concentration is about 10–12% (Bai et al. 2008, Catsberg et al. 1997). The fermentation industry, therefore, applies a tanks-in-series system to alleviate product inhibition. Today, it can reach a yield of 90–93% of the theoretical value of glucose to ethanol (Rosillo-Calle et al. 2007).

Oilseed crops, such as rapeseed, soybean, and jatropha, are used to produce straight vegetable oil or biodiesel. Straight vegetable oil is oil extracted from an oilseed crop and directly applied for energy purposes (Worldwatch Institute 2007). An example is olive oil for lighting. Due to its chemical properties, such as the high viscosity at low temperatures, it is often difficult to use straight vegetable oil as a biofuel in diesel engines. In countries with warm climates, the relatively high temperatures prevent the oil from thickening and the straight vegetable oil is a viable fuel. In countries with temperate climates, the oil needs additional treatment to manufacture a biodiesel less sensitive to lower temperatures. Biodiesel is manufactured in a chemical reaction termed transesterification in which oil reacts with an alcohol resulting in an alkyl ester of the fatty acid with glycerine molecules as the primary co-product. In Europe, rapeseed oil is the dominant feedstock for biodiesel, with some sunflower oil also used. In the US, the main feedstock is soybean oil; in tropical and subtropical countries, palm, coconut, and jatropha oil is used (Worldwatch Institute 2007).

When calculating natural resource use, the whole life cycle of a product should be taken into account. The use of water, however, is dominated by the first link of the production chain—agriculture. Ethanol production, for example, requires about 21 litres of water per litre of ethanol. Moreover, this water is often reused (Institute for Agriculture and Trade Policy 2007). This study therefore only took water requirements in agriculture into account and neglected water use in industrial links of the production chain.

The ethanol-energy yield of a crop (in GJ/ton) has been calculated as follows:

$$E_{\text{ethanol}}(c) = DMF_y(c) \times f_{\text{carbohydr}}(c) \times f_{\text{ethanol}} \times HHV_{\text{ethanol}} \qquad \text{Eq. 8-4}$$

where $DMF_y(c)$ is the dry mass fraction in the crop yield (gram/gram), $f_{\text{carbohydr}}(c)$ the fraction of carbohydrates in the dry mass of the crop yield (gram/gram), f_{ethanol} the amount of ethanol obtained per unit of carbohydrate (gram/gram), and HHV_{ethanol} the higher heating value of ethanol (kJ/gram). For the amount of ethanol per unit of sugar, we assumed the theoretical maximum value of 0.51 g/g; for starch, 0.53 g/g (Bai et al. 2008).

The biodiesel-energy yield of a crop (in GJ/ton) has been calculated as follows:

$$E_{\text{diesel}}(c) = DMF_y(c) \times f_{\text{fat}}(c) \times f_{\text{diesel}} \times HHV_{\text{diesel}} \qquad \text{Eq. 8-5}$$

where $DMF_y(c)$ is the dry mass fraction in the crop yield (gram/gram), $f_{\text{fat}}(c)$ the fraction of fats in the dry mass of the crop yield (gram/gram), f_{diesel} the amount of biodiesel obtained per unit of fat (gram/gram), and HHV_{diesel} the higher heating value of biodiesel (kJ/g). For the fraction biodiesel per fat weight, we assumed the value 1. The fractions of carbohydrates and fats in the dry mass of crop yields are given in Table 8-1. The higher heating value for ethanol is 29.7 kJ/g and for biodiesel, 37.7 kJ/g.

The WF of ethanol energy from a crop *c* (m^3/GJ) is calculated by dividing the WF of the crop (m^3/ton) by the ethanol-energy yield of the crop (GJ/ton). The WF of biodiesel energy from a crop *c* (m^3/GJ) is calculated in a similar way:

$$WF_{\text{ethanol}}(c) = \frac{WF(c)}{E_{\text{ethanol}}(c)}; \qquad WF_{\text{diesel}}(c) = \frac{WF(c)}{E_{\text{diesel}}(c)} \qquad \text{Eq. 8-6}$$

For the calculation of the WF of first-generation biofuels, this study fully allocated the WF of the crop to the biofuels derived, assuming that the value of the residues of production is much lower than the value of the biofuel.

Calculation of the WF of Next-Generation Biofuels

Biomass not only contains starch, sugar, and oil that can be processed into biofuels, it also contains large amounts of cellulosic biomass. So far, the cellulosic fraction could be used for energy by burning it to provide heat and produce electricity. It is expected that these cellulosic fractions will form an attractive source for the production of liquid, next-generation biofuels. For next-generation biofuels, industry can apply total biomass, including wastes. It is not yet clear what efficiency will be achieved in converting total biomass into biofuel. It is safe to assume that the WF of next-generation biofuels will never be lower than the WF of the crop (m^3/ton) divided by the energy content of the crop (GJ/ton), where the latter is expressed in terms of its higher heating value (HHV).

THE WATER FOOTPRINT OF BIO-ELECTRICITY AND BIOFUELS

Crop Production, Crop Water Requirements, and Irrigation Requirements

Some countries have a large contribution to global production. For example, Brazil produces 27% of globally available sugar cane; the United States almost half of global soybean production, 40% of the maize, and one-quarter of the sorghum; China 18% of all wheat, one-third of the paddy rice, one-fifth of the potatoes, and 27% of the rapeseed. Half of the global production of rye takes place in Russia and Germany, while Nigeria shows the largest contribution to cassava production. For other crops, such as sugar beet and barley, production is distributed more evenly.

On almost every crop location, irrigation is required. Exceptions are sugar beet grown in Japan; maize from South Africa; wheat from Australia; cassava from Nigeria, Angola, Benin, Guinea, the Philippines, Vietnam, and India; potato from Bangladesh, Peru, and Japan; sorghum from Nigeria, Ethiopia, Chad, and Venezuela; and rapeseed from Bangladesh. In some countries, crop water requirements are completely or almost completely covered with irrigation water. These crops and countries are: sugar cane from Argentina (96%) and Egypt (92%); wheat from Argentina (100%), Kazakhstan (98%), and Uzbekistan (98%); potato and barley from Kazakhstan (100%); sorghum from Yemen (100%); and soybean from Brazil (95%). For the other crops and production locations, irrigation requirements find themselves in between the two extremes.

The WF of Biomass

The WFs of biomass (m^3/ton) show large variations among crops. For each specific crop, WFs vary among countries, dependent on agricultural production systems applied and climate conditions. Most WFs show a variation of a factor of 4 to 15, with two exceptions. These are the values for wheat and sorghum, with a difference of a factor of 20 and 47, respectively.

The WF of Heat and Electricity from Biomass

Table 8-2 shows the total weighted global average WF for 13 crops providing electricity. It is assumed that not only crop yields, but total biomass yields are applied for the generation. The largest difference is found between jatropha and sugar beet: beets are 10 times more water efficient. The WF of heat is always 59% of the WF of electricity, as shown in the table, based on the energy efficiency assumed in this study.

The WF of First-Generation Biofuels

Biofuel Energy Production per Crop Unit

Table 8-3 shows energy provided by ethanol (HHV ethanol in MJ/kg fresh weight of the crop) from two sugar- and eight starch-providing crops included in this study. It shows three categories: sugar-providing crops and one starch-providing crop with relatively low values for energy provided by ethanol (sugar beet, sugar cane, and potato); the category of starch-

Table 8-2. Total weighted global average water footprint for 13 crops providing electricity (m^3/GJ). It is assumed that not only crop yields, but total biomass yields are applied for the generation of electricity.

Crop	Total Water Footprint	Blue Water Footprint	Green Water Footprint
		m^3 per GJ Electricity	
Sugar beet	46	27	19
Maize	50	20	30
Sugar cane	50	27	23
Barley	70	39	31
Rye	77	36	42
Paddy rice	85	31	54
Wheat	93	54	39
Potato	105	47	58
Cassava	148	21	127
Soybean	173	95	78
Sorghum	180	78	102
Rapeseed	383	229	154
Jatropha[a]	396	231	165

[a]Average numbers for five countries (India, Indonesia, Nicaragua, Brazil, and Guatemala).

Table 8-3. Energy provided by ethanol from the two sugar- and 10 starch-providing crops that were included in this study, as well as the energy provided by oil from the three oil providing crops.

Crop	MJ of Biofuel per kg Fresh Weight Crop
Ethanol from sugar	
Sugar cane	2.3
Sugar beet	2.6
Ethanol from starch	
Potato	3.1
Cassava	5.2
Sorghum	10.0
Maize	10.0
Wheat	10.2
Barley	10.2
Paddy rice	10.5
Rye	10.5
Biodiesel from oil	
Soybean	6.4
Rapeseed	11.7
Jatropha	12.8

providing crops with relatively large values for energy provided by ethanol (sorghum, maize, wheat, barley, paddy rice, and rye); and one category in between (cassava). These differences are caused by differences in the water contents of the crops, where a large water content relates to relatively low energy values provided by ethanol. Table 8-3 also shows the energy provided by oil from the three oil-providing crops included in this study. The HHV of oil from soybean is smallest, about half the value of rapeseed or jatropha.

The WF of Bio-ethanol

Figure 8-2 shows the lowest value, the highest value, and the weighted average global value of the WF for energy for 10 crops providing ethanol showing the enormous variation in the total WF among crops. Especially sorghum shows large variation, mainly caused by unfavourable conditions in Niger and high production efficiency in Egypt. Figure 8-3 shows weighted global average green and blue WFs for 10 crops providing ethanol. It shows that differences among crops are large.

At present, sugar beet is the most favourable crop, sorghum the most unfavourable with a difference of a factor of seven. When data for the two main ethanol- producing countries, Brazil and the United States are compared, in Brazil ethanol from sugar cane is more efficient than maize (99 versus 140 m^3/GJ ethanol), while in the United States, maize is more attractive than sugarcane (78 versus 104 m^3/GJ ethanol). Figure 8-3 shows the distinction between green and blue water. As a global average, the blue WF of cassava is smallest. Other favourable crops are sugar beet, potato, maize, and sugar cane. In terms of blue water, sorghum is unfavourable.

Table 8-4 shows the total weighted global average WF for 10 crops providing ethanol, as well as the blue and green WF. The table also shows the amount of water needed for a specific crop to produce one litre of ethanol.

On average, it takes 1,400 L of water to produce 1 L of ethanol from sugar beet, 2,400 L for 1 L of ethanol from potato, 2,500 L from sugar cane, and 2,600 L from maize. Sorghum is the most inefficient crop, 9,800 L for 1 L ethanol. Irrigation is smallest for cassava, 400 L of blue water for 1 L of ethanol, followed by 800 L for sugar beet, and 1,000 L for maize. Sorghum is the crop showing the largest WF, 4,250 L per litre ethanol. As one can see from Tables 8-2 and 8-4, sugar beet is most efficient in terms of ethanol and electricity. The other crops show a different order for the efficiency in which electricity and ethanol are produced. In general, the production of ethanol of only part of the crop is less water efficient than the production of electricity from total biomass.

The WF of Biodiesel

The WF of biodiesel derived from soybean, rapeseed, and jatropha shows differences among the main producing countries. For rapeseed, western Europe shows the smallest WFs, Asia the largest. Especially in India, rapeseed has a large blue WF. For soybean, Italy, Paraguay, and Argentina have the smallest WFs, India the largest. Biodiesel from jatropha is produced in the most water efficient way in Brazil, and inefficiently in India. Table 8-4 shows the total weighted global average WF for biodiesel (soybean and rapeseed) and the average WF for jatropha, as well as the blue and green WF. It also shows the amount of water needed to produce 1 L of biodiesel. On average, it takes 14,000 L of water to produce 1 L of biodiesel from soybean or rapeseed, and 20,000 L for 1 L of biodiesel from jatropha.

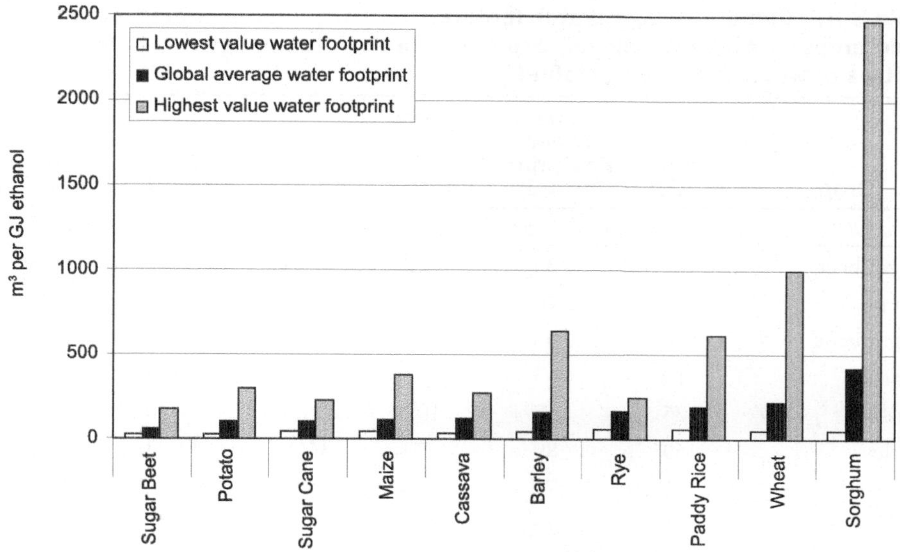

Figure 8-2. Lowest value, highest value, and weighted average global value of the water footprint for energy for 10 crops providing ethanol

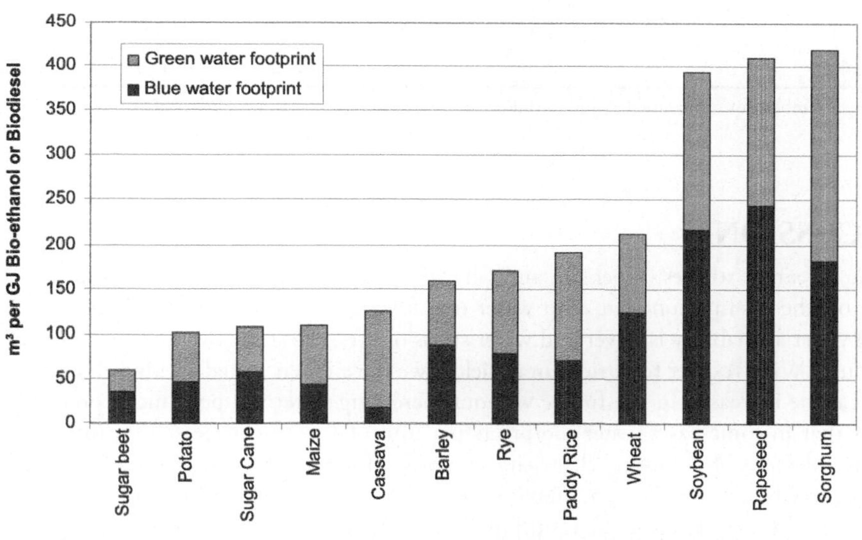

Figure 8-3. The weighted global average water footprint for 10 crops providing ethanol and for two crops providing oil for biodiesel

The WF of Next-Generation Biofuels

For next-generation biofuels, total biomass of a crop can be applied. When we optimistically assume that the production of next-generation biofuels will be as efficient as the productionof electricity from biomass (in terms of GJ/ton), the results shown in Table 8-2 form a lower limit for the WF of these next-generation biofuels.

Table 8-4. Global average water footprint for 10 crops providing ethanol and three crops providing biodiesel, expressed both in terms of m³/GJ and in terms of litres of water per litre of biofuel

Crop	Total Water Footprint (m^3/GJ)	Blue Water Footprint (m^3/GJ)	Green Water Footprint (m^3/GJ)	Total Water (L/L)	Blue Water (L/L)	Green Water (L/L)
Ethanol	m^3 per GJ ethanol			litres water per litre ethanol		
Sugar beet	59	35	24	1,388	822	566
Potato	103	46	56	2,399	1,078	1,321
Sugar cane	108	58	49	2,516	1,364	1,152
Maize	110	43	67	2,570	1,013	1,557
Cassava	125	18	107	2,926	420	2,506
Barley	159	89	70	3,727	2,083	1,644
Rye	171	79	92	3,990	1,846	2,143
Paddy rice	191	70	121	4,476	1,641	2,835
Wheat	211	123	89	4,946	2,873	2,073
Sorghum	419	182	238	9,812	4,254	5,558
Biodiesel	m^3 per GJ Biodiesel			Litres Water per Litre Biodiesel		
Soybean	394	217	177	13,676	7,521	6,155
Rapeseed	409	245	165	14,201	8,487	5,714
Jatropha[a]	574	335	239	19,924	11,636	8288

[a] average numbers for five countries (India, Indonesia, Nicaragua, Brazil, and Guatemala).

DISCUSSION

Similar to earlier studies (Hoekstra and Chapagain 2007, 2008), the calculations have been based on the assumption that crop water use is equal to crop water requirements. When actual water availability is lower and water stress occurs, the study overestimates actual crop water use. With respect to agricultural yields, we have taken actual yields, which in many cases can be increased in the future without increasing water use per unit of product. This means that in some cases water footprints per unit of energy can be significantly lowered. For the efficiency of obtaining electricity or biofuels from biomass, we have taken optimistic assumptions by taking theoretic maximum values or values that refer to the best available technology. This means that the resulting water footprint figures are conservative.

The results of this study are based on rough estimates of freshwater requirements in crop production and on theoretically maximum conversion efficiencies in the production of bio-electricity and biofuels. For the assessment of the WF of bio-energy, the study integrated data from several sources, each of which adds a degree of uncertainty. For example, the calculations using the CROPWAT model (FAO 2007) required input of meteorological data that are averages over several years rather than data for a specific year. The data presented do thus not reflect annual variations. Estimated crop water requirements are sensitive to the input of climatic data and assumptions concerning the start of the growing season. In the most extreme cases, this study found crop water requirements that were a factor two different from earlier studies (Hoekstra and Chapagain 2007, 2008). Other times results were similar. The factors mentioned imply that results presented here are indicative. However, the differences in calculated WFs are so large that general conclusions with respect to the water footprint of bio-ethanol versus the water footprint of biodiesel can be drawn, and that also conclusions can be drawn about the relative water footprints of different crops.

There is a distinction between gross and net production of bio-energy (Giampietro and Ulgiati 2005, Pimentel and Patzek 2005). In assessing the WF of heat, electricity, and fuels from biomass, we looked at the WF of the gross energy output from crops. We did not study energy inputs in the production chain, like energy requirements in the agricultural system (e.g., energy use for the production of fertilizers and pesticides) or the energy use during the industrial production of the biofuel. This means that this study underestimates the WF of bio-energy, especially in cases in which agricultural systems have a relatively large energy input. As an example, in a case where energy input equals 50% of the energy output – a case common in bio-energy production systems (Pimentel and Patzek 2005)—the water footprint of the net bio-energy production would be twice the water footprint of gross energy production.

CONCLUSIONS

The water footprint of bio-energy is large if compared to other forms of energy. In general, it is more efficient to use total biomass, including stems and leaves, and generate electricity than producing a biofuel. For most crops, the WF of bio-electricity is about a factor of two smaller than for bio-ethanol or biodiesel. The difference is caused by the crop fraction that can be applied. For electricity, total biomass can be used; for bio-ethanol or biodiesel, only the starch or oil fraction of the yield. In general, the WF of bio-ethanol is smaller than of biodiesel. The WF of bio-energy shows a large variation, depending on three factors: (1) the crop used, (2) the climate at the location of production, and (3) the agricultural practice.

For electricity generation, sugar beet, maize, and sugar cane with WFs of about 50 m^3/GJ are the most favourable crops, followed by barley, rye, and rice with WFs of about 70–80 m^3/GJ. Rapeseed and jatropha, typical energy crops, showing WFs of about 400 m^3/GJ are the most unfavourable crops. For the production of ethanol, two crops grown in a temperate climate, sugar beet and potato, with WFs of 60 and 100 m^3/GJ, respectively, are most favourable, followed by a crop typical for a warm climate, sugar cane, also showing a WF of about 110 m^3/GJ. Values for maize and cassava show the same order of magnitude. With a WF of 400 m^3/GJ, sorghum is by far the most unfavourable crop. For biodiesel production, soybean and rapeseed, crops mainly grown for food, show the most favourable WF of 400 m^3/GJ; jatropha has the most unfavourable WF of about 600 m^3/GJ.

Results show large differences in crop water requirements among countries caused by differences in climate. The crop water requirement of sugar beet grown in Iran, for example, is twice the weighted global average value.

Agricultural practice determines yields and thus differences among WFs of crops even in cases with a similar climate. When yield levels are relatively low, WFs are high and vice versa. For example, in Kazakhstan, yields of barley, potato, and wheat are relatively low. In combination with unfavourable climatic factors, this results in high values for the WFs. Conditions in Denmark are favourable, resulting in relatively low crop water requirements for wheat.

Theoretically, all crops can be applied for energy, including crops like rice and rye that are currently mainly applied for food. Water use for a specific crop does not depend on whether the crop is applied for energy or for food. Some food crops, including rice, are more water efficient in producing a unit of ethanol, biodiesel, or electricity than some typical energy crops, such as rapeseed or jatropha. The ethical discussion whether food crops can be used for energy should be extended to the discussion whether we should apply our limited water resource base for food or for energy.

The scientific and the international political communities promote a shift toward renewable energy sources, such as biomass, to avoid emissions of greenhouse gases. This study has shown that biomass production goes along with large water requirements. Already there are reasons for profound concern in several regions and countries with limited water resources if food and fibre needs of future generations can be met. If a shift toward a larger contribution of bio-energy to total energy supply takes place, results of this study can be used to select the crops and countries that (under current production circumstances) produce bio-energy in the most water-efficient way.

REFERENCES

Akhtar, N. 2004. Agro-physiological response of spring sown sunflower (*Helianthus annuus L.*) to various management practices. PhD thesis, University of Agriculture, Faisalabad, Pakistan.

Allen, R.G., Pereira, L.S., Raes, D., and Smith, M. 1998. Crop evapotranspiration: Guidelines for computing crop water requirements. FAO Irrigation and Drainage Paper 56, FAO, Rome.

Bai, F.W., Anderson, W.A., Moo-Young, M. 2008. Ethanol fermentation technologies from sugar and starch feedstocks. *Biotechnology Advances* 26: 89–105.

Banerji, R., Chowdhury, A.R., Misra, G., Sudarsanam, G., Verma, S.C., and Srivastava, G.S. 1985. Jatropha seed oils for energy. *Biomass* 8: 277–282.

Berndes, G. 2002. Bioenergy and water—the implications of large-scale bioenergy production for water use and supply. *Global Environmental Change* 12: 253–271.

Blok, K. 2006. *Introduction to Energy Analysis.* Techne Press, Amsterdam, the Netherlands.

Catsberg, C.M.E., and Kempen-van Dommelen, G.J.M. 1997. *Levensmiddelenleer.* Uitgeverij Intro, Baarn, the Netherlands.

Chapagain, A.K., and Hoekstra, A.Y. 2004. *Water Footprints of Nations,* Value of Water Research Report Series No.16, UNESCO-IHE, Delft, the Netherlands.

Daey Ouwens, K., Francis, G., Franken, Y.J., Rijssenbeek, W., Riedacker, A., Foidl, N., Jongschaap, R., and Bindraban, P. 2000. Position paper on *Jatropha curcas.* State of the art, small and large scale project development. Results of the seminar held in March 2007, Wageningen University, the Netherlands.

Faay, A.P.C. 1997. Energy from biomass and waste. Thesis University of Utrecht, Utrecht, the Netherlands.

FAO. 2003. *World Agriculture Towards 2015/2030.* An FAO Perspective (ed. J. Bruinsma). Earthscan, London, UK.

FAO. 2007. CROPWAT 4.3 decision support system, Food and Agriculture Organization, Rome, Italy. www.fao.org/nr/water/infores_databases_cropwat.html.

FAO. 2008a. Food outlook. Global market analysis. www.fao.org.

FAO. 2008b. The state of food and agriculture 2008. Biofuels: prospects, risks and opportunities. Rome.

Fresco, L.O. 2006. Biomass for food or fuel: Is there a dilemma? The Duisenberg Lecture Singapore, September 17, 2006.

Gerbens-Leenes, P.W., Hoekstra, A.Y., and van der Meer, Th.H. 2009. The water footprint of energy from biomass: A quantitative assessment and consequences of an increasing share of bio-energy supply. *Ecological Economics* 68(4): 1052–1060.

Giampietro, M., and Ulgiati, S. 2005. Integrated assessment of large-scale biofuel production. *Critical Review in Plant Sciences* 24 (5): 365–384.

Goudriaan, J., Groot, J.J.R., and Uithol, P.W.J., 2001. Productivity of agro-ecosystems. In: *Terrestrial Global Productivity*, pp. 301–304. Academic Press.

Habekotté, B. 1997. Identification of strong and weak yield determining components of winter oilseed rape compared with winter wheat. *European Journal of Agronomy* 7: 315–321.

Hoekstra, A.Y., and Chapagain, A.K. 2007. Water footprints of nations: Water use by people as a function of their consumption pattern. *Water Resources Management* 21: 35–48.

Hoekstra, A.Y., and Chapagain, A.K. 2008. *Globalization of Water. Sharing the Planet's Freshwater Resources.* Blackwell Publishing, Oxford, UK.

Hoekstra, A.Y., Hung, P.Q. 2002. Virtual water trade: A quantification of virtual water flows between nations in relation to international crop trade. Value of Water Research Report Series No.11, UNESCO-IHE, Delft, the Netherlands.

Hughes, S., Partzch, L., and Gaskell, S. 2007. The development of biofuels within the context of the global water crisis. *Sustainable Development Law & Policy* 62:58–62.

Institute for Agriculture and Trade Policy. 2007. Biofuels and global water challenges. www.iatp.org.

Madison Center for Sustainability and the Global Environment, University of Wisconsin, 2008. www.sage.wisc.edu/download/majorcrops/majorcrops.html.

Minnesma, M., and Hisschemöller, M. 2003. Biomassa – een wenkend perspectief. Instituut voor Milieuvraagstukken (IVM), Free University, Amsterdam, the Netherlands.

Mitchel, D. 2008. A note on rising food prices. The World Bank Development Prospects Group. Policy Research Working Paper 4682.

Müller, M.J., and Hennings, D. 2000. *Climate 1, the Global Climate Data Atlas.* University of Flensburg, Inst. F. Geografie, Flensburg, Germany.

Nonhebel, S. 2002. Energy use efficiency in biomass production systems. In: *Economics of Sustainable Energy in Agriculture* (eds. E.C. van Ierland and A. Oude Lansink), pp. 75–85. Kluwer Academic Publishers, the Netherlands.

Penning de Vries, F.W.T., Jansen, D.M., Ten Berge, H.F.M., and Bakema, A.L. 1989. *Simulation of Ecophysiological Processes of Growth in Several Annual Crops,* pp. 63–64. Centre for Agricultural Publishing and Documentation (Pudoc), Wageningen, the Netherlands.

Pimentel, D., and Patzek, T.W. 2005. Ethanol production using corn, switch grass, and wood: biodiesel production using soybean and sunflower. *Natural Resources Research* 14(1): 65–76.

Postel, S.L., Daily, G.C., and Ehrlich, P.R. 1996. Human appropriation of renewable freshwater. *Science* 271 (9 February): 785–788.

Rosillo-Calle, F., De Groot, P., Hemstock, S.L., and Woods, J. 2007. *The Biomass Assessment Handbook,. Bioenergy for a Sustainable Environment.* Earthscan, London, Sterling, VA.

Verkerk, G., Broens, J.B., Kranendonk, W., Puijl van der, F.J., Sikkema, J.L., and Stam, C.W. 1986. Binas, informatieboek vwo-havo voor het onderwijs in de natuurwetenschappen, Tweede druk. Wolters-Noordhof, Groningen, the Netherlands.

Worldwatch Institute. 2007. Biofuels for Transport. *Global Potential and Implications for Sustainable Energy and Agriculture.* Earthscan, London, UK.

Mitigation

INTRODUCTION

Water utilities can choose from a wide range of mitigation strategies, both internal and external to the utility. These strategies include reduction of energy use through modification of existing operations, construction of renewable energy projects to reduce energy and GHG emissions usage at the utility, or participation in or construction of carbon offset projects. Customer water conservation can reduce energy use through reducing uses that require heating of water. Water utility policy and programs can encourage customer water conservation, and utilities can provide the education and incentives needed. Carbon offset projects and customer water conservation reduce emissions not directly under the control of the utility; however, they provide potential opportunity to achieve substantial emission reductions.

There is a growing awareness of the links between energy, water supply, and climate. Retamal, Turner, and White explore the need to break the link between climate change and supply capacity, and to move to a systems approach to water management that focuses on demand-side options. They explore the advantages and disadvantages of centralized versus distributed systems and suggest the use of an integrated resource planning framework for decision making.

Understanding of energy use across the urban water cycle begins with the ability to take action to reduce impact. Kenway et al. take a comprehensive look at the energy use of 10 water utilities (water and wastewater) in seven cities in Australia and New Zealand, including the energy needs for residential hot water systems. Intercity comparisons and future energy use scenarios are given. A main conclusion of the paper is that a targeted 20 percent reduction in residential hot-water use would completely offset the energy used for water service provision.

WATERGY is a term used by the Alliance to Save Energy to describe the strong connection between water and energy in distribution systems. Setty takes a practical look at the actions that utilities can take to improve energy efficiency in water supply and wastewater treatment operations, including how to generate political will, improve technical management, and implement efficiency measures.

Wastewater industry GHG emissions come primarily from two sources: digestion or decomposition of sewage and the decomposable fraction of sludge, and energy use associated with lift stations and other pumps used to convey wastewater. Eckhard's paper concentrates on energy use reductions. Utilities can increase pump efficiencies and lower energy use through actions such as replacing worn parts, strategically installing upgrades, matching pump size to the job, and improving pump management. Monitoring the system and collecting data can help utilities understand which actions to select and configure replacement hardware.

Lane reviews some of the climate change–related legislation in the U.S. Congress, reviews climate change–related programs implemented by U.S. federal agencies, and addresses potential impacts to drinking water utilities from GHG mitigation technologies, including geologic sequestration and use of biofuels, which are implemented outside of the water industry.

The Kyoto Protocol has influenced how climate change mitigation activities are viewed by water utilities in the developing world. Broeck and Nolasco's paper uses case studies of companies that are implementing programs to reduce emissions and trading "certified emissions reduction credits (CER)" under Clean Development Mechanism projects. Methods and examples for calculating methane recovery, avoidance of methane emissions and production, and mitigation of GHG emissions are discussed. The case studies show that significant CERs can be achieved but with a concluding caution that methods for calculating emission reductions need further research.

Paper 9. **The Water–Energy–Climate Nexus—Systems Thinking and Virtuous Circles**

Paper 10. **Energy Use in Urban Water**

Paper 11. **WATERGY: Energy and Water Efficiency in Municipal Water Supply and Wastewater Treatment: Cost-Effective Savings of Water and Energy Leading to Climate Change Mitigation**

Paper 12. **Station Efficiency Reduces Greenhouse Gas Emissions**

Paper 13. **Climate Change—Mitigation Policy Issues**

Paper 14. **Climate Change Mitigation Strategies in the Water Sector in Developing Countries**

– 9 –

The Water–Energy–Climate Nexus—Systems Thinking and Virtuous Circles

Monique Retamal, Andrea Turner, and Stuart White

INTRODUCTION

In recent years, there has been a growing awareness of the links between energy and water supply and the interrelationships between climate change and the water cycle. This nexus has been examined from various perspectives in the literature. Within the water industry, studies have largely focused on the energy used for water service provision and the potential impacts of long-term climate change on the security of water supply. However, the link between these two issues is not sufficiently recognised or considered in current water planning or decision making. The twin pressures of increasing water and energy demands need to be considered concurrently, to ensure that changes to one system do not adversely affect the other. This is particularly important at this point in time, especially in Australia, as climate change impacts are being felt and water service infrastructure is undergoing unprecedented investment.

Many countries, especially Australia, are currently being affected by unusual weather patterns and extended drought conditions. During this period there has been a trend to respond to drought and potential long-term climate change by investing in additional supply capacity. This supply-side approach has concentrated predominantly on nonrainfall dependent sources such as desalination and large-scale reuse schemes in addition to other solutions such as intercatchment transfers. However, such approaches often have the paradoxical effect of increasing energy use in the water sector, thereby increasing greenhouse gas emissions (GHG) and ultimately contributing to long-term climate change. This negative feedback loop or vicious circle is illustrated in Figure 9-1.

This problem can be characterised by the lack of application of a systems perspective. That is, the issue is dealt with one problem at a time (in this case climate change–induced drought), without appropriate identification of the consequences of the solution (energy-intensive centralised water supply infrastructure) in terms of generating further problems. These consequences are often not identified due to this compartmentalisation, and lack of a systems perspective.

As infrastructure providers strive to develop more sustainable systems in the face of climate change, it is essential that the augmentation of supply systems does not place greater strain on other limited resources. In the case of water infrastructure, it is imperative that new water supplies have minimal impacts on increasing energy demand and do not foreclose opportunities to improve the sustainability of both water and energy systems.

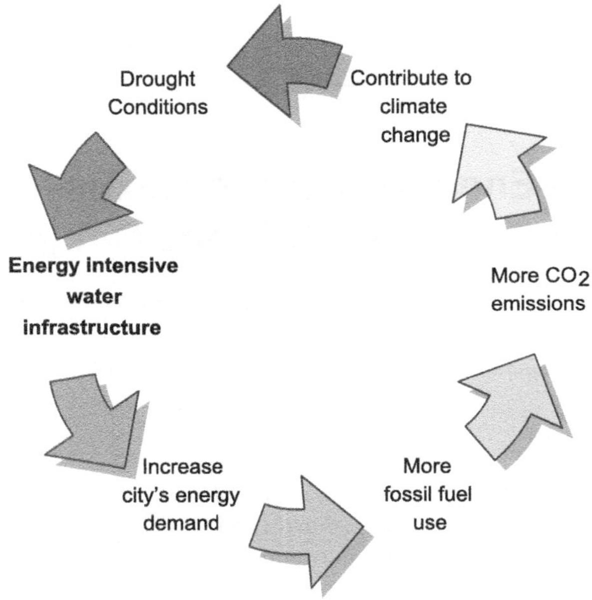

Figure 9-1. The vicious circle of solutions and responses that reinforces the problem

THE CURRENT SITUATION

Centralised Large-Scale Water Supply Systems

Over the last five years a number of studies have examined the energy intensity of existing water supply systems, including the Pacific Institute's 2004 report "Energy Down the Drain: The Hidden Costs of California's Water Supply" (Cohen et al. 2004) which calculated the average energy intensity for each step in the water cycle. In California, the average energy intensity for surface water treatment was found to be 0.55 kWh/kL and was dominated by the energy required to pump treated water. The delivery of water in local distribution systems used an average 0.21 kWh/kL, but was as high as 1.16 kWh/kL in some areas, depending on the topography. Wastewater treatment varied between 0.28 and 1.1 kWh/kL, depending on the type of technology used. This brings the total energy consumed per unit of water passing through the system to between 1.0 and 1.5 kWh/kL on average. However, in southern California, the energy intensity of water supplied is much higher as water is sourced from either San Francisco (2.43 kWh/kL) or the Colorado River (1.6 kWh/kL). In this area, the energy used to transfer water between catchments is several times greater than the energy used for water servicing elsewhere in California (Cohen et al. 2004). The example of southern California demonstrates the impact that large-scale water supply augmentation can have on energy consumption.

In Australia, the recent drought has brought about significant changes in water service provision, which has seen the energy intensity of water multiply in several cities. In a recent study by Kenway et al. (2008) the energy consumption of water infrastructure was examined for all major cities in Australia. The study found that the energy intensity of water supplied varied between 0.1 and 0.7 kWh/kL for those cities currently without major supply augmentation, and ranged between 1.0 and 1.8 kWh/kL for those cities that received additional water supplies through inter-catchment transfers and desalination plants. An intercatchment transfer from the Shoalhaven River has effectively quadrupled the energy intensity of water supply in Sydney from 0.25 to 1.03 kWh/kL over the past seven years. In Adelaide, an increase in pumping from the Murray River due to drought more than doubled the energy intensity of water supplied from 0.85 to 1.84 kWh/kL. In Perth, the energy intensity of water supplied almost doubled from 0.56 to 0.98 kWh/kL with the introduction of a desalination plant in 2007 (Kenway et al. 2008).

Even though significant investment has already been made in recent years on supply-side options in major urban centres across Australia, further augmentation is planned due to the continuing drought conditions. Many dams have recently been at their lowest levels in decades, putting into question the reliable yield estimates for these systems. Hence, desalination plants are currently being built or are planned for Sydney, Melbourne, and Adelaide and a second plant is being planned for Perth (Knights et al. 2007). In addition, in South East Queensland (SEQ), one of the fastest growing urban areas in Australia, major investment is being made into a desalination plant, a reuse scheme (including indirect potable reuse) and a new dam (over 100 km north of Brisbane) as well as a suite of other supply and demand-side options. These large supply-side solutions, which are being networked through a grid system over hundreds of square kilometres, will have a major impact on the energy intensity of water service provision in SEQ for decades to come (Turner et al. 2007a). These examples demonstrate an alarming trend toward more energy-intensive water supply infrastructure in Australia.

Distributed Small-Scale Systems

In addition to this trend toward major supply-side augmentation, drought in Australia has spurred the proliferation of smaller lot- and estate-scale water supply systems. These include household rainwater harvesting and greywater reuse systems as well as precinct and estate-scale wastewater reuse and stormwater harvesting schemes. From a water savings perspective, these options have the potential to save millions of litres of water if connected to multiple end uses (both indoor and outdoor), thereby optimising the system. However, very limited evaluation of the savings obtained from rainwater tank installations has been undertaken in Australia despite the significant government investment in rebates (for example, see Commonwealth of Australia 2009) and despite the imposition of regulations in various states requiring new homes to install such devices. A recent study undertaken by the Institute for Sustainable Futures (ISF) is one of the few studies that has investigated ex-post savings from rainwater tank installation stimulated by rebate programs. The study found that, based on theoretical analysis, rainwater tanks could save on average approximately 70 kL/household/annum in a particular region of SEQ. However, following statistical evaluation it was found that the savings were closer to 20 kL/household/annum because only a small proportion of households connected the rainwater tanks to indoor end uses that optimise the water savings potential (Turner et al. 2007b).

From an energy perspective, little is known about the energy intensity of these systems. One of the first studies to be conducted in Australia into the energy consumption of smaller-scale systems was carried out by Gardner et al. (2006) and Beal et al. (2008). This study monitored the energy intensity of a small estate-scale rainwater supply system in SEQ. This system is unusual in that it has individual rainwater harvesting systems for each house with a centralised overflow tank to provide extra storage and top up when individual tanks run low. The supply system operates through the use of individual household pumps plus a larger pump that recirculates water from the overflow tank. The energy intensity of this water supply was found to be 2.6 kWh/kL (Gardner et al. 2006), which is very high compared to the energy intensity of other centralised supply systems. In Sydney, preliminary monitoring studies carried out by ISF during 2008–2009 on the energy consumption of individual household rainwater supply systems found wide variations in system configurations and corresponding energy intensities. Typical rainwater tank configurations in a small sample of households monitored in Sydney were found to have an energy intensity of between 1.0 and 1.7 kWh/kL (ISF, forthcoming), which is much lower than the estate-scale system investigated by Gardner et al. (2006) for the reasons outlined above. The results do, however, raise questions regarding the increasing energy intensity of distributed water supply systems which, depending on their configuration, could actually be more energy intensive than large-scale systems. This will merely add to the demands on an already constrained energy system. The constraints on the energy system have been leading to major investments, which in New South Wales (the largest Australian state) alone are expected to run to approximately AUD $18 billion over the next five years (approximately 9.2 billion euro) for augmentation of the transmission and distribution network (i.e., poles, wires, transformers, and substations) for the current population of approximately 6.8 million (AER 2008).

The Costs of New Infrastructure

The examples of the large-scale, supply-side water system augmentation (e.g., desalination, major reuse and intercatchment transfer); smaller-scale water systems (e.g., rainwater tanks) and their subsequent impact on the energy system also call into question the wisdom of a fragmented approach to investment in water and energy infrastructure (i.e., the economic perspective). For example, in SEQ, major supply-side options such as the desalination plant, a major reuse scheme, and the Traveston dam (associated with intercatchment transfer) all have a unit cost of AUD $3.00/kL (1.56 euro/kL) or more. Similarly, the rainwater tank rebate program driving the adoption of rainwater tanks in existing households also has a unit cost well in excess of $3.00/kL (Turner et al. 2007a). Such high unit costs will have a significant impact on the price of water paid for by the community in SEQ. This will occur in parallel with the increasing price of electricity paid by customers, which in part will be due to the need to upgrade the electricity system due to the additional load being required for the more energy-intensive water and wastewater systems (both large- and small-scale). This situation will therefore drive yet another vicious circle of infrastructure investment that will be locked in for decades to come.

Greenhouse Gas Emissions

Another example of the lack of careful consideration of the link between the water and energy sector and greenhouse gas emissions is clearly demonstrated by concerns surrounding the practice of dedicating green power to service high energy–using technology, such as desalination plants. For example, at Perth's first desalination plant in Western Australia, green electricity is purchased from the grid to offset the GHG produced by the plant. This is in turn being used by the power utility to claim renewable energy certificates (RECs), which are included as part of a mandatory renewable energy target (MRET). This means that the renewable energy being generated is not reducing the amount of fossil fuel being burned in the city, but is merely offsetting the additional demand from the plant (Knights et al. 2007). Figure 9-2 illustrates this energy/GHG vicious circle.

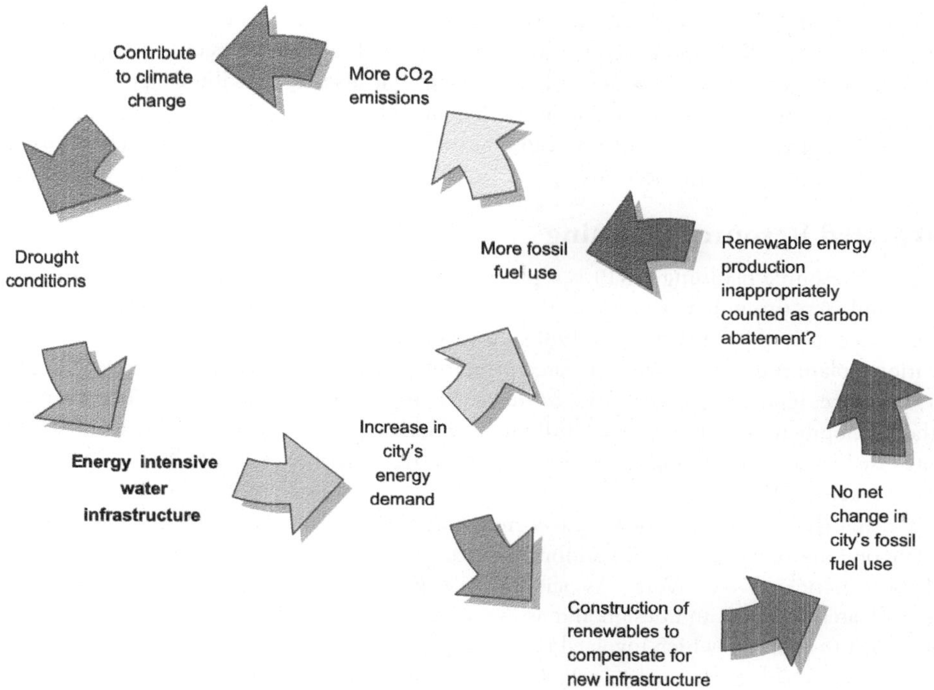

Figure 9-2. Renewable energy vicious circle

A WAY FORWARD?

We are currently at a critical juncture where we need to take stock of:

- The potential impacts of climate change (more extreme weather patterns which will not only increase the potential for drought but in some case increase the potential for flooding)
- How we should design our water systems to be more resilient
- How we can reduce energy usage and ultimately GHG emissions
- How we can invest in solutions that have multiple benefits for multiple physical systems (i.e., water, wastewater, stormwater, energy), the environment, and society as a whole.

A first step in the process of seeing these various layers and their complex interdependencies is to use a systems approach. This perspective has been applied in some cases from the water perspective for more than a decade (i.e., consideration of the water cycle within an urban catchment). However, the energy system and its impact on GHGs and ultimately climate change are seldom adequately integrated into water planning. An example of a systems perspective was used by the California energy commission (Klein et al. 2005) "*When energy initiatives reduce the air conditioning load in large buildings using cooling towers, every unit of heat that does not need to be removed means that less water is used in the cooling towers for evaporation. The cooling water saved in the building means less energy consumption by the water supply system; this energy saving also means less cooling water needed at thermal power plants—the action of multiple feedback loops.*"

It is precisely this systems perspective that is now required in the analysis and implementation in the urban water sector. This perspective will allow planners to recognise the potential for water efficiency to not only save water but also to contribute significantly to energy saving. It will also allow a recognition that an end use approach and the application of integrated resource planning principles can point to strategies which minimise societal costs, as indicated in the following sections.

Integrated Resource Planning

Integrated resource planning (IRP) is a process that enables resource managers to conceptualise and step through a structured process that incorporates systems thinking. IRP was originally conceived in the US in the 1980s. At this time, the energy industry recognised that traditional planning methods concentrated primarily on the construction of electricity supply infrastructure, ignoring opportunities to make existing networks more productive. This led to the development of IRP in which a full range of both supply-side and demand-side options are assessed against a common set of planning objectives (Swisher et al. 1997, Tellus Institute 2000, Vickers 2001).

During the 1990s, these new ideas were incorporated into water industry thinking. In the US, organisations such as the California Urban Water Conservation Council (CUWCC) and the American Water Works Association (AWWA) developed methodologies to better forecast water demand, and design and assess water conservation options. These approaches used "end-use analysis" which involved the disaggregation of demand into end uses (e.g., toilets, showers, washing machines) as the focal point of the analysis in order to better forecast demand and to develop strategies to reduce demand appropriately at the level of end use. Similar approaches were adopted by the UK Environment Agency (UKEA) to ensure water utilities manage both water demand and supply by considering the supply–demand balance and associated supply- and demand-side options in the same framework (UK Environment Agency 2003). In Australia, ISF has assisted in further refining the IRP process for both the Australian and international context (Turner et al. 2006). In Australia, an IRP approach has been adopted by the peak water body, the Water Services Association of Australia (WSAA), which represents the major water utilities across the country (Turner et al. 2008). In addition, it is being adopted by the National Water Commission for broader water industry application. Figure 9-3 illustrates an overview of the process.

The IRP framework has been used extensively in Australia (Turner et al. 2008), which has led to significant investment in demand management in recent years, thereby helping to actually reduce the water needed by the community. However, facing the potential impact of climate change, decision makers have often opted for familiar large-scale supply-side solutions instead of grappling with the complexity of the situation and assessing the impacts from a whole-of-system perspective. However, these large supply-side solutions may be forcing the water industry into a vicious circle of energy-intensive infrastructure that will become problematic as resources become more limited in the future.

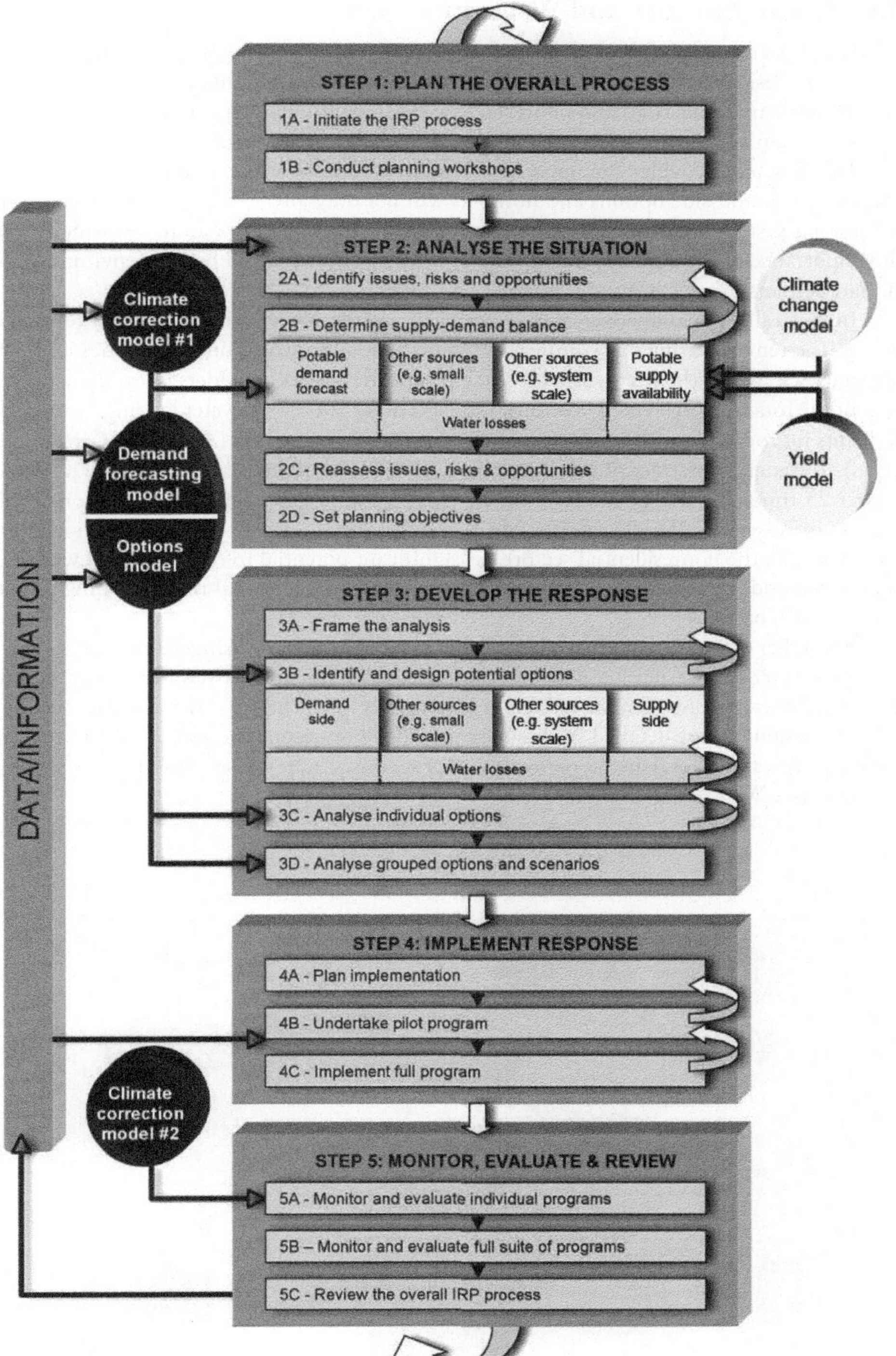

Figure 9-3. Integrated resource planning framework for decision making (Turner et al. 2006, 2008)

The Role of End Use and Water Efficiency

As indicated, end-use analysis is fundamental to the IRP process. By disaggregating water usage into the residential sector (single- and multidwelling houses); nonresidential (commercial, institutional, and industrial) and nonrevenue water (including leakage in the system), a greater appreciation of how water is used and should be used (the reference case) is obtained. This detailed view also provides the opportunity to determine how water can be saved through the use of demand-side options and how this will not only affect the water, wastewater, and stormwater system but also the energy system. This is illustrated in Figure 9-4 which shows the importance of the end user in the water and energy systems and broader environmental/climate systems.

In Australia, generally over 50% of water in an urban centre is used by the residential sector. The remainder goes to commercial, industrial, and institutional properties and nonrevenue uses. While the energy used by the major water service providers is less than 1% percent of the total national energy consumption (Kenway 2008), hot water heating in Australia accounts for roughly a quarter of all residential energy consumption (Australian Government 2008). The energy intensity of this hot water is 47 kWh/kL if heated from 18 to 60°C, which is over 25 times the energy intensity of the current Perth water supply system—one of the highest in Australia. Hence, efficiency in the use of water and hot water in the home (and more broadly, the nonresidential sector) has significant potential to reduce overall water and energy demand, GHG emissions, and ultimately to perform a pivotal role in helping to mitigate climate change.

A number of studies have investigated the potential benefits of using demand-side options to reduce water and wastewater demand, the associated water and wastewater energy required by the water service provider, the reduction in hot water by the householder, and a reduction in GHG emissions. These detailed studies have provided both economic and social and environmental justifications for demand management programs, e.g., for an example in Alice Springs, Australia (see White et al. 2006).

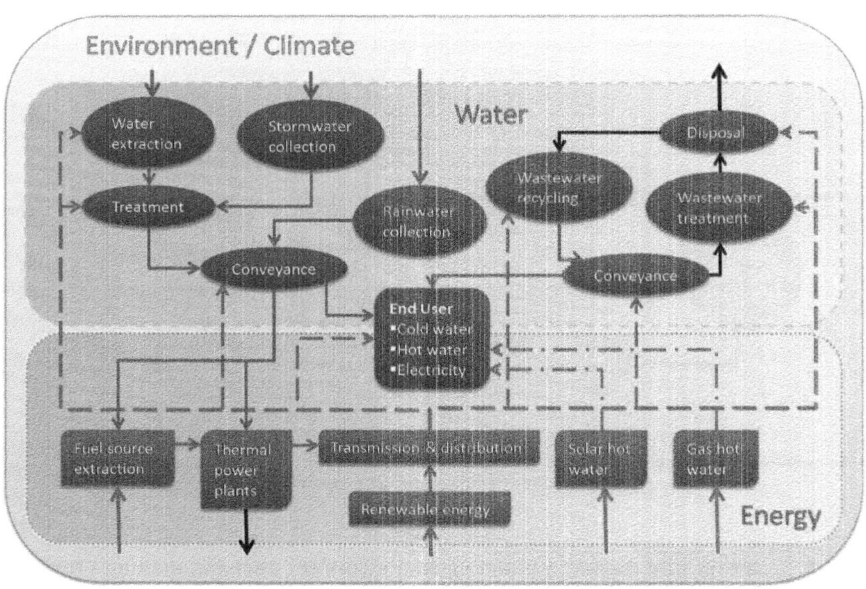

Figure 9-4. Simplified view of the inter-relationship between the water, energy, and climate systems

For example, a study by Turner et al. (2007a) for SEQ, Australia, compares a more traditional supply-side option (new dam and intercatchment transfers) with a unit cost of $3.00/kL (1.56 euro/kL) to a well-designed diverse portfolio of options concentrating on demand management with a unit cost of approximately AUD $1.00/kL (0.55 euro/kL). This analysis, developed as part of an IRP process, and illustrated in Figure 9-5, illustrates that not only does the demand-side suite of options cost less to society, but it has a net reduction in energy equivalent to the GHG reduction which would result from taking 230,000 vehicles off the road, approximately 15% of SEQ vehicles (Turner et al. 2007a).

The study by Kenway et al. (2008) investigated the energy intensity of centralised water supply systems in Australia. As part of the study, scenarios were examined to help mitigate increasing demands for water and energy. One of the scenarios modelled for improved household water efficiency compared the use of efficient (8.1 L/min) and nonefficient showerheads and found that a high water–using household could be expected to save 4,981 MJ/year. A low water–using household could be expected to save 1,609 MJ/year. This example illustrates not only the difference that efficient appliances can make, but also how water use behaviour can affect overall water and energy consumption (Kenway et al. 2008).

Flower et al. (2007) examined the GHG emissions related to domestic water end uses in different appliances (such as showers, toilets, dishwashers, etc.), and the impact of demand management. The Melbourne-based theoretical case study compared several scenarios of structural and nonstructural demand management with a base case using the least efficient appliances currently available. A Life Cycle Analysis approach (LCA) was used to determine water savings and GHG emissions for each appliance in each scenario, assuming a gas hot water system. The results showed that the scenario using the most efficient appliances reduced GHG emissions by 53% (1,618 kg CO_2-eq/year) when compared to the least efficient scenario. The scenario that combined structural and behavioural change with the most efficient appliances and reduced water use saved 63% (1 927 kg CO_2-eq/year) of GHG emissions produced in the least efficient scenario (Flower et al. 2007).

The Virtuous Circle

The water and energy implications of new large-scale centralised water supply projects need to be considered and questioned. While the motivation for constructing these projects has been based on concerns regarding climate change–induced reductions in supply yield, steps need to be taken now to ensure that water planners are not locking into a vicious circle of increased energy intensity of water supply. This will merely fuel the climate change problem and generate high costs that the community will need to pay for over decades to come. In addition, with the proliferation of smaller-scale lot and estate-scale distributed systems, it is now also important to evaluate their actual water savings and energy intensity to ensure that these systems are not resulting in suboptimal energy consumption outcomes which will be difficult to manage in the future.

With water use and hot water heating and usage being such significant contributors to household water and energy demand, it is vital to acknowledge the important role that energy and water efficiency play in reducing overall energy demand (in both the water and energy sectors) and that they can play a pivotal role in helping to mitigate climate change.

We are currently at a critical juncture where we need to take stock of the potential impacts of climate change; how we should design our water systems to be more resilient; how we can reduce energy usage and ultimately, GHG emissions, and how we can invest in "solution-multiplying solutions," rather that "problem-multiplying solutions."

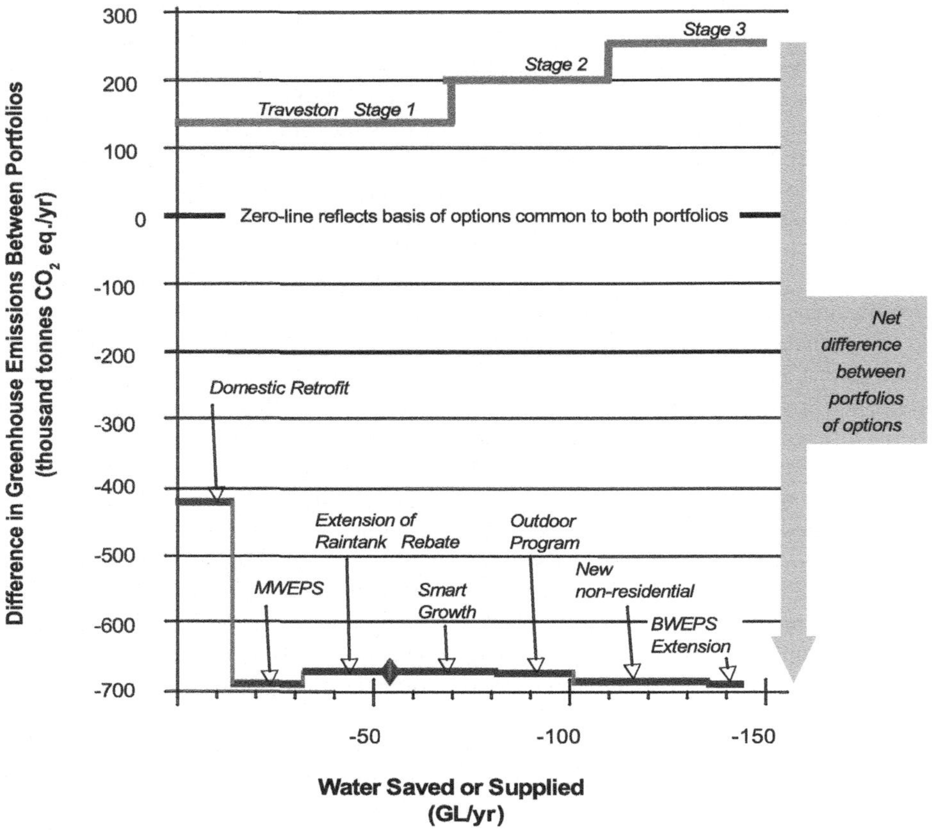

Figure 9-5. Net difference in GHG emissions between two portfolios of options providing the same volume of water in the future

The water–energy–climate nexus is complex. As researchers and water and energy practitioners, we need to act now to unveil this complex picture so that water and energy service providers and decision makers can make well-informed decisions that will help mitigate climate change and not fuel it.

REFERENCES

Australian Energy Regulator. 2008. *Draft Decision New South Wales Draft Distribution Determination 2009–10 to 2013–14*, 21 November 2008.

Beal, C., Hood, B., Gardner, T., Lane, J., and Christiansen, C. 2008. Energy and water metabolism of a sustainable subdivision in South East Qld: a reality check. Paper presented to the ENVIRO'08.

Klein, G., Krebs, M., Hall, V., O'Brien, T., and Blevins, B. 2005. *California's Water–Energy Relationship*, California Energy Commission.

Cohen, R., Nelson, B., and Wolff, G. 2004. *Energy Down the Drain: the Hidden Costs of California's Water Supply*. Natural Resources Defense Council and Pacific Institute, Calif.

Commonwealth of Australia. 2008. Energy Use in the Australian Residential Sector 1986–2020. Report prepared by Energy Efficient Strategies for Department of Environment, Water, Heritage and the Arts.

Commonwealth of Australia. 2009. *National Rainwater and Greywater Initiative.* Department of Environment, Water, Heritage and the Arts. http://www.environment.gov.au/water/programs/nrgi/index.html.

Flower, D.J.M., Mitchell, V.G., and Codner, G.P. 2008. The potential of water demand management strategies to reduce the greenhouse gas emissions associated with urban water systems. 1st Conference on Sustainable Urban Water Management & 9th Conference on Computing and Control in the Water Industry, Leicester, UK, 3–5 September, 2007.

Gardner, E.A., Millar, G.E., Christiansen, C., Vieritz, A.M., and Chapman, H. 2006. Energy and water use at a WSUD subdivision in Brisbane, Australia. *Australian Journal of Water Resources,* 10, No. 3.

ISF (forthcoming). Water energy nexus: final report [prepared for CSIRO]. Institute for Sustainable Futures, UTS, Sydney.

Kenway, S.J., Priestley, A., Cook, S., Seo, S., Inman, M., Gregory, A., and Hall, M. 2008. *Energy use in the provision and consumption of urban water in Australia and New Zealand.* CSIRO Water For a Healthy Country National Research Flagship.

Knights, D., MacGill, I., and Passey, R. 2007. The sustainability of desalination plants in Australia: is renewable energy the answer?, OzWater Conference, 5th March 2007, Sydney.

Swisher, J.N., Jannuzzi, Gilberto de Martino, and Redlinger, R.Y. 1997. *Tools and Methods for Integrated Resources Planning: improving energy efficiency and protecting the environment.* UNEP Collaborating Centre on Energy and Environment, Riso National Laboratory, Roskilde, Denmark.

Tellus Institute. 2000. *Best Practices Guide: Integrated Resource Planning for Electricity.* The Energy Group, Institute of International Education, Washington D.C.

Turner, A.J., Willetts, J.R., and White, S. 2006. *The International Demand Management Framework Stage 1* [prepared for Canal de Isabel II]. Institute for Sustainable Futures, UTS, Sydney.

Turner, A.J., Hausler, G., Carrard, N.R., Kazaglis, A., White, S., Hughes, A., and Johnson, T. 2007a, *Review of Water Supply-Demand Options for South East Queensland,* [prepared for Mary River Council of Mayors], Institute for Sustainable Futures, UTS and Cardno (Brisbane), Sydney.

Turner, A.J., White, S., Kazaglis, A., and Simard, S. 2007b. Have we achieved the savings? The importance of evaluations when implementing demand management. *Water Science and Technology,* 7, no. 5–6, 203–210.

Turner, A.J., White, S., and Edgerton, N. 2007c. Alice Springs Water Efficiency Study Stage III—Implementation of the Alice Springs Water Efficiency Program—Feasibility Study—Final Report [prepared for Northern Territory Government]. Institute for Sustainable Futures, UTS, Sydney.

Turner, A.J., Willetts, J.R., Fane, S.A., Giurco, D., Kazaglis, A., and White, S. 2008. *Guide to Demand Management* [prepared for Water Services Association of Australia]. Institute for Sustainable Futures, UTS, Sydney.

UK Environment Agency. 2003. Water Resources Planning Guideline. Version 3.3.

US DoE. 2006. Energy Demands of Water Resources—Report to Congress on the interdependency of energy and water. US Department of Energy.

Vickers, A. 2001. *Handbook of Water Use and Conservation.* Water Flow Press, Amherst, Mass.

– 10 –

Energy Use in Urban Water

S.J. Kenway, A. Priestley, S. Cook, A. Gregory, A. Lovell, and N. Smith

INTRODUCTION AND BACKGROUND

This paper considers energy use through the urban water cycle, including water consumption. Many cities worldwide are evaluating the implications of climate change on water resources for urban, industrial, and agricultural uses. This is particularly the case in Australia, where sustained below-average rainfall and above-average temperatures over the last decade have devastated agricultural production in many inland areas and reduced water availability in major and regional cities to levels well below historical experience. There is growing consensus that these climatic variances are at least partly due to climate change and that a return to climatic conditions typical of recorded history is unlikely.

All Australian cities are responding to this through strategies to improve the efficiency of water use and diversify supplies, with major investments in more energy intensive manufactured water supplies such as desalination and water recycling in its various forms. Understanding the relationship between water and energy use across the urban water cycle is increasingly important to inform the adaptation of water systems, address escalating energy costs, and minimise the greenhouse emissions related to water supply and end use.

The paper draws on a project undertaken collaboratively by CSIRO and the Water Services Association of Australia (Kenway et al. 2008). Sustainability Victoria, and representatives of 10 water utilities also contributed to the project and are acknowledged here.

Management of urban water in Australia is diverse. In some systems, such as Western Australia, a single entity (The Water Corporation) manages all water and wastewater management for the entire state, including the city of Perth. In other cities, very different arrangements exist. In some cases, this makes the task of sourcing and forecasting energy data more complex. For example, Sydney Water manages water treatment, distribution, and retail water services as well as all wastewater transport and treatment. Sydney Catchment Authority provides bulk water supplies. In Melbourne, three entities (Yarra Valley Water, South East Water, and City West Water) manage retail functions and minor wastewater treatment plants, while Melbourne Water provides bulk water and wastewater services. In other systems, such as Brisbane, a number of local government entities have traditionally managed water and wastewater; however, this is now in the process of centralisation. Private operators are increasingly running water and wastewater treatment and transport operations and desalination facilities in most cities.

Utilities provided detailed data on energy use to a pro-forma designed by CSIRO. These data underpinned current energy use estimates. Water use projections and a range of future water sources were used to project energy use out to 2030. Hot water energy estimates were derived from water usage data provided, coupled with literature review for end-use breakdown and first-principles modelling of energy required to heat water. In all cases, offsets such as

Figure 10-1. Location of cities

greenpower purchase or sequestration are ignored to avoid obscuring the real energy consumption in each system.

The aim of the study was to improve the understanding of energy use by Australian and New Zealand water utilities and place that in context with energy needs for residential hot water and total urban systems. It was also undertaken to meet an increasing need for information on options that improve energy efficiency and mitigate greenhouse gas emissions in urban water into the future. The work was intended to help find improved approaches and policies to enable greater influence on energy use and greenhouse gas emissions abatement.

The project focused on seven cities in Australia and New Zealand: Sydney, Melbourne, Brisbane, Gold Coast, Adelaide, Perth, and Auckland, shown in Figure 10-1. The climatic conditions vary considerably between these cities, with Brisbane and Gold Coast influenced by more tropical rainfall patterns, Sydney and Melbourne have relatively even temperate rainfall, and Perth and Adelaide experience marked seasonal differences in rainfall.

The analysis included extensive industry survey for energy use through the water cycle; modelling of residential hot water use, projecting future energy use, and placing this in context of total urban systems energy use. Energy use was the focus due to the relative paucity of data sets for urban water cycle greenhouse gas emissions and the large general contribution to greenhouse gas emissions that energy use currently makes in Australia. Consequently, this paper largely uses units of Petajoules (PJ). Use of 0.36 PJ electrical energy (100 GWh) from Australia's coal-fired electricity grid contributes approximately 100,000 tonnes of CO_2-eq. in indirect emissions factors as full fuel cycle including Scope 2 and 3 emissions (Department of Climate Change 2008).

Data from 10 utilities surveyed for energy use through the water cycle (Table 10-4, Appendix) were aggregated to give city-wide energy use (Table 10-1). Pro-rata estimates where necessary, for example, where only one retail entity (of several) were surveyed.

DETAILS AND CONTEXT FOR INDIVIDUAL CITIES

Supply to Sydney comes mainly as gravity feed from its major Dam at Warragamba. However, in drought periods (e.g., 2006/07), there is extensive pumping, which sharply lifts Sydney's

Table 10-1. Energy and water use by city (2006/07)

	Sydney	Melbourne	Perth	Brisbane	Gold Coast	Adelaide	Auckland
Population served	4,300,000	3,621,000	1,538,000	1,006,000	492,000	1,095,000	1,232,000
Water supplied (GL)							
Total	507	412	235	113	65	159	136
Residential water	315	257	170	61	40	112	83
Indoor water use (%)	65	84	53	—	—	—	—
Wastewater collected (GL)	508	296	119	86	47	89	104
Total energy (GJ)							
Water supply							
Pumping	1,687,960	125,355	423,000	28,245	39,416	1,041,901	44,460
Treatment	186,009	12,860	409,000	246,337	9,234	55,418	56,749
Wastewater							
Pumping	119,916	459,713	92,800	39,726	50,030	32,064	42,697
Treatment	698,205	739,243	213,000	138,028	119,389	185,194	273,593
Other energy demand	250,838	131,728	162,700	49,070	39,461	123,240	23,157
Total energy demands	2,942,929	1,468,900	1,300,500	501,406	257,530	1,437,817	430,504
GHG emissions for energy-related sources (k t CO_2-eq.)	774	302	313	138	75	392	31

water supply energy consumption. Sydney also has 14 water filtration and/or chlorination plants, of which four are privately operated. The city has three major coastal plants processing about 75% of the total volume of wastewater. These plants provide primary treatment, with deep ocean outfall disposal. One of these plants, North Head Sewage Treatment Plant (STP), requires all wastewater to be lifted 50 m to the top of the headland because the ocean outfalls require sufficient pressure to enable dispersal of the effluent. North Head STP accounts for approximately 15% of the total electricity consumption of Sydney Water. Energy requirements for both water treatment and wastewater pumping are relatively low. The energy intensity of water supply in Sydney in 2006/07 was 3,696 GJ/GL—an increase of 300% since 2000/01 when the energy intensity was 915 GJ/GL (Figure 10-2).

Melbourne Water provides bulk water and wastewater services for Melbourne. Demand management and enhanced public awareness of decreasing rainfall patterns, declining inflows to storages, and subsequent water restrictions have played a significant role in this reduction. The energy requirement for water supply fluctuated significantly during this period due to changes in the amount of pumping from the Yarra River into Sugarloaf Reservoir. In 2006/07, the energy intensity of Melbourne's water supply was only 335 GJ/GL. In contrast, the city used 4,051 GJ/GL to treat and dispose of its wastewater. This relationship is because most of Melbourne's water is gravity fed from protected mountain catchments and only a small percentage is treated, while the wastewater is pumped long distances and requires extended levels of treatment.

Until recently, water that was supplied to Perth had a relatively consistent energy intensity of around 2,000 GJ/GL. This jumped dramatically to 3,540 GJ/GL when the desalination plant was commissioned in 2006. The dominant source of energy for Perth's water system is electricity (1.15 × 106 GJ), with a small amount (57,200 GJ) being generated from biogas produced in wastewater treatment plants. The additional demand for energy created in 2006/07 for the desalination plant means that now the power requirements for water supply pumping and treatment are two to three times those for wastewater—attributed to the relatively low pumping energies required and relatively small volumes of wastewater flow compared with water supply.

Brisbane Water's energy use profile has declined despite a slow but steady (1.2%) increase in population. This is likely due to the declining volume of water demand from 2004/05 in response to the severe water restrictions imposed as a result of very low storage levels. Brisbane Water has relatively high energy intensity for water supply at 2,431 GJ/GL, mainly due to the need to pump water to the Mt. Crosby and North Pine treatment plants. This energy value for pumping is included with the total energy value reported for water treatment. The requirement for tertiary treatment of wastewater before discharge to Moreton Bay is also a significant driver of energy use, resulting in an energy intensity of 2,069 GJ/GL for wastewater disposal and a contribution of approximately 40% of total energy requirements. The Gold Coast displays similar trends.

For Adelaide, the total power requirements for water supply depends significantly on the proportion of supply pumped from the Murray River at Mannum. In 2006/07, for example, the power requirements for pumping jumped to a high of 1,042,000 GJ as an extra 48 GL were pumped from the Murray River to provide extra storage during drought conditions. The energy demand for water supply pumping increased by 117% between 2005/06 and 2006/07 and this is reflected in the energy intensity of water supply of 6,901 GJ/GL in 2006/07. Energy consumption for wastewater pumping and treatment in Adelaide was about 20% of that for water supply pumping and treatment, at 217,258 GJ. This is largely due to a very low pumping energy requirement of only approximately 32,000 GJ/year. (Note: there is a 20 m fall between Adelaide city and Bolivar Wastewater Treatment Plant.)

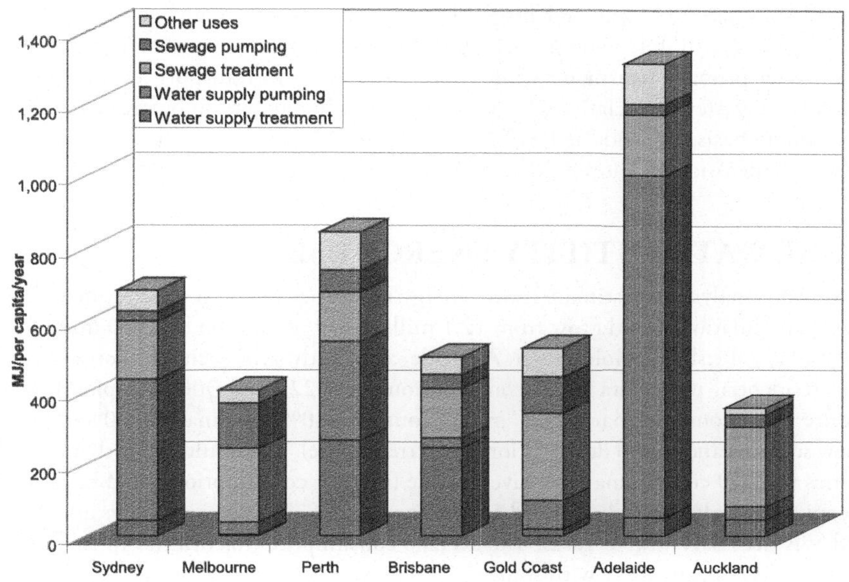

Figure 10-2. Energy use for water and wastewater services per capita (2006/07)

INTERCITY COMPARISON

For Adelaide, Perth, and Sydney, energy input was highest for water supply. In Melbourne and the Gold Coast, wastewater disposal used more energy. Adelaide and Perth use significantly more energy per customer than Melbourne and Auckland. A major point is that fragmentation of the water cycle can obscure energy use patterns. For example, Sydney Catchment Authority used nearly 500,000 GJ in 2006/07. Consequently, considering only data from Sydney Water (which used approximately 1,200,000 GJ) would only provide part of the energy use picture.

Collectively, Australian utilities used 7.1 PJ energy (590 MJ/capita national average) in 2006/07. The majority of this energy (76%) was sourced from coal-fired power plants via the electricity grid. Local conditions, including water consumption patterns, topography, and water sources, have a major influence on energy use. For example, in some systems, large amounts of energy are used to pump raw and treated water to elevated treatment plants.

Energy intensities for water supply (the amount of energy needed to deliver water) ranged from 335 GJ/GL for Melbourne to 6,901 GJ/GL for Adelaide. This wide disparity can be explained in part by extreme drought conditions and atypical water shortages requiring increased long-distance pumping for Adelaide and Sydney in 2006/07. In contrast, most of Melbourne's water supply was gravity fed from elevated catchment storages. Significant energy is required to lift water against gravity—typically needed for long distance water transport. Adelaide's energy use for water supply (6,901 GJ/GL) is more than half of the energy intensity of seawater desalination (approximately 13,000 GJ/GL), based on various estimates including Gardner et al. (2006).

For wastewater treatment and disposal, energy intensities ranged from 1,610 GJ/GL (Sydney) to 4,051 GJ/GL (Melbourne). Sydney discharges most of its wastewater directly to the ocean after primary treatment, while Melbourne needs to transport its secondary- and tertiary-treated wastewater relatively long distances over higher terrain before ocean disposal. On a per-capita basis, Adelaide at 1,313 MJ/(capita*year) is the most energy-intensive supply per person, while Auckland at 349 MJ/(capita*year) is the least intensive supply.

FUTURE WATER UTILITY ENERGY USE

Scenarios were analysed to estimate future energy use associated with urban water. These assumed that population would grow from 12.1 million to 15.8 million by 2030 (total for Sydney, Melbourne, Brisbane, Gold Coast, Adelaide, and Perth). The scenarios considered three levels of residential per capita water consumption—150,225 and 300 L/(capita*day)—and two future water source scenarios—a "mix of sources" (40% desalination, 40% re-use, and 20% new sources) and 100% desalination (an extreme case). The results show that energy use by utilities in 2030 could remain relatively stable (if water consumption could be reduced to 150 L/capita*day); alternatively it could grow to around 26–36 PJ/year if water use of 300 L/capita*day is necessary (Table 10-2). The average consumption for utilities in this study for 2006/07 was 217 L/capita*day with some cities on severe water restrictions.

ENERGY USE FOR RESIDENTIAL HOT WATER

Energy required to heat residential hot water was calculated using data derived from utilities (refer above) and using the equation $E=CM\Delta T$ (based on <http://hyperphysics.phy-astr.gsu.edu/hbase/thermo/spht.html>). In this equation, E is the energy required to heat the water to the desired temperature; C is the specific heat for water (amount of heat per unit of mass required to raise temperature 1°C is 4.186 J); M is the volume of water to be heated; T is the increase in temperature required. A temperature end point of 60°C and start point of 18°C was assumed for all cities. The estimate of energy use for residential hot water of 46 PJ (Table 10-3) is an order of magnitude higher than the energy used directly by utilities. Energy use associated with industrial and commercial water use (e.g., water heating) is anticipated to be of similar magnitude to the energy for residential water heating. However, only minimal data could be found to verify this and consequently this information should be sourced as a priority.

Residential hot water uses on average 6.5 times the energy that is used to deliver urban water services, this ratio ranging from 4.7 in Adelaide to 11.2 in Melbourne. This means that at a national level, a 15% reduction in the use of residential hot water or an equivalent increase in the efficiency of residential hot water systems would completely offset the total energy used by the utilities providing water to those households in 2006/07. However, care must be taken interpreting this for any particular city.

The results suggest that residential water demand management strategies should be targeted at energy-intensive end-uses, such as showers and washing machines, as these can significantly reduce household energy demand and associated greenhouse gas emissions. Analysis showed that shifting to a WELS 3-star shower rose would decrease energy consumption for hot water by approximately 50% for households with considerably greater-than-average water use.

Table 10-2. Energy demand in 2030 for water at three residential water consumption levels considering two supply scenarios

	Mix of Supply Sources (40% desalination, 40% re-use, 20% new sources)			100% Desalination (extreme case)		
Residential Water Consumption Levels L/((cap*d))	Total Energy Required (PJ)	% Increase for Utilities (from 2006/07 use of 7.1 PJ)	% of Total Urban System	Total Energy Required (PJ)	% Increase for Utilities (from 2006/07 use of 7.1 PJ)	% of Total Urban System
300	26	260	0.5	36	400	0.7
225	16	130	0.3	21	200	0.4
150	7	0	0.1	7	0	0.1

Note: Estimates are based on current yields and assume the yield and energy intensity of sources providing water in 2006/07 remain constant. Under these assumptions, an additional 1,400, 700, and 0 GL respectively would be required to meet residential consumption levels of 300, 225, and 150 L(cap day) respectively for 15.8 million people. Energy use for the total urban system is a pro-rata use of state-wide energy use (including agriculture, mining, and electricity generation).

Table 10-3. Residential hot water – volume and energy (2006/07)

	Residential Water Supplied (GL/year)	Total Indoor (GL/year)	Volume of Residential Hot Water (GL/year)	Energy for Residential Hot Water (PJ/year)
Sydney	315	205	82	14
Melbourne	257	216	86	15
Brisbane	61	45	18	3
Gold Coast	40	30	12	2
Perth	170	87	35	6
Adelaide	112	83	34	6
Total	955	665	267	46

INDUSTRY IMPLICATIONS AND ACTIONS

Water utilities use less than 15% of the energy used for residential water heating (estimated as 46 PJ in 2006/07) and about 0.2% of total urban energy use. A 20% reduction in the residential hot water use or an equivalent increase in the efficiency of residential hot water systems would completely offset the total energy currently used for water service provision. It is critical that appropriate policies be developed to ensure that energy efficiency (and greenhouse gas reduction) opportunities at the household level are maximised to ensure the whole water cycle is considered, rather than simply energy use in the provision of water services.

Local conditions significantly influence energy use through the water cycle, including the relative consumption of energy in water versus wastewater components of the cycle. The work could only consider one year of data in detail. Consequently, many seasonal or one-off trends could not be considered or analysed. Understanding the true trends and variances of energy use through urban water will be critical to minimising its use.

There are many areas that water utilities are progressing to improve management of their direct energy use. Improved definitions of "treatment" and "transport" will help in these efforts. Industry is also improving monitoring and operational energy-efficiency programs. Other schemes are developing and implementing schemes for the internal generation of energy such as biogas generation or mini-hydro schemes.

Other Recommendations Include:

- Analysis of data from a larger number of utilities, particularly to characterise energy used with different supply and wastewater treatment processes (e.g., primary, secondary).
- The influence of increased uptake of decentralised water supply options on energy consumption warrants further analysis. Due to lack of data, decentralised systems were not considered in this project.
- Characterisation over a longer period of time would help detect longer-term drivers and also smooth results for local influences on particular systems that may have affected the 2006/07 result. Analysis over a longer time period would be essential to identify underlying trends and to remove variability potentially influencing some cities in the particular year analysed.
- Analysis needs to extend beyond individual water utilities. Where more than one utility is involved through the supply chain for water and wastewater services, it becomes increasingly difficult to obtain a true picture of the total energy required to provide water and wastewater services. It also becomes increasingly difficult to estimate the impacts of alternative strategies for future water provision. Alignment of strategies through the water cycle will be necessary to ensure that the best "whole-of-system" outcomes or that the lowest possible "whole-of-system" greenhouse gas emissions rates are achieved.
- Private participation in water management is increasing and in places, water cycle management is being fragmented and decentralised systems are being increasingly used. System-wide information will be necessary in the future to enable decision making that provides the most cost- and energy-efficient solutions for urban communities.
- Improved monitoring of energy associated with water end use is needed to help quantify the relationship between energy and use, and to better inform management options.
- The heating of water for industrial and commercial uses is expected to represent a similarly significant water-related energy use to the residential sector. Improved data are necessary to characterise this.
- Improved tracking of greenhouse gas emissions within utilities (particularly Scope 3 and Scope 2 emissions) is necessary, as relatively little data or knowledge about this issue currently exist.
- Public reporting of energy use associated with water use is recommended. Public reporting can help communities understand their energy use. It would also help improve associated monitoring, auditing, and accounting processes. Some utilities are already reporting the estimated greenhouse gas savings attributable to water conservation measures (Sydney Water 2008). It is suggested that WSAA voluntarily

adopt such reporting measures. This would require development of appropriate reporting indicators, data definitions, monitoring programs to acquire data, and development of an agreed-upon energy calculation methodology. Such information would help confirm the magnitude of current energy use associated with water use and thereby improve estimates of the influence of water supply options on energy use and associated greenhouse gas emissions. It is possible that collaboration with energy supply companies (e.g., electricity and/or gas service providers) could improve the quality of information in this effort.
- Financial analysis is necessary to ensure that least-cost solutions are found, not just those that save the greatest amount of greenhouse gas emissions. Consideration of future carbon prices will also help water service providers (and others) identify the most cost-effective options.

This study has provided a new understanding of the direct energy used by water utilities and energy use associated with residential water heating. While water utilities have a direct responsibility for a small portion of total urban energy (0.2%), urban water choices—including the use of water—have a much wider opportunity to impact energy use. When all energy use is considered (including 1.2% for residential water heating, approximately 1.0% for nonresidential water heating, energy "embedded" in nutrients in wastewater that have to be subsequently synthesised, etc.) it is possible that water policy choices could influence 3–6% of total urban systems energy use.

BROADER CONSIDERATIONS FOR WATER AND ENERGY POLICY

The study has also highlighted the increasing need to understand the broader implications of water and energy management choices right across the urban water cycle. These choices are often beyond the control of an individual water utility but can have significant implications for city sustainability. A few key considerations include:

- Further analysis is warranted to characterise the direct and indirect water and energy linkages across the whole urban water system including water use, energy demand, and greenhouse gas emissions. This will enable improved understanding of the water, energy, and greenhouse gas emissions implications associated with water management strategy choices.
- Water is a key input to energy production and energy a major input to water production and distribution. For example, in some Australian cities large volumes of manufactured water are being pumped uphill for significant distances to supplement water needed for energy production. Similarly, policy decisions regarding distributed energy production versus more power stations will impact on water supplies and associated energy use. Improved integration of water and energy planning has the potential to generate city-wide co-benefits for both systems and the community.
- Managing demand—the demand for water and energy water and energy distribution infrastructure—is sized to meet the peaks in demand. The higher probability of extreme climatic events due to climate change will place even greater peak demand pressures on water and energy networks; indeed, these extremes are already being experienced in Australia. Pursuing water and energy demand-side options that help to flatten demand peaks as well as reduce base-load demands can return benefits across to both systems. For example, the major residential showerhead retrofit program delivered in Sydney since 2000 has reduced city wide water demand by 10 GL/year with a co-benefit of reducing household hot water-related greenhouse emissions by an

estimated 290,000 tonnes per year (Sydney Water 2008). This effectively offsets 80% of the water utilties' average annual emissions footprint. Similarly, as demonstrated in this paper, pursuing a widespread residential solar hot water policy has the potential to provide a five-fold offset of all water service-related emissions.

- Hybrid systems—The above study highlighted the large differences in energy used for treatment and pumping in different cities, largely driven by local conditions and the technologies historically available for large centralised supply networks. In combination with water efficiency initiatives, decentralised water systems for rainwater harvesting, and water reuse recycling offer the potential to significantly reduce water demands on central networks. There is a need for improved data on actual performance of these systems, including their energy use. However, considering the rapidly increasing energy footprint of centralised systems, further research is warranted to clarify the potential benefits of these alternatives if adopted at local- and citywide scale. This should be analysed in consideration of emerging distributed energy generation alternatives that can power these alternatives and the citywide implications for water and energy systems.
- Resource recovery—urban wastewater has traditionally been viewed as a pollutant requiring sufficient nutrient removal to minimise or avoid environmental degradation upon discharge to receiving waters. In light of increasingly stringent environmental water quality standards, rising water system energy costs, the desire to reduce greenhouse emissions, and diminishing worldwide phosphorous reserves for fertiliser production, urban wastewater may provide a significant source of required nutrient and energy resources. The realisation of this potential will require the development of new recovery technologies, possibly in conjunction with reconfiguration of wastewater systems.

REFERENCES

Department of Climate Change. (2008). National Greenhouse Accounts (NGA) Factors, Australian Government, Canberra.

Kenway, S.J., Priestley, A., Cook, S., Seo, S., Inman, M., and Gregory, A. (2008). Energy use in the Provision and Consumption of Urban Water in Australia and New Zealand. Water Services Association of Australia.

Sydney Water Corporation. (2008). Water Conservation and Recycling Report 2007-08.

Table 10-4. Raw data from water utilities (2006/07)

	Sydney Water	Sydney Catchment Authority	Melbourne Water	Yarra Valley Water	Water Corporation	Brisbane Water	Gold Coast Water	SA Water	Watercare (NZ)	Metrowater (NZ)
Population	4,300,000	N/A	N/A	1,571,650	1,538,000	1,006,000	492,000	1,095,000	1,232,000	431,000
Water supplied (GL)	509	507	412	160	235	113	65	159	136	54
Residential water supplied (GL)	315	N/A	—	107	170	61	40	111	N/A	—
Wastewater collected (GL)	508	N/A	296	108	119	86	47	88	104	53
Total Energy (GJ)										
Water supply—pumping	476,298	1,211,662	86,185	16,321	423,000	28,245	39,416	1,041,901	39,327	1,711
Water supply—treatment	186,009	N/A	12,860	Not reported	409,000	246,337	9,234	55,418	56,749	—
Wastewater—pumping	119,916	N/A	436,467	9,686	92,800	39,726	50,030	32,064	37,978	1,573
Wastewater—treatment	698,205	N/A	645,715	38,970	213,000	138,028	119,389	185,194	273,293	100
Other energy demand	220,522	30,316	67,000	26,970	162,700	49,070	39,461	123,240	23,157	—
All demands	1,700,950	1,241,979	1,248,227	91,947	1,300,500	501,406	257,530	1,437,817	430,504	—
Greenhouse gas emissions—total reported (or FFC) for energy-related sources (t CO_2-eq.)	448,282	326,110	233,633	28,482	312,850	138 000	74,505	392,486	31,883	—

N/A = not applicable; – = no data provided; Data provided by Water Utilities in May 2008.

– 11 –

WATERGY: Energy and Water Efficiency in Municipal Water Supply and Wastewater Treatment

Cost-Effective Savings of Water and Energy Leading to Climate Change Mitigation

Alliance to Save Energy[1]

INTRODUCTION

The term "Watergy" was coined by the Alliance to Save Energy to describe the strong link between water and energy in municipal water distribution systems. The Watergy approach helps cities realize significant energy, water, and monetary savings through technical and managerial interventions in water and wastewater systems, providing consumers with quality water while using a minimum of water and energy. In areas where either resource is constrained, Watergy is a valuable resource waiting to be tapped, and is so cost-effective that the time it takes for improvements to pay for themselves generally ranges from a few months to three years. The resulting savings leave more funds for critical public services. For those bearing the financial responsibility for local public services, efficiency in the provision of energy and water is one of the few cost-effective options available for meeting growing demands for vital services such as electricity, water supply, and wastewater treatment. In addition to direct savings in energy and water, efficiency improvements in municipal water utilities generate impressive reductions in operating costs and often improve service. These efficiency improvements also result in reduced emissions of carbon dioxide.

> ***Watergy makes the best use of two valuable, limited resources: water and energy.***

The integral relationship between water and energy is not widely understood or sufficiently exploited through coordinated, holistic efficiency approaches. The water–energy relationship is based on the reality that treating water for human consumption and moving treated water to the consumer is an extremely energy-intensive undertaking. Every liter of

1 The original version of this paper was written by Judith Barry, a former employee of the Alliance to Save Energy, with input from the Alliance's International Team. Sudha Setty condensed the paper for the current publication and added a section on incorporating energy efficiency at the design stage of new water systems.

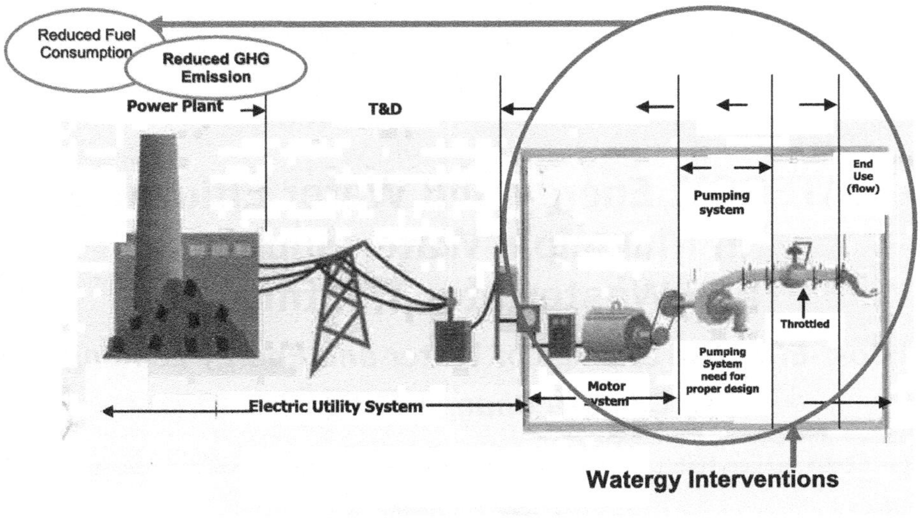

Figure 11-1. Emission reductions in municipal water supply system

water that passes through a system represents a significant energy cost. Figure 11-1 illustrates the major steps involved with providing municipal citizens with clean water and treating wastewater for safe discharge, each of which requires a significant and quantifiable amount of energy. The relative energy importance of the different steps depends on factors like the topography between the water source and its destination (especially elevation change), distance from the bulk water supply, and the integrity of the primary mains (supply pipes) and secondary mains (distribution pipes). Globally, energy is among the top three cost items to water utilities, often coming second after labor costs. In developing countries, energy is usually the highest cost associated with water supply.

Water supply and wastewater treatment are also infrastructure intensive, requiring expensive pumping, treatment, and conveyance systems. Water access can be expanded much more quickly and inexpensively through efficiency than new infrastructure, deferring the need for additional infrastructure investment. Also, water and wastewater investment decisions that neglect energy efficiency have a domino effect that increase investments in other sectors, such as power plants, investments to extract and transport the additional fuel, and the environmental costs associated with air emissions and declining water and hydrocarbon reserves. A development agenda that maximizes the capacity of existing infrastructure through efficiency before encouraging new construction is the most cost-effective and sustainable way to meet the growing need for clean water. By incorporating efficiency into existing and planned infrastructure systems, costs can be controlled, service delivery improved, and access expanded without necessarily adding to the cost of the service.

Part of what makes water supply such an energy-intensive process, as well as one that squanders precious fresh water, is that so many distribution systems around the world are leaky. Even many cities in fully industrialized countries have high leakage rates due to aging infrastructure. Whenever water is lost to leaks, also lost are the energy and cost of energy embodied in that water, from pumping, treating, and conveying the water. When water is lost before it reaches the end user—which is where the majority of losses occur in many countries—the associated water production costs cannot be recovered.

Common Barriers to Energy and Water Efficiency

For some, the case for efficiency seems sufficiently compelling that its universal adoption should simply be a matter of course. In reality there are serious obstacles to the widespread adoption of more efficient practices and technologies, which can be grouped into five main categories:

1. *Lack of Awareness.* People will not make changes toward efficiency unless they are aware of the cost–benefit arguments for doing so. This is especially true in the case of applying energy efficiency to water supply, since those who operate day to day in the water sector are not accustomed to focusing on energy.
2. *Aversion to Risk.* Deviating from the usual routine is associated with risk, real or perceived, such as added burden on staff or financial risk. Fear of change has a rational basis and breaking through it requires that the fears be addressed and that the benefits of change clearly outweigh risks.
3. *Change May Imply a Problem with the Status Quo.* It is not uncommon for staff to be resistant to new ideas and procedures due to a feeling that suggestions for change imply criticism of their performance and ability.
4. *Subsidies.* Although subsidies have a role in providing essential services to the poor, when they are poorly planned or implemented they often greatly reduce the cost incentives inherent in efficiency. Some subsidies are unofficial, such as the tacit approval by the authorities of water or electricity theft.
5. *Financing Efficiency.* Quite a few Watergy measures cost little if anything. For those that do require capital outlays, performance contracting approaches pay for project costs from the cost savings on water and energy. Those contemplating efficiency improvements often lack an understanding of performance contracting mechanisms, especially the awareness that they can be applied to the water sector. In some countries, financing issues are compounded by an insufficient supply of service providers capable of performance contracting; or even where suppliers exist, the industry is often so nascent that confidence in them is lacking. This lack of confidence usually translates into an inability of these firms to provide the project financing, since their unproven creditworthiness either denies them access to loans altogether, or gives them access only to loans whose terms are poor. On the flip side, in some countries municipal governments do not have a good track record of sound financial management or honoring contracts, making efficiency service providers reluctant to enter into contracts with municipalities.

In developing countries, often one-third to one-half of the volume of water produced is lost to leaks and system inefficiencies. In Latin America, for example, one-third of the water that enters distribution systems in Mexico is lost before it reaches the consumer; the losses average 44% in Brazil. Indian cities lose about half of their water to leakage on average, with actual figures ranging from 40% in large metropolitan cities to 50–60% in smaller cities. In South Africa, most big cities also average about 40% loss. Watergy can produce the dramatic reductions in such losses needed to bridge the growing gap between the need for water and the capacity to provide it. For example, the water currently lost to leaks in Mexico would cover the growth in demand expected there for the next six years, meeting future needs and diminishing current shortfalls so that investments in infrastructure could be postponed, and stretching limited water and energy resources.

TECHNICAL AND MANAGERIAL APPROACHES TO WATERGY

To the extent that municipalities and water utilities consider efficiency in the water sector, it is most often limited to demand-side management or conservation *after* the water arrives at end user properties. While there are tremendous efficiency gains to be made downstream of the water meter, there are numerous fruitful efficiency opportunities in the various stages preceding the arrival of the water at its end use destination. Table 11-1 summarizes the many components involved in supplying water and treating wastewater that use energy, apart from end use.

In the experience of the Alliance to Save Energy, which has helped more than 40 water utilities implement Watergy programs, no two Watergy projects are alike. The specific form that a Watergy project will take depends on a myriad of local factors such as water scarcity, geography, tariff structures for water and electricity, culture, and the way that water supply is governed and administrated. However, over the years the Alliance has used its practical experience facilitating Watergy projects around the world to forge a general approach to creating lasting water efficiency improvements. This approach is illustrated in Figure 11-2 and the three steps in the process—generating political will, technical management and analysis, and implementing efficiency measures—are described in this section.

Table 11-1. Measures to improve efficiency in water supply and wastewater treatment and illustrative payback periods

Area	Function	Typical Payback Period (years)
Electricity rates	Reduce demand during periods of peak electricity rates	0–2 depending on storage capacity
Electric installations	Power factor optimization with capacitors	0.8–1.5
	Reduction in voltage imbalance	1–1.5
Operations and maintenance	Routine pump maintenance	2
	Deep well maintenance and rehabilitation	1–2
Production and distribution	Use automation (such as telemetry, SCADA, and electronic controllers on modulating valves), for example to control pressure and output in the networks, and to optimize the operation of pumping equipment	0–5
	New efficient pump	1–2
	New efficient motor	2–3
	Replace impeller	0.5
	Optimize the distribution network (e.g., by removing unnecessary valves, sectoring, and installing variable speed drives and regulating valves)	0.5–3
End use	Incentive program for the use of efficiency technologies	1–3
	Effective metering of consumption	1–2

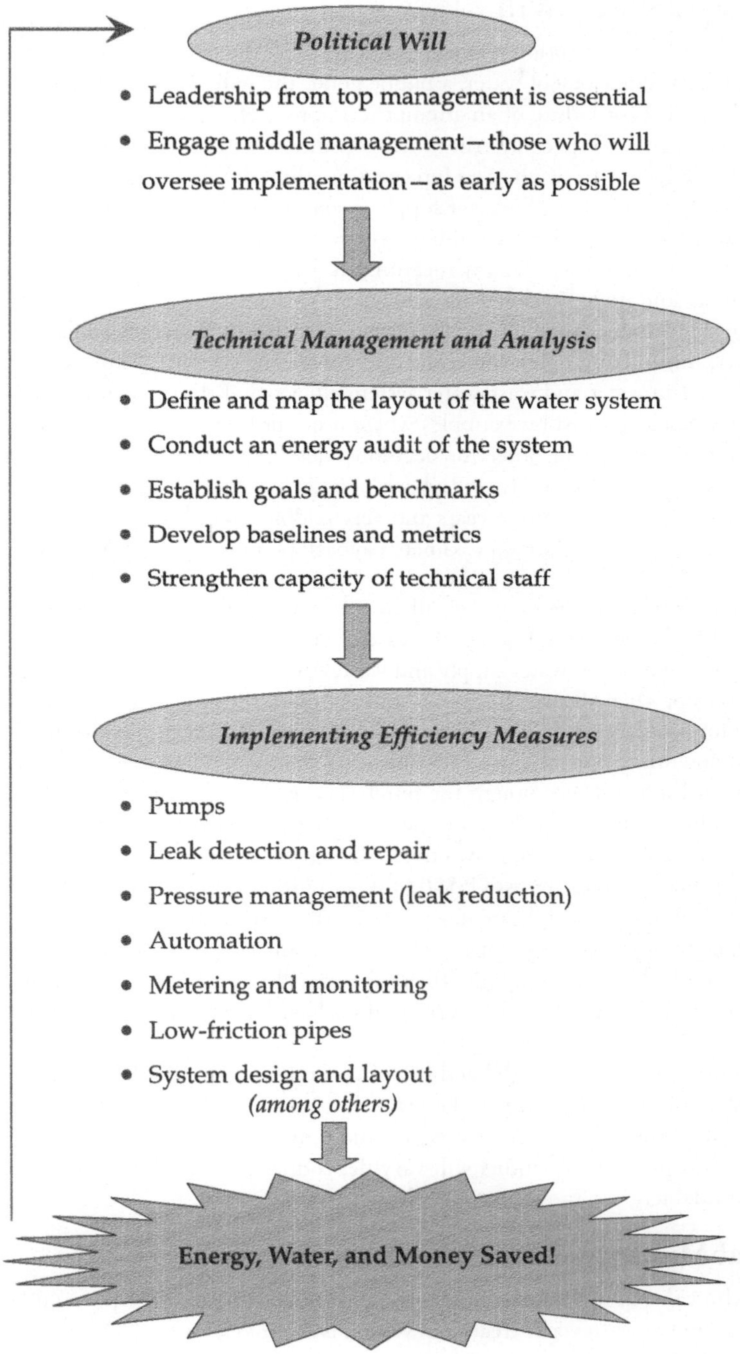

Figure 11-2. Stages of the Watergy Approach

Generating Political Will

Lasting change in any institution requires political will at the senior management level. In the context of end use efficiency of water, a public good, the will of the community can also be critical to the success or failure of an attempt to improve efficiency of water supply or wastewater treatment. This is especially true when consumers will be asked to take on additional costs or responsibilities as a result of an intervention, though this will generally not be the case for improvements focused on the water supply system itself.

The water sector languishes in a climate where few efficiency examples serve as models for implementing these initiatives on any meaningful scale, although Watergy is rapidly changing that as more and more successes are achieved.

Management understands all the concepts embodied in Watergy: efficiency, productivity, reliability, minimizing operating costs, and generating revenue. Managers will generally become believers in water and wastewater system efficiency if the opportunities are presented in these terms, accompanied by examples where other utilities have been successful. A cost–benefit analysis is needed in order for decision makers to know how long it will take for improvements to pay for themselves. Payback periods for Watergy measures, summarized in Table 11-1, are often less than two years and very seldom exceed three years. Rates of return depend on local conditions, but approximate payback periods for some Watergy measures are also given.

Another key component to an overall cost–benefit analysis is an understanding of life-cycle costs. This is especially critical in the case of pumps since pumping systems account for most of the energy usage in water supply and wastewater treatment systems. Over the lifetime of a typical pump, when all costs to operate and maintain the pump are considered, 3% of the total cost is for purchase and 74% is for energy! A more efficient pump also has lower maintenance and downtime costs. Figure 11-3 illustrates the tremendous cost advantage of buying energy-efficient pumps. Even though the purchase price of an efficient pump can be significantly higher than an inefficient one, the energy savings make the efficient pump less expensive even in the year when the purchase was made. In subsequent years, the lower operating costs go right to the bottom line, almost US$50,000 per pump in this example, year after year.

It is best to engage middle management in the discussions as early in the process as possible. Once middle management receives word from the top to proceed, they will be the ones overseeing the Watergy program. The program will run much more smoothly if they are convinced of its merits and if their expert opinions have been incorporated into the planning process.

A preliminary "walk-through" audit is often necessary to get commitment from management because it generates a list of the major interventions that can be made and at least a rough estimate of the potential energy, water, and monetary savings.

The cost for preliminary audits will vary depending on the level of detail, but they are generally affordable.

Technical Management and Analysis

The first technical step toward an efficient system is to ensure that the physical layout of the water distribution or wastewater treatment system is accurately captured both in graphic form (some form of schematic layout) and descriptively. Ideally, the system layout will be entered into a geographic information system (GIS), and linked to the engineering database and the database of billing information for the system. This level of sophistication will not be possible in all locations, but in any case, a system cannot be managed effectively unless its boundaries are defined and the technical parameters and location of all of its components are described.

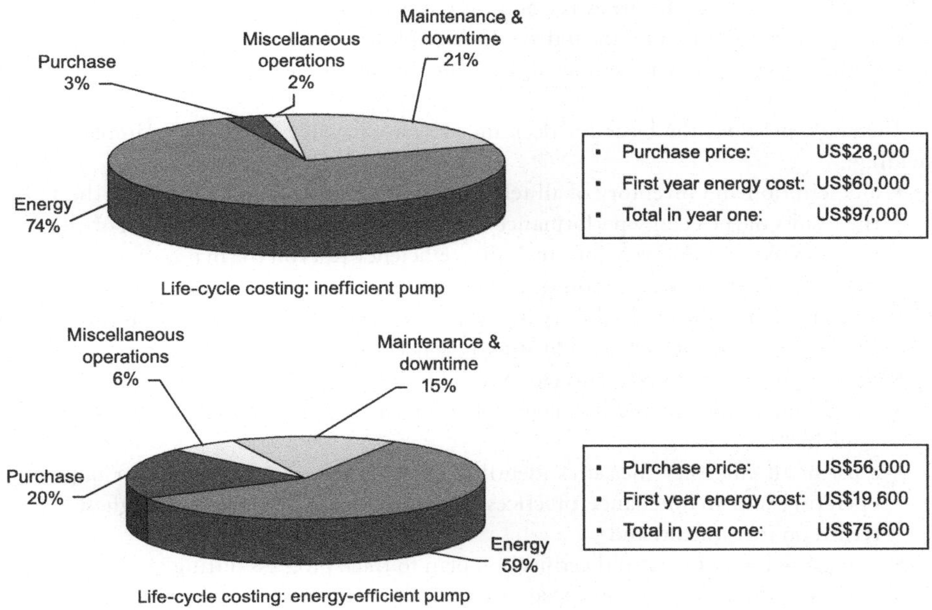

Figure 11-3. Lifecycle costs of an efficient versus inefficient pump

If the system has not been well defined, or an existing schematic layout has not been updated recently, this can be done as part of the detailed efficiency audit on the system.

The next step is to have a qualified efficiency service provider perform an investment-grade audit on the system to determine what measures can be taken to improve energy and water efficiency, and their payback periods. The main purpose of this step is to translate technical findings into the financial terms needed for decision makers. In cases where external financing is sought, banks need this financial information to assess whether the project is a bankable one capable of servicing a loan.

The audit begins with discussions between the auditor and key personnel from both the managerial and technical levels of the system. The purpose of these meetings is two-fold. One is to ensure that the decision makers thoroughly understand and are on board with the process. The other is to ensure that relevant facility staff has an adequate understanding of the process since they will be providing the auditor with data and specifications about the

Examples of measurements made in an audit of water supply or wastewater treatment system

Pumping systems

- Total head (water pressure in the system)
- Pumping system efficiency

Electrical systems

- Power factor
- Losses in transformers and cables
- Analyzing electricity bills
- Peak and off-peak operating loads

Electrical motors

- Motor losses and loads
- Potential savings from improvements like variable speed drives and capacitors

facility essential to the audit. Next, the auditor will visit all facilities involved in the project to ascertain the availability of data and system complexity, formulate a data collection strategy, and identify the proper personnel at different facilities with whom to coordinate on the audit.

The audit provides the basis for deciding which efficiency measures to implement by providing:

- a description and inventory of all relevant systems and facilities covered by the audit,
- the results of the energy performance and system optimization evaluation of all facilities targeted for improvements, including efficiency tests on the major energy consuming equipment and energy consumption details,
- a description of the methodology followed in establishing the baseline parameters and the criteria to be followed in adjusting it as needed over time,
- the various types of data collected and their sources,
- system maps and the methodology followed for mapping the system (if not already available),
- a list of all efficiency measures identified—including any suggestions for improving operating and maintenance practices—prioritized according to the highest rate of return on investment, and
- a suggested monitoring and verification plan to track progress during implementation.

In addition, the following three steps are critical to maximizing the savings and system improvements that are made:

- establishing performance targets and benchmarks against which to evaluate progress,
- developing metrics relative to the baseline, as well as a monitoring system to measure them, and
- strengthening the capacity of technical staff in new procedures and technologies.

Implementing Efficiency Measures

The four most fruitful areas of intervention for water and wastewater efficiency improvements are:

- Pumping systems
- Leak management
- System automation
- Metering and monitoring

The intention is to provide an overview of the most common interventions, rather than to delve into operational details on implementing them.

Pumping Systems

One of the most critical elements to improving water or wastewater system efficiency is to optimize energy consumption by the pumping systems. Optimizing the system includes improvements such as matching the pump to requirements, optimizing the distribution piping, eliminating unnecessary valves, controlling pump speed where appropriate, and institutionalizing improved operations and maintenance (O&M) practices. For example, according to a study by the U.S. Department of Energy,[2] matching pumps to the system requirements so no more flow and pressure are pumped than needed generally saves between

2 United States Industrial Motor Systems Market Opportunities Assessment, December 1998, www1.eere.energy.gov/industry/bestpractices/pdfs/mtrmkt.pdf.

10 and 30%, while the appropriate use of variable speed drives[3] to adjust pump speeds saves on average 10.5%. Pump system optimization can readily result in 20% energy savings, with savings of 30–40% often feasible.

The pumping system should be analyzed to identify the opportunities for improvements. A common problem in pumping systems is for pumps to be too large for the needs of the system. All pumps have a best efficiency point (BEP)—the flow rate at which the pump operates with the lowest energy intensity (volume moved per kWh)—and a pump operating at a flow rate significantly lower than its BEP is wasting a lot of energy and wearing out more quickly. Besides switching to a smaller pump, other possible approaches are to trim the impeller[4] (that is, to reduce the length of the blades that move the water), install a smaller impeller, reduce pump speed, and—for a multi-stage pump—remove one or more stages.

Aside from an initial set of efficiency measures, it is critical to institute improvements in the routine operations and maintenance protocol. Managers should develop a facility layout showing the location of all critical pumps and use it as a road map for the maintenance technicians to follow for troubleshooting, preventive inspection, cleaning, and minor adjustments. Some routine operations and maintenance procedures that increase the life of equipment as well as efficiency are to:

- Monitor the electrical system, including motors (such as motor operating parameters, power factors, peak and off-peak loads, and electricity consumption).
- Optimize pump variables (pressure, flow, peak load, and motor starting and stopping).
- Clean the impellers and replace as needed.
- Periodically run the manufacturer's field test on the pumps and check the packing and lubrication of the bearings.
- Check pumps for excessive heat, leaks, vibration, and noise.
- Maintain uniform flow profiles at pump inlets and outlet.

Automated Controls

Automation in a water supply or wastewater treatment system monitors—and if sophisticated enough, also controls—various system components for optimal system performance and efficiency. Automation varies in complexity but it always has at its core sensors that measure system parameters such as pressure, water level, and flow rates. The most basic and inexpensive form of automation consists of stand-alone devices that act on the information from the sensor to perform simple actions only at the site where they are placed, such as an automatic shut-off valve responding to water level indicator. Next in line of complexity is telemetry, which automatically measures and transmits measurements from sensing devices located throughout the network to a central receiving station.

In the absence of control devices, telemetry requires an operator to analyze the input and act upon it. A control system not only monitors variables in the water distribution or wastewater treatment system, but it automatically maintains them at a desired range of

3 Electronic variable speed drives (VSD) adjust the speed of the motors they control by varying the voltage and frequency of the electric power supplied to the motor. They sense the load required of the motor and change its power and speed accordingly. VSDs save energy because pumps often do not need to operate at their maximum load, so the motor driving the pump can run with less power whenever the application requires less than a full load.

4 The impeller is basically a propeller inside the pump, powered by the shaft upon which it is mounted. The torque power of the engine is translated from the shaft to the blades of the impeller to move water.

values. A set of controlled devices, such as control valves, motors, pumps, and fans, act in response to a set of controllers, such as an automatic power factor controller. For example, a motor is a controlled device that responds to a variable speed drive, a controller. One of the major benefits offered by control systems is the ability to adjust the scheduling and operation of component devices to changing demand and conditions, allowing the performance of various system functions to be continually and automatically optimized.

The most common type of control automation used in water and wastewater systems is the supervisory control and data acquisition (SCADA) system. SCADA can remotely monitor and control *in real time* a large array of components spread over large geographic areas, while also providing performance information on individual sites. The basic SCADA system consists of one central computer (the master terminal unit, MTU) that communicates with and controls a number of remote terminal units (RTUs) at key control points such as pumping and metering stations, plus the equipment to communicate between the MTU and RTUs. The RTU serves two functions: it uses information from those sensors connected to it to control system components such as pumps and valves, and it enables the user to access performance information on the site.

A properly designed SCADA system for water pumping saves time and money while improving service. Some of the many benefits include:

- Complete information on pumping operation provided in real time
- Pump operation adjusted automatically as needed to ensure reliable water supply
- Reduced need for service personnel to visit sites for inspection, data collection, or adjustments
- Alarms sent to the central control location in case of emergency
- Optimal pressure maintained in the water supply network, minimizing service interruptions and eliminating water hammers
- Electricity consumption reduced and pump productivity increased
- Troubleshooting facilitated
- Equipment life increased
- Reports generated automatically

Although the exact numbers will vary greatly depending on conditions, the following are typical saving ranges from a control system:

- water savings: about 10%
- electricity savings: 12–30%
- reduced downtime: up to 30%
- reduced maintenance and personnel costs: 15–30%

Metering and Monitoring

There is a longstanding saying in business that "you can't manage what you don't measure." In water systems around the world, only about 35% of consumption and 50% of supply is metered, making it difficult to improve system performance. Any Watergy approach must include the establishment of a system to regularly monitor the various components and locations within the water or wastewater system. The basic steps for putting in place a successful monitoring and metering system are to:

- Create a system for water metering and monitoring if there is none, or expand and upgrade the existing system.
- Develop baselines and metrics for regular monitoring.
- Create targets and gauge success toward achieving them against baselines and benchmarks.
- Procure the proper measurement instrumentation.

Experience shows that it is cost effective to regularly replace water meters. This is because the working parts in a meter wear out over time, causing the meter to slow down and lose accuracy. In addition, older meters generally do not register water passing through the meter resulting from a leak on the property, and resulting water loss becomes the liability of the municipality. A 10-year replacement cycle is typical because—assuming that water is priced to fully recover the cost of providing it—meters older than 10 years are losing the utility more money than it costs to replace them.

The lists below summarize the parameters typically monitored in a regular monitoring protocol. In order to isolate problem areas within the system, measurements should be made on discrete system components such as specific equipment and areas of facility buildings.

Examples of ways to use a control system in a water supply system

- Regulate the use of pumps based on the water level in storage tanks.
- Use frequency converters to control pumps according to changes in demand and to control daytime and nighttime pressure in the network, saving energy when demand is lower.
- Use variable speed drives and automated pressure controls to stabilize water pressure in the system and reduces downtime.
- Save energy costs by scheduling reservoir filling to coincide with low electricity rates and low demand).

Pump system

- Valves—Inspect for leaks, clogs, or an open or closed stationary position.
- Flow—Compare water flow rates in different parts of the system to help pinpoint leaks and other problem areas needing attention.
- Pressure—Monitor working head and suction and delivery pressure (meter/kg-cm^2), not only to identify leaks but to indicate whether there is excessive pumping, enabling pressure, and therefore service, to be kept constant. Monitor network pressure at the critical point in the distribution zone to ensure sufficient supply pressure.
- Rotating speed—Measure pump speed with a strobe light to determine whether a motor is operating at optimal efficiency.
- Power consumption—An increase indicates a loss of pump efficiency.
- Volume of water—pumped per day

Electrical systems

- Motors—Under different flow conditions, measure operating parameters such as input power (to indicate whether motor is operating at optimal efficiency), the current upon start-up and while running, frequency, and speed.
- Power factors
- Operating loads—peak and off-peak
- Energy consumption—track and analyze electricity bills.

Incorporating Energy Efficiency at the Design Stage of New Water Utilities and Wastewater Systems

In addition to the three steps in the process described above—generating political will, technical management and analysis, and implementing efficiency measures—there is a fourth component that is key to water system efficiency in the case of newly planned systems.

"Prevention is better than a cure" goes the age-old adage, and so is the case with energy conservation. Energy efficiency aspects incorporated into water systems at the design stage can ensure substantial energy savings from the commencement of operation.

The design teams at water utilities should incorporate standard energy efficiency guidelines at the design stage of the work, and during the tendering stage of any procurement. Many players are involved in the selection of equipment such as pumps and motors, starting at the specification and design stage, through installation and operation. Safety margins (in terms of size) are added at each of these stages, primarily due to the high degree of uncertainty in pinpointing design parameters. However, this practice leads to considerable loss of efficiency and energy in pumps. The energy-efficient operation of any water supply system or equipment depends on the right specification, selection, and installation at the design/retrofit/project expansion stages. If empowered, a design team can devise and enforce guidelines to assist the various divisions developing new water systems and oversee incorporation of energy efficiency practices.

The first step during the design stage is to identify specific energy consumption targets for the:

- Overall systems/process
- Individual equipment/sections

All subsequent actions should be designed to meet these specific energy consumption targets.

Overall Approach

A multipronged approach, as described in Figure 11-4, is advocated for successfully achieving energy efficiency at the design stage.

Installation of Energy Efficient Technology

For achieving significant increases in energy efficiency, the use of efficient technology is vital. Energy efficient technologies can be adopted most easily at the design stage, when the additional costs involved are only marginal, in comparison to the benefits of long-term energy efficient operations. Retrofitting facilities with efficient technologies after the fact has many disadvantages, including operations down time, increased investment, and resistance of operating personnel.

Usage of Energy Saving Devices

While incorporation of the latest technologies sets the stage for energy conservation at the overall facility, energy saving devices such as automatic power factor controllers, variable frequency drives (VFDs) for motors, and remote control systems such as SCADA ensure that these technologies are operated at their maximum efficiencies. As with efficient equipment, the incorporation of energy-saving devices at the design stage avoids the many disadvantages of later retrofitting. In some cases, the retrofitting of energy-saving devices may even call for replacement of the original major equipment, underscoring the value of their inclusion from the beginning.

System Flexibility

During the design stage, designers and planners should suggest a modular approach to allow flexibility in addressing future water supply needs. A well-planned system design can provide flexibility for meeting future demands through options such as increases in the size of transmission mains and adding booster pumping stations for higher head requirements.

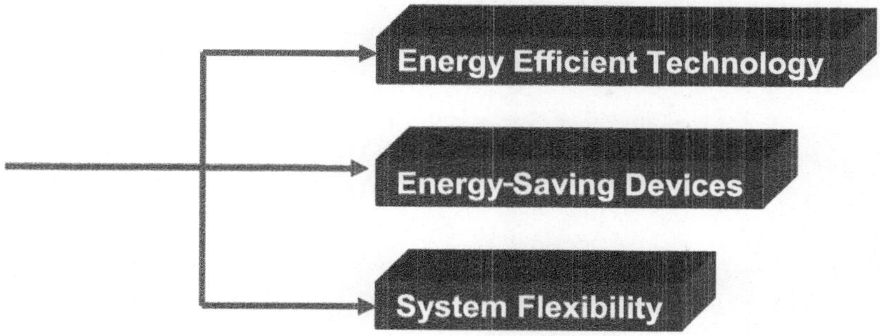

Figure 11-4. Overall approach

Provision should be made within the pumping stations for additional pumps and incorporating the new pumps to the header line. Although the flexibility of pumping systems does not have an immediate direct bearing on energy efficiency, a flexible system can increase long-term efficiency by enabling appropriate changes to be made to the running capacity of the pumping stations throughout the facility's lifetime.

Energy Efficiency Design Considerations

The parameters a design team should consider during the selection or design of a pumping system include:

- Minimizing over-sized pumps
- Selection of the right types of efficient pumps and motors
- Installation of variable-speed drives (VSDs)
- Segregation of high-head and low-head pumps
- Utilization of gravity flow
- Unitization of pumps
- Parallel operation of pumps
- Piping design

RESULTS

Over more than a decade, the Alliance to Save Energy's Watergy Program has helped more than 100 cities in nine countries realize annual savings of more than 20.8 million kWh of electricity and US$5.3 million in operating costs.

– 12 –

Station Efficiency Reduces Greenhouse Gas Emissions

Robert Eckard

Scientists recently published strong evidence that climate warming trends observed around the world are a result of human activity (IPCC 2007a). Their studies indicate that temperature increases observed since the mid-20th century have resulted from concurrent increases in greenhouse gas (GHG) emissions, which include carbon dioxide (CO_2), methane (CH_4), and other gases emitted as a result of human activities (IPCC 2007b).

GLOBAL WARMING AND GREENHOUSE GAS EMISSIONS

Anthropogenic greenhouse gas emissions result primarily from burning of fossil fuels for electricity, heat, and transportation. For instance, US industrial activities, including wastewater treatment handling and processing, generate approximately 1.75 billion tons of CO_2 equivalent GHG emissions per year (to provide a single metric, the effects of other GHGs, such as methane, are presented as CO_2 equivalents), or approximately 25% of net US annual GHG emissions (USDOS 2007).

Of that amount, it is estimated that approximately 17 million tons of CO_2 equivalent GHG emissions per year are released during wastewater treatment. This figure does not include GHG emissions resulting from sludge degradation and methane outgassing, which accounted for an additional 37 million tons of CO_2 equivalent GHG emissions in 2006 (USEPA 2008). These emissions result primarily from wastewater pumping as well as other treatment operations (CSS 2007).

The US federal government has yet to take significant action directed explicitly toward reducing GHG emissions. Instead, many state and regional organizations have taken steps to reduce and regulate GHG emissions. Regional entities including the Western Governors' Association, the Eastern Climate Registry, Powering the Plains, the US Mayors' Climate Protection Agreement, and several other groups have implemented GHG reduction targets and cap-and-trade programs, and programs for increased energy efficiency, and to provide clean and diversified energy. Additionally, state initiatives including climate action plans (29 states), renewable energy portfolio standards (23 states), climate advisory boards (12 states), GHG emissions targets (12 states), energy-efficiency portfolio standards (10 states), mandatory CO_2 reporting for stationary sources (5 states), and GHG emission caps or registries (3 states) specifically target reduction of GHG emissions.

The current US political climate, outside of the Bush administration, appears to be favorable to GHG regulation. All three front-runners in the 2009 presidential campaign support adoption of a cap-and-trade system, which is set in place when regulators (typically government) set a maximum emissions "cap," then issue emission permits to GHG-emitting companies or entities. The companies or entities may not exceed the cap without purchasing additional "credits" from a company or entity that is not using its full share of permitted emissions (the "trade"). The current legislature has also shown substantial support for a cap-and-trade system for increases in energy-efficiency standards and for additional goals or standards for use of renewable energy in place of fossil fuels. In general, US states, counties, cities, and other local government entities are also initiating GHG emissions analyses and passing measures that support or require reductions in GHG emissions and increases in energy efficiency.

WHAT DOES EMISSIONS REDUCTION MEAN FOR THE WASTEWATER INDUSTRY?

Because wastewater industry GHG emissions represent less than 1% of the US total, do wastewater treatment managers and administrators really need to be concerned about GHG emissions? For the time being and under the current federal administration, the answer is, for most parts of the country, yes. Wastewater treatment operations in Nebraska, North Dakota, and South Carolina probably have little to worry about. At the time this article went to press, these states had yet to conduct GHG inventories or set relevant state standards for GHG emissions. However, all other states are taking substantial initiatives, with California and many northeastern states leading the way toward stringent GHG regulations. Many cities and local jurisdictions such as Portland, Ore., and Sonoma County, Calif., are implementing regulations above and beyond those required at the state level. Over the past year and a half, even federal lawmakers' opinions about climate change and GHG emissions have come to align with the stringent regulatory environments of Europe, Japan, and Australia. In 5 to 10 years, it is reasonable to predict that the US wastewater industry as a whole will be subject to substantial GHG regulations.

SOURCES OF GHG EMISSIONS IN THE WASTEWATER INDUSTRY

Wastewater industry GHG emissions come from two primary sources: digestion or decomposition of sewage and the decomposable fraction of sludge, which releases CO_2 and CH_4; and consumption of energy used to power lift stations and other pumps used to convey wastewater. Some GHG emissions result from wastewater treatment processes such as dewatering of biosolids or from maintenance and operation of office buildings and other support facilities. However, these sources are generally minor compared with digestion and decomposition and pump power usage.

GHG emissions associated with digestion and decomposition of sewage and sludge can be reduced by collecting and disposing of (generally via burning) emitted methane. Additionally, sewage and sludge can be treated to induce release of methane gas for collection using a methane digester. The methane digester creates an anaerobic environment in which microbial decay is promoted, resulting in the generation of methane. The methane is then collected and flared or, more typically, purified and used as fuel to generate heat or electricity. This process has been well-researched and documented; methane digesters and other facilities are available from various companies throughout the United States to reduce GHG emissions and provide low-cost heat and electricity.

Less widely discussed are contributions of GHG emissions associated with the energy use of pumps. Wastewater pumps housed in lift stations and elsewhere within a water treatment

Common Causes of Centrifugal Pump Efficiency Loss

- Bent shaft
- Buildup that may rub along mechanical seals
- Cavitation
- Clogged pipelines or pumps
- Degradation of wear rings
- Eroded or corroded internal pump passages
- Foreign objects
- Harmonic vibration
- Impeller and casing wear causing increased clearances
- Impeller imbalance or wear
- Improperly managed hourly, daily, or monthly increases/reductions in flow/pressure
- Increased internal recirculation
- Loose hardware
- Loss of proper impeller clearance
- Operating at critical speed
- Operating too far from the best-efficiency point of the pump
- Overlubricated or overloaded bearings
- Overtightening of packing or improper seal installation
- Pressure surges
- Shape of the impeller
- Solids rubbing against components, especially the seal
- Use of vortex pumps
- Water hammer

Source: McNally Institute (2008)

facility require substantial amounts of energy to operate. Water is intrinsically heavy; consequently, pumping it uphill or against a pressure gradient consumes energy and results in GHG emissions.

For example, the Inland Empire Utilities Agency operates a moderate-sized, 8.5-mgd wastewater treatment plant in Ontario, Calif. The facility's total combined 1,195 hp for all pumps as of 2001 consumed approximately 4,750,000 kW·h of electricity annually. Assuming 1.345 lb of CO_2 emitted per kW·h of electricity during electricity generation (USDOE 2000), operation of the pumps in this single facility for one year resulted in the emission of approximately 3,200 tons of CO_2, which is equivalent to traveling about 7 million miles in a passenger vehicle that has a fuel efficiency of 20 mpg.

Obviously, pumps use energy to lift or move water; however, pumps also use considerable amounts of energy to overcome frictional and vibrational forces within the motor and pump mechanism and to overcome the turbulence and friction associated with the water as it passes through the pump and pipelines. The wire-to-water efficiency of a pump describes the amount of energy that a pump uses specifically to move water versus energy used to overcome friction, vibration, and turbulence. As efficiency increases, more of the energy applied to a pump is transferred into moving water. Therefore, by increasing pump efficiency, less energy can be used to perform the same amount of work, thereby reducing GHG emissions and energy costs.

As an example, the wastewater treatment plant in Ontario was part of a joint study by the California Energy Commission and the US Department of Energy (DOE). By studying

State and Regional Entities Taking Steps to Reduce and Regulate Greenhouse Gas Emissions

Western Governors' Association, Clean and Diversified Energy Initiative
www.westgov.org/wga/initiatives/cdeac/index.htm

Eastern Climate Registry
www.easternclimateregistry.org/

Great Plains Institute, Powering the Plains
www.gpisd.net/resource.html?Id=61

U.S. Conference of Mayors, U.S. Mayors Climate Protection Agreement
www.ci.seattle.wa.us/mayor/climate/default.htm#what

US Environmental Protection Agency (USEPA), Combined Heat and Power Partnership
epa.gov/chp/markets/wastewater.html, and epa.gov/chp/documents/wwtf_opportunities.pdf

USEPA Wastewater Fact Sheet
www.epa.gov/CHP/markets/wastewater_fs.html

USEPA Funding Database—Biomass/biogas (list of potential state grants and tax breaks for adding a methane digester to a wastewater treatment plant)
www.epa.gov/CHP/funding/bio.html

the energy efficiency of the plant's 19 water pumps, the plant's operators found that only three of the most frequently maintained pumps had good wire-to-water efficiencies (near 70%); all others operated between 47 and 60% efficiency. After replacing worn parts, including eddy current clutches and old motors, the plant reduced its energy consumption by about 475,000 kW•h/year, saving enough money on electricity to repay the repair cost in two years and reduce annual GHG emissions by about 320 tons (the equivalent of driving 700,000 miles at 20 mpg; CEC & USDOE 2002).

This example is not unique. DOE studies indicate that by increasing pump efficiency, wastewater treatment facilities can realistically save up to 20% of the total energy consumed during wastewater pumping (DEP 2002). Several case studies also support this claim for water pump stations in general. An efficiency study and retrofit at Austin Energy (Texas) revealed that strategic upgrades annually saved 43,000 tons of CO_2 emissions and $1.2 million in energy costs for pumping cooling water (USDOE 2006a). In addition, a study and upgrades for a pumping system at Boise Paper in Wallula, Wash., saved an estimated 330 tons of CO_2 emissions and $17,500 in energy costs annually (USDOE 2006b).

WHY DO PUMPS LOSE EFFICIENCY?

Losses in pump efficiency occur over time as parts wear, degrade, or are damaged during pump operations. The sidebar on page 139 lists potential causes of pump efficiency losses. These causes fit into two general categories: mechanical degradation and ineffective operation.

Mechanical degradation, such as degradation of wear rings, loss of impeller clearance, or enhanced wear from buildups, requires ongoing pump and facility maintenance to correct and ensure maximum efficiency. Perhaps most critical to reducing the effects of wear is maintaining the internal wearing ring clearance and the smoothness of impeller and casing waterways. In addition, by maintaining pumps and all system components in virtually new condition, losses in efficiency that result from worn parts and equipment will be avoided (Hydraulic Institute 2008).

Ineffective pump operation may be the result of

- pumps that are oversized for the job,
- misapplied pumps,
- operation of pumps in a throttled condition to compensate for use of oversized pumps,
- careless management of system flows and pressure,
- pumping systems with bypass flow,
- use of multiple pump systems where excess pressure is generated,
- use of high system pressure instead of a booster pump, or
- changes from initial design conditions, such as cross-connections, parallel main lines, or changes in pipe diameter or material (USDOE 2005).

EFFICIENCY AND GHG REDUCTION SOLUTIONS

To most effectively increase pump efficiency, reduce electricity use, and reduce GHG emissions, it is useful to ask these two questions:

- What physical changes need to be made—maintenance, replacement of worn or inappropriately sized pumps, removal of internal buildups on pipes and pumps, impeller replacement?
- What ongoing monitoring is needed to identify problems and substandard conditions that cause or could cause reduced efficiency?

Unfortunately, the complex network of pumps and lift stations, pipelines, and other collection and treatment facilities associated with wastewater treatment begs yet another question—where do I start?

In wastewater facilities, it is common for several pumps or pump stations to operate with high or relatively high efficiency, whereas the remainder (sometimes only a handful) operate at low or extremely low efficiency. This type of scenario results in excess power consumption and excess GHG emissions for the facility as a whole and raises the problem of how to identify the problematic components. Without efficiency data for each pump station or pump in a facility, it is nearly impossible to make a cost-effective or time-effective decision regarding maintenance or replacement.

A second, longer-term challenge is identifying the most appropriate pumps for each pump station or pump application to ensure that ongoing electricity-use reduction, GHG emissions reduction, and efficiency are optimized. Selecting the most appropriate pumps for ongoing and future uses requires a significant amount of data on historic flow rates, water volumes, and efficiency problem areas. Without this type of information, it is not possible to select the most efficient pumps and pump configurations in the future.

Perhaps the easiest and most cost-effective way to fill in these data gaps is to monitor the system on an ongoing basis. For example, parameters such as flow rate, energy consumption, and pump efficiency, if monitored consistently, will provide the data needed to identify inefficient or problem equipment. Over time, data collected by ongoing pump and pump station monitoring can be used to select and configure replacement hardware.

Monitoring Technology

Fortunately, advances in flow and efficiency monitoring technology can significantly streamline the monitoring process (Figure 12-1). For example, MultiTrode Inc.'s MultiSmart

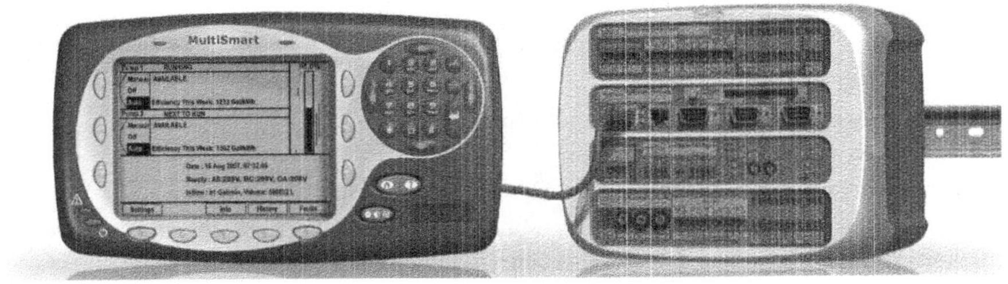

Figure 12-1. Advances in flow and efficiency monitoring technology can significantly streamline the monitoring process.

Table 12-1. Comparison of annual energy, cost, and GHG emissions after implementing efficiency improvements

Annual Energy Requirement, Costs, and Emissions	Energy Price—kW·h $0.06	$0.08	$0.10	$0.12
	100-hp Pump			
kW·h	726,000	726,000	726,000	726,000
Electricity—$	43,600	58,000	72,600	87,120
GHG—tons	488	488	488	488
	100-hp Pump with 20% Efficiency Increase			
kW·h	580,800	580,800	580,800	580,800
Electricity—$	34,880	46,400	58,080	69,696
GHG—tons	391	391	391	391
	Savings			
kW·h	145,200	145,200	145,200	145,200
Electricity—$	8,720	11,600	14,520	17,424
GHG—tons	98	98	98	98

Source: RSE Consulting 2008; USDOE 2005;
GHG—greenhouse gas, hp—horsepower, kW·h—kilowatt hours

pump station managing system includes built-in features that datalog level, pressure, pump energy usage, flow rates, and system efficiency. Algorithms include running the most efficient pump. This system also includes complete integration with any supervisory control and data acquisition system to facilitate easy data acquisition and real-time supervisory control. This capability allows comprehensive reporting and availability of real-time data, empowering operations staff to optimize the network on a short-term and a long-term basis.

Although pump station managers require some up-front investment, studies have shown that for wastewater pumps and lift stations, costs for efficiency improvements are quickly offset by energy savings (Asdal 2007, USDOE 2005). As shown in Table 12-1, at only \$0.08/kW•h, increasing the efficiency of a single 100-hp pump by 20% results in an annual savings of 145,000 kW•h of electricity, \$11,600 in electricity cost, and 98 tons of CO_2 equivalent GHG emissions, or the emissions equivalent of driving 215,600 miles at 20 mpg. These numbers speak for themselves: investing in pump efficiency reduces GHG emissions and saves money.

REFERENCES

Asdal, R. 2007. The Time Has Come: Pump Systems Do Matter. *Pumps and Systems*, November.

CEC and USDOE (California Energy Commission & US Department of Energy). 2002. Energy Solutions for California Industry: Ways to Improve Operations and Profitability: Case Study. USDOE, Washington.

CSS (Center for Sustainable Systems). 2007. Factsheets: US Wastewater Treatment. css.snre.umich.edu (accessed April 2008).

Hydraulic Institute. 2008. Energy Savings: 7 Ways to Save Energy. www.pumps.org/content_detail.aspx?id=2438 (accessed April 2008).

IPCC (Intergovernmental Panel on Climate Change). 2007a. Climate Change, 2007, the Fourth IPCC Assessment Report; AR4 Synthesis Report. www.ipcc.ch/ipccreports/assessments-reports.htm (accessed April 2008).

IPCC. 2007b. Climate Change, 2007, the Fourth IPCC Assessment Report; The Physical Science Basis. www.ipcc.ch/ipccreports/assessments-reports.htm (accessed April 2008).

McNally Institute. 2008. What Do We Mean by Pump Efficiency? www.mcnallyinstitute.com/06-html/6-01.html (April 2008).

DEP (State of Maine Department of Environmental Protection). 2002. Energy Conservation in Wastewater Treatment Facilities: A Monthly Newsletter for Wastewater Discharge Licensees, Treatment Facility Operators and Associated Persons. Maine DEP, Augusta.

USDOE (US Department of Energy). 2006a. Pumping System Improvements Save Energy at Texas Power Plant. *Energy Matters*, Spring.

USDOE. 2006b. Performance Spotlight: Proven Tools and Practices to Increase Industrial System Energy Efficiency: Boise Paper: Process Pumping System Optimization Saves Energy and Improves Production. USDOE, Washington.

USDOE. 2005. Energy Tips—Pumping Systems; Pumping Systems Tip Sheet Number 1: Conduct an In-Plant Pumping System Survey. USDOE, Washington.

USDOE. 2000. Carbon Dioxide Emissions From the Generation of Electric Power in the United States. USDOE, Washington.

USDOS (US Department of State). 2007. Fourth Climate Action Report to the UN Framework Convention on Climate Change. www.state.gov/g/oes/rls/rpts/car/ (accessed Apr. 4, 2008).

USEPA (US Environmental Protection Agency). 2008. Draft Inventory of Greenhouse Gas Emissions and Sinks: 1990–2006. www.epa.gov/climatechange/emissions/usinventoryreport.html (accessed Apr. 3, 2008).

Originally printed in Journal AWWA *June 2008.*

– 13 –

Climate Change—Mitigation Policy Issues

Cynthia Lane

INTRODUCTION

The topic of climate change has been a concern for Congress and various federal agencies in the past few years, and future decades will provide many opportunities for involvement and action by the drinking water industry. From the cap and trade bill to USEPA research and policy needs to the increasing recognition of the interdependence of water and energy, much attention is being given to climate change mitigation. This article explores some of the climate change related efforts underway in Congress, reviews the actions taken by federal agencies on climate change, and addresses potential impacts to drinking water utilities from greenhouse gas (GHG) mitigation technologies that are implemented outside of the water industry.

CONGRESSIONAL ACTIVITIES

Cap and Trade Legislation

In October 2007, America's Climate Security Act of 2007 (S.2191) was introduced into the United States Senate by Senator Joseph Lieberman. This bill established a cap-and-trade program that requires a 70 percent reduction in GHG emissions from covered sources, which represent more than 80 percent of total US emissions (Thomas 2008a). Also included in the bill were requirements for the development of a low carbon fuel standard and provisions aimed at enhancing energy efficiency. The provisions included in this bill were expected to reduce overall GHG emissions from the United States by approximately 63 percent by 2050 (Pew 2008).

In May of 2008, after various negotiations and revisions, the bill was reintroduced as the Lieberman-Warner Climate Security Act of 2008 (S.3036), with amendments by Senator Barbara Boxer. The revised bill includes many different provisions and the following ones would be of interest to the drinking water industry (Thomas 2008b):

- Administrator of the US Environmental Protection Agency (USEPA) will establish:
  - A federal greenhouse gas registry, for which certain facilities must report information regarding the amount of fossil fuels and GHGs produced and consumed; and,
  - Specified quantities of GHG emission allowances, which decline for each year of 2012 to 2050.
- Administrator of U SEPA will establish a GHG emission allowance transfer system for coal-fired power plants and other entities that use more than 5,000 metric tons of coal, natural gas processors and importers, petroleum processors and refiners, manufacturers and importers of more than 10,000 metric tons of GHGs, and any entity that emits

more than 10,000 metric tons of hydrofluorocarbons (HFCs) as a byproduct of the manufacture of hydrochlorofluorocarbons (HCFCs).
- Several different programs will be established including a domestic offset program to sequester GHGs and the Bonus Allowance Account for carbon capture and sequestration projects.
- The allocation and distribution of emission allowances are provided to states and facilities on the basis of emission reductions, energy savings, state programs that exceed federal emission reduction targets, state expenditures for low-income home energy assistance programs, and relative potential carbon dioxide emissions and also specifies how states shall use such allowances.
- The deposit of auction proceeds to, and allocations from, various programs are established and provided.
- USEPA must establish new qualifying levels and requirements of Energy Star certification for retail carbon offsets.
- The use of auction proceeds, including for programs concerning zero- or low-carbon energy technologies, advanced coal and sequestration technologies, incentives for production of fuel from cellulosic biomass, advanced technology vehicles manufacturing incentives, and sustainable energy is provided.
- The Energy Policy and Conservation Act is amended to set forth provisions concerning appliance energy efficiency requirements and state building energy efficiency code updates.
- The Safe Drinking Water Act is amended to require the Administrator to permit commercial-scale underground injection of carbon dioxide for purposes of geological sequestration.

Of special interest to drinking water utilities are the amendments put forward by Senator Boxer that contain provisions creating programs of research into, among other things, the effects of climate change on drinking water utilities. Ten areas of research are listed including

- Impacts of climate change on water quality and potential solutions;
- Impacts on groundwater supplies from carbon sequestration, including research to evaluate potential water quality consequences of carbon sequestration in various regional aquifers, soil conditions, and mineral deposits;
- Impacts of climate change on drinking water quantity and potential solutions;
- Potential climate-change-related infrastructure impacts and solutions for water treatment facilities and underground pipelines;
- Desalination, water reuse, and alternative supply technologies;
- Efficiency and the minimization of greenhouse gas emissions;
- Regional and hydrological basin cooperative water management solutions;
- Utility management, decision support systems, and water management models;
- Water conservation and demand management; and,
- Communications, education, and public acceptance related to improving strategies and approaches for communicating with customers, decision makers, and other stakeholders about the implications of climate change regarding water supply.

Although there was a great deal of momentum in 2008 regarding the cap-and-trade legislation, it is not known when the legislation would be introduced for debate again. House Speaker Nancy Pelosi has stated that the European model is currently being studied to determine how to best execute a cap-and-trade system for the United States (EED 2009). As a result of changes in committee leadership, incoming House Energy and Commerce Chairman Henry Waxman will be leading the efforts in 2009 on a climate cap-and-trade bill. To date, Waxman has not provided his plans for that legislation.

Water Efficiency Legislation

In July of 2008, the U.S. House of Representatives passed *the Water Use Efficiency and Conservation Research Act (H.R.3957)*. The purpose of this Act is to require the USEPA Office of Research and Development (ORD) to establish a research and development program to promote water use efficiency and conservation. Research would be performed in the following areas (Thomas 2008c):

- Technologies and processes that enable the collection, storage, treatment, and reuse of rainwater, stormwater, and greywater;
- Water storage and distribution systems; and
- Behavioral, social, and economic barriers to achieving greater water use efficiency.

The Act also requires the development of a strategic research plan for the water use efficiency and conservation research and development program. ORD is directed to facilitate the adoption of technology and processes to promote water use efficiency and conservation, and to collect and disseminate information on technologies and processes to promote water use efficiency and conservation, including

- Information on incentives and impediments to development and commercialization;
- Best practices; and,
- Anticipated increases in water use efficiency and conservation resulting from the implementation of specific technologies and processes.

In January of 2009, this Act was re-introduced into the House as bill H.R. 631 (Thomas 2009a). The bill had bi-partisan support in 2008 and could be similarly supported in the 111th Congress. Drinking water utilities would benefit from this legislation, as it would provide for technologies that would reduce the energy and water consumption of their customers, thereby reducing demand on the utilities.

ACTIVITIES BY FEDERAL AGENCIES

Several federal programs and agencies are performing work in the area of climate change mitigation specifically, the U.S. Climate Change Science Program and the Environmental Protection Agency. This section discusses the current and upcoming efforts of each group.

U.S. Climate Change Science Program

In February 2002, the U.S. Climate Change Science Program (CCSP) was created with the purpose of integrating federal research on climate and global change, as sponsored by thirteen federal agencies and overseen by the Office of Science and Technology Policy, the Council on Environmental Quality, the National Economic Council and the Office of Management and Budget. The goal of the CCSP is to provide the best possible scientific knowledge to manage climate variability and other related aspects of climate change. The various options that are proposed for society to mitigate and adapt to climate change may have substantial environmental, economic, and societal consequences. It is extremely important that policy makers have reliable information on which to base their decisions as unintended consequences can be a side-effect of hasty judgments. As such, the CCSP "...generates science in support of decision making related to the management of risks and opportunities of climate variability and change, including adaptive management and mitigation efforts, with an increased emphasis on generating scientific results at regional and local scales." (CCSP 2003)

There are seven interdisciplinary research elements in which the participating agencies coordinate and support scientific research. The following elements relate to major components of environmental and human systems that are undergoing changes caused by a variety of natural and human-induced causes (CCSP 2003).

Atmospheric composition—Activities in this area focus on how human and natural activities are altering the composition of the global atmosphere. Scientists are also exploring how these changes might influence climate, pollutant exposure, ecosystems, and human health.

Climate variability and change—This research looks at the climate elements that are particularly important to human systems, such as temperature and precipitation, to determine how they would be affected by changes in the Earth system.

Global water cycle—This is one of the more important research areas for a drinking water utility. Research is focused on how the distribution and quality of water is influenced by natural processes and human activities. Scientists are trying to determine if the changes are predictable as well as how variability and change in the water cycle would affect human systems.

Land-use/land-cover change—Scientists are examining the processes that determine the distributions of land cover and land use at local, regional, and global scales. They will be focusing on how, and to what extreme, land use and land cover can be predicted over various time scales; and how any changes in land usage may affect environmental and socio-economic conditions at any level.

Global carbon cycle—This research will identify the size, variability, and potential future changes to carbon reservoirs within the Earth system; and will provide data that will assist policy-makers in evaluating options to manage carbon sources and sinks.

Ecosystems—Research on this topic focuses on the interactions between natural and human-induced environmental changes and how they impact the structure and functional ability of existing ecosystems. Scientists are also examining the options that society may have to ensure the sustainability of these ecosystems in the face of uncertain regional and global environmental changes.

Human contributions and responses—The current focus of this research is on the potential effects of climate variability and change on human health and welfare. Studies are also being performed to determine if society is vulnerable or resilient to global climate change as well as how decision making is conducted when significant complexity and uncertainty exists.

Drinking water utilities will benefit from the research performed in almost all of these research elements. A better understanding of the global water cycle, and the potential impacts from climate change, allows for a focused approach to water management. The approach will be built on information provided from both the land-use/land-cover change and human contributions and responses elements. Research in these areas can provide utilities with a better idea of where to focus their mitigation and adaptation activities to realize the best return on their investment.

USEPA National Water Program Strategy: Response to Climate Change

In 2007, the USEPA acknowledged two major issues associated with climate change. One, that climate change will substantially impact water resources and watersheds across the United States and, two, that USEPA efforts to ensure progress according to the Clean Water Act, Safe Drinking Water Act, and various ocean and coastal laws could also be affected. In an effort to respond to these concerns, USEPA created a National Water Program Climate Change Workgroup that was tasked with assessing the relevant information related to climate change, evaluating potential impacts on water programs, and identifying appropriate response actions for the NWP. As part of this effort, the Workgroup developed a formal strategy to respond to climate change, which was issued in the fall of 2008 and is entitled *National Water Program Strategy: Response to Climate Change.* Contained within the strategy is an overview of the likely effects of climate change on water resources. The strategy also addresses the potential impacts of climate change on the nation's clean water and safe drinking water programs and provides recommendations for more than 40 key action items that can be undertaken by the NWP to respond to the challenges posed by climate change. It was also developed to complement related efforts by the USEPA Office of Air and Radiation in climate change policy and program development and the Office of Research and Development's management of climate change related research.

The strategy goals are based on projected impacts of climate change, providing for response actions that would directly address the needs of the water industry. The five general goals are (USEPA 2008):

- **Goal 1: Water Program Mitigation of Greenhouse Gases:** use water programs to contribute to greenhouse gas mitigation;
- **Goal 2: Water Program Adaptation to Climate Change:** adapt implementation of core water programs to maintain and improve program effectiveness in the context of a changing climate;
- **Goal 3: Climate Change Research Related to Water:** strengthen the link between USEPA water programs and climate change research.
- **Goal 4: Water Program Education on Climate Change:** educate water program professionals and stakeholders on climate change impacts on water resources and programs; and
- **Goal 5: Water Program Management of Climate Change:** establish the management capability within the National Water Program to engage climate change challenges on a sustained basis.

This section of the article will focus on Goal 1 (mitigation of greenhouse gases). With respect to this goal, the strategy puts forward several general recommendations as to how USEPA can reduce the severity of climate change impacts on water resources by addressing and reducing greenhouse gas emissions. These recommendations include (USEPA 2008):

- Expand existing programs that result in greenhouse gas mitigation;
- Expand efforts related to geologic and biological sequestration of carbon dioxide;
- Use the best available science and technology to support responsible operation of water treatment and delivery systems through water conservation and energy efficiency; and
- Support carbon sequestration related to energy production and industrial processes.

To support these general recommendations, the strategy highlights eleven (11) key actions to be performed by the NWP by 2011. The NWP will either expand existing efforts to better address climate change or take new actions specifically focused on climate change issues. More specifically, the key actions were identified with the hope of defining areas of uncertainty, evaluating proactive and reactive actions, and guarding against unintended consequences. It is through these key actions that drinking water utilities anticipate climate change mitigation policy and regulatory decisions to be made.

Water-Related Energy Conservation and Production

The first area for action identified in the strategy is water-related energy conservation and production. While energy usage by drinking water and wastewater facilities only accounts for approximately three percent of the United States' energy consumption, it is a major component of the everyday operations of the facilities. Water conveyance, including pumping water to treatment plants and distributing treated water to customers, is generally acknowledged to be one of the most energy intensive components of drinking water systems. Drinking water treatment facilities are consuming larger and larger amounts of energy. This trend is not expected to change over the upcoming decades as new and/or revised drinking water treatment regulations that consume more energy are implemented. Pumping needs will also increase as utilities face reducing drinking water supplies and increasing demand.

Key Action #1: Improve Energy Efficiency at Water and Wastewater Utilities

The first key action identified in the strategy is to improve energy efficiency at water and wastewater utilities. In the strategy, USEPA states that the National Water Program will "...continue to work with the Office of Air and Radiation to promote energy performance benchmarking programs, use of energy audits and energy tracking systems, use of alternative energy sources within plants (e.g., solar, wind, hydro), installation of combined heat and power systems for heat and energy generation in facilities that use anaerobic digesters, and will provide state and local governments information on available and emerging treatment technology."

One of the main components of this key action is a focus on the drinking water industry within USEPA's ENERGY STAR program. USEPA created this program to focus on the development of tools for utility managers that will enable them to successfully implement energy management programs. The National Water Program will be working with the ENERGY STAR program on the development of these tools that will allow for the benchmarking of energy performance among public water utilities.

In addition to the development of these tools, the strategy states that the NWP will also focus on encouraging water utilities to produce energy to offset their energy purchases from local power utilities. To accomplish this, the NWP will work with the Office of Air and Radiation to promote the installation of alternative energy power production facilities, such as solar, wind, and hydro, for heating and electricity generation at the treatment facilities. Some utilities may find it difficult to implement the alternative energy production technologies, caused by financial restrictions or logistical issues. These utilities can realize immediate energy savings by installing energy efficient lighting and adequate installation in the buildings used for operations and maintenance. The NWP will also be encouraging utilities to do these simple energy saving measures as these can be the stepping stones to the next step of alternative energy production. It is important to note that alternative energy power production facilities produce "green power." Because several states are requiring water utilities to have alternative or "green" power as a part of their overall production, a water utility that installs adequate facilities would be in compliance with these laws.

Water Conservation

Water conservation is another area in which the NWP will be focusing its efforts to mitigate climate change. Through the more efficient use of water, several benefits can be realized including the extension of the water supply and the reduction of greenhouse gasses. Because water usage by customers is reduced, less water will be pumped and treated. Also, when energy usage is reduced, water savings are realized because less is needed to operate power plants.

The strategy identifies many areas of opportunities for energy savings on both the supply side and the demand side of a drinking water utility. On the supply side, better planning, maintenance, and operation of water delivery systems can provide for substantial water and energy savings. On the demand side, the promotion of conservation programs can also effectively increase water and energy savings. Five key actions are identified in the strategy that will be implemented by the NWP to mitigate climate change.

Key Action #2: Implement the WaterSense Program

One of the current efforts that will continue is the implementation and expansion of the WaterSense program. Because the conservation of water results in a conservation of energy, USEPA plans to expand the labeling guidelines for WaterSense products to include information on the reduction in energy consumption that is realized by the reduction in water consumption. Water utilities will be able to use this information to promote the WaterSense program to their residential customers and to demonstrate the opportunity to save both water and energy by installing water-efficient fixtures and appliances, including toilets, showerheads, faucets, clothes washers, dishwashers, and irrigation equipment.

Key Action #3: Water Conservation and Management for Drinking Water Systems

The strategy puts forward key action #3 as "...explore opportunities with states and drinking water systems to better address expected impacts of climate change on water supply and water usage rates through water conservation and water resources management." By promoting water and energy efficient fixtures and appliances, drinking water utilities will see a reduction in water demand, resulting in climate change mitigation opportunities through lower energy usages for water treatment and distribution and fewer greenhouse gas emissions. Key action #3 allows for a wide variety of involvement by USEPA, which might manifest as research prioritization workshops, the development of strategies for water utilities to implement conservation programs, or guidelines for the development of alternative sources of water.

Key Action #4: Water Conveyance Leak Detection and Remediation

Activities by the NWP in this area would include the promotion of available technologies to identify and address leakage from water pipes and other means of conveyance. Repairing leaking distribution systems is a major opportunity to recover water that would otherwise be lost from the system. Water utilities that invest in leak correction and detection technologies can realize significant savings in water. The NWP will be encouraging water utilities to implement these technologies during the initial project design or pipe replacement projects. There may also be some partnerships with other entities to further develop the existing technologies into ones that are more precise and adaptable to utilities of all sizes.

Key Action #5: Industrial Water Conservation Reuse and Recycling Technology Transfer

The NWP expects to "...identify industries and facilities that best maximize their water efficiency and develop a technical guide for control authorities and industry for promoting water minimization, re-use, and recycling." As most water utilities know, industrial facilities consume large amounts of water. The companies that manage these facilities are becoming more aware of the importance of measuring, managing, and controlling water use because some are finding water scarcity to limit growth. An economic incentive exists for facilities to use as little water as possible in their industrial operations; however water reuse and recycling have not been widely promoted. The NWP is going to focus on changing this with the goal of reducing the amount of potable water that utilities will have to provide to industrial facilities. By reducing water usage in the industrial sector, savings in treatment costs, including chemical consumption energy usage, could be realized by water utilities. Reductions in both of these areas contribute to the overall goal of mitigating climate change.

Key Action #6: Federal Agency Water Conservation Guidance

In 2008, Executive Order 13423, Section 2 (c) was issued requiring "...Federal agencies reduce water consumption intensity, relative to the baseline of the agency's water consumption in fiscal year 2007, through life-cycle cost-effective measures, by 2 percent annually through the end of fiscal year 2015, or 16 percent by the end of fiscal year 2015." Under this order, the NWP is responsible for developing a Water Efficiency Implementation Guidance for all federal agencies.

The Guidance will cover the development of the baseline water consumption, efficiency opportunity identification/implementation, and reporting. The usage of WaterSense products and services in implementation strategies is strongly encouraged. While this is currently only a federal requirement, it is possible that state governments that do not already have a similar policy may adopt it at some point in the future. If they do, the local water utilities that serve the state buildings may consider working with the State administration to better understand the energy and water savings that could be realized through the implementation of this guidance.

Promote "Green Building" Design and "Smart Growth"

The previous six key actions were focused on mitigating greenhouse gases by increasing the water and energy efficiency of water utilities. Sometimes, this only has short term effects as customers revert to previous behaviors after a water crisis has passed. Sustaining the water and energy efficiencies over the long-term is an extremely important component of the mitigation toolbox. The commitment to water and energy efficiency must be incorporated into building codes and community design, thereby allowing for "green building" principles and "smart growth" policies to multiply the energy and water savings that can be realized by water utilities.

The U.S. Green Building Council's Leadership for Energy and Environmental Design (LEED) program and the American National Standards Institute (ANSI) are currently working to promote the "green buildings" concept and LEED rating systems within state and local governments. Rating systems are built on the idea that a developer will follow "green" building practices during construction and will receive recognition for doing so. Practices that are recognized as "green" within the various rating systems include

- Reduced use of energy and water;
- On-site (decentralized) energy generation (e.g., solar power, geothermal); and,
- Water retention (e.g., green roofs).

LEED has begun to expand their traditional ratings to include other characteristics of sustainable cities such as smart growth and low impact design. A new rating system for neighborhoods now evaluates site selection and design, infrastructure linkages, and provides credits for onsite stormwater management practices such as green roofs and rain gardens. While the main area of focus for the NWP would be on stormwater management, the strategy provides for the promotion of "green" facilities within all aspects of the community, including water treatment and distribution systems.

Key Action #7: Promote Energy Saving/Generating "Green Buildings" and "Green Infrastructure" Including Provisions Allowing Such Practices in Stormwater Permits

The strategy states that the NWP will "...work with other USEPA offices to support states, tribes, and local governments and the private sector in promoting the "green buildings" rating systems, with a focus on saving water and energy and will work to integrate provisions allowing "green infrastructure" practices into stormwater permits." As previously stated, much of the NWP's efforts as part of this key action will be on allowing "green building" standards and "green infrastructure" practices to be used during the design and construction of stormwater management structures. USEPA seem to recognize that this may require greater flexibility in permitting the use of such standards and practices and acknowledges that allowing part of stormwater permits could encourage their adoption. USEPA will be promoting general "green buildings/infrastructure" concepts over the next few years. While not directly impacting water utilities, the implementation of these concepts will reduce water demands for utilities and provide for reductions in energy usage and greenhouse gas emissions.

OTHER POLICY ISSUES

Energy/Water Connection

When considering possible climate change mitigation technologies, it is important to remember the inherent connection between energy and water. Much of the focus of climate change mitigation has been on changing the types of energy and energy generation methods that the United States uses. These changes can pose several challenges for water utility managers and regional environmental departments including

- Increased water use and withdrawals by power plants will require expanded efforts to assure that drinking water availability is not compromised;
- Increased attention to potential nonpoint pollution impacts of expanded agricultural production may be needed;
- Increased monitoring of discharge permit conditions to address increased temperature and concentration of pollutants caused by low flows;
- Need for more efficient use of electrical energy at water facilities and production of power from methane at some wastewater treatment facilities; and,
- Recognition that effective, market-based practices with water quality benefits could be a source of revenue.

Many of the decisions that will be made related to the type of energy generation used as a nation will undoubtedly affect drinking water utilities. It will be important for drinking water utilities to become part of the decision-making process on energy and energy generation methods at both the local and regional level. These new mitigation technologies, such as geologic carbon sequestration or increased use of biofuels, can make a significant difference in GHG emissions; however, there is the potential for unintended consequences that could adversely impact drinking water sources.

Consequences of Mitigation Technologies

As previously mentioned, there is the potential for unintended consequences with the various technologies that are proposed for large-scale GHG mitigation. Significant adverse impacts to drinking water sources can be associated with two of the most high profile technologies, geologic carbon sequestration and biofuels. These potential impacts are discussed in more detail the following sections.

Geologic Carbon Sequestration

Geologic carbon sequestration consists of the injection of supercritical carbon dioxide into deep underground geological formations such as saline aquifers, unreachable coals seams or basalt formations. Theoretically, physical and geochemical trapping mechanisms would prevent the carbon dioxide from escaping to the surface. The sequestration of carbon dioxide is currently being proposed by the Department of Energy as a viable method to curb carbon emissions into the atmosphere. The drinking water industry has expressed many concerns regarding carbon dioxide sequestration technology and its potential impacts on underground sources of drinking water (USDWs). These potential impacts are as follows:

- **Water chemistry:** Several carbon sequestration references discuss changes in the carbonate cycle resulting in lowered pH conditions and iron, manganese, arsenic, and possibly other inorganics released into groundwater surrounding the injection zone. Additionally, silica and boron, depending on aquifer composition, can dissolve into the groundwater.
- **Acidic reactions:** Carbon dioxide in either a gas or supercritical liquid state is very acidic and could react with the surrounding strata. Without neutralization, there is the possibility that the carbon dioxide could change the equilibrium state for the sediments within those strata and result in the contamination of USDWs.
- **Other contaminants:** There is the potential for contamination of USDWs because of the presence of other compounds, such as nitrogen, carbon monoxide, sulfur dioxide and possibly mercury, in the carbon dioxide injection stream.
- **Injection well construction:** Because the injection wells are encased in cement, the long term integrity of the cements that are used during construction will need to be extensively tested under real-world conditions.
- **Injection well depth:** There is the potential for USDWs to be displaced both vertically and laterally because of changes in pressure gradients resulting from the injection of carbon dioxide into deep saline aquifers.
- **Future USDW classification:** As the demand for water increases during the 21st century and changes in climate impact traditional water supplies, utilities will look for new sources of drinking water. The possibility exists that utilities may use some of the injection site aquifers as new potable sources.

It should also be noted that groundwater storage of water resources may become a favorable adaptation strategy for water management under climate change. This could be especially true for areas in the United States that will lose storage (e.g. decreasing snowpack) and/or require more storage (e.g. increasing population). Groundwater storage may be a better option than surface water storage options such as dams and reservoirs, which have high costs, environmental opposition, and potentially higher evaporation rates as a result of climate change. There could be potential conflicts between sequestration sites and aquifer storage sites if they are located in the same region.

It is important that drinking water utilities become aware of this technology and participate in any pilot projects that are located nearby. DOE has created seven Regional Carbon Sequestration Partnerships (RCSPs) that will assist with technology development through various research and outreach projects. Each of the RCSPs is involved in a number of geologic carbon sequestration projects, ranging from small-scale tests to validate geologic formations injection possibilities, containment, and storage effectiveness to large-scale sequestration projects that will attempt to validate the reservoir models against field data and verify the fate of the injected CO_2 using the most advanced monitoring networks applied to date. Drinking water utilities are encouraged to participate in these RCSPs through the stakeholder outreach process that is a major component of each project. Utility involvement in sequestration projects will help the industry protect USDWs and will raise awareness of the potential for unintended consequences that could result from this GHG mitigation technology.

Biofuels

Increased use of biofuels as a GHG mitigation technology could have a significant impact on the water resources of the country. Changing agricultural practices, increased corn production, and growth in the number of biorefineries result from an increase in the amount of biofuels produced in the US. These changes could add pressure to the water quantity challenges currently faced by drinking water utilities. A Report in Brief issued by the National Resource Council (NRC) in 2007 raises another important issue related to biofuels production. It is possible that an increase in the production of biofuels could drive the expansion of agriculture into regions that do not currently support much agriculture. Depending on the specific type of crops grown, the location of the fields, and whether there is an increase in overall agricultural production, significant acceleration of biofuels production could cause much greater water quantity problems than are currently experienced (NRC 2007).

With respect to water quality, a shift in agricultural practices to incorporate more biofuel crops could cause degradation in water quality. Increased problems associated with fertilizer runoff and soil erosion may be observed when pastures or woodlands are converted into cornfields (NRC 2007).

With the increased focus on climate change in the next few years, it is extremely likely that these policy topics will be discussed as part of the bigger picture of large-scale mitigation of GHGs. Drinking water utilities should be aware of these policies and determine how their water management practices could be impacted.

CONCLUSIONS

In summary, mitigation of greenhouse gases is being investigated by Congress as well as many different federal agencies and associations. While there is currently a great deal of focus on adaptation techniques for the drinking water industry, attention is also concentrating on the development of technologies for GHG mitigation within water utilities. It is important that water managers are aware of the potential federal resources for information, such as the CCSP or USEPA, as well as the policy issues associated with mitigation technologies for other industries. Energy and water are interconnected, and any change in either industry has the potential to significantly impact the other.

REFERENCES

CCSP. 2003. *Strategic Plan for the U.S. Climate Change Science Program.* Climate Change Science Program. Washington, D.C.

CCSP. 2008. *Revised Research Plan for the U.S. Climate Change Science Program.* Climate Change Science Program and the Subcommittee on Global Change Research. Washington, D.C.

DOE. 2009. *Regional Carbon Sequestration Partnerships.* United States Department of Energy. Washington, D.C. http://www.netl.doe.gov/technologies/carbon_seq/partnerships/partnerships.html.

Samuelsohn, Darren. 2009. "Pelosi says House can pass cap and trade, but timetable is uncertain." *Energy and Environment Daily.* January 6, 2009. Washington, D.C.

NRC. 2007. *Water Implications of Biofuels Production in the United States.* Committee on Water Implications of Biofuels Production in the United States, National Research Council, National Academy Press, Washington, D.C.: National Academy Press.

Pew. 2008. *Summary of the Boxer Substitute Amendment to the Lieberman-Warner Climate Security Act.* Pew Center on Climate Change. Washington, D.C.

http://www.pewclimate.org/docUploads/L-WFullSummary.pdf. June 4, 2008.

Thomas. 2008a. *S. 2191—America's Climate Security Act of 2007.* Library of Congress. THOMAS. Washington, D.C.
http://thomas.loc.gov/cgi-bin/bdquery/z?d110:SN02191:@@@L&summ2=m&.
Thomas. 2008b. *S.3036—Lieberman-Warner Climate Security Act of 2008.* Library of Congress, THOMAS. Washington, D.C.
http://thomas.loc.gov/cgi-bin/bdquery/z?d110:SN03036:@@@L&summ2=m&.
Thomas. 2008c. *H.R.3957—Water Use Efficiency and Conservation Research Act.* Library of Congress, THOMAS. Washington, D.C.
http://thomas.loc.gov/cgi-bin/bdquery/z?d110:HR03957:@@@L&summ2 =m&
Thomas. 2009a. H.R.631—*Water Use Efficiency and Conservation Research Act.* Library of Congress, THOMAS. Washington, D.C.
http://thomas.loc.gov/cgi-bin/bdquery/z?d111:HR00631:@@@L&summ2 =m&
USEPA 2008. *National Water Program Strategy: Response to Climate Change.* Office of Water, U.S. Environmental Protection Agency. EPA 800-R-08-001. Washington, D.C.

– 14 –

Climate Change Mitigation Strategies in the Water Sector in Developing Countries

Juan van den Broeck and Daniel A. Nolasco

INTRODUCTION

In the developing world, particularly in Latin America, climate change mitigation activities in water utilities are mostly fostered by the Kyoto Protocol.

Based on the fact that greenhouse gas (GHG) emission reductions (ERs) have the same beneficial effect on global warming disregarding where the reduction takes place, the Kyoto Protocol includes three market-based mechanisms to assist developed countries (Annex I countries) meet their targets in a more cost-effective way:

- Emissions Trading (within one Annex I country),
- Joint Implementation (one country of Annex I finances a project in another Annex I country and gets the credits), and
- Clean Development Mechanism (CDM, mitigation in a non-Annex I country generates credits—certified emission reductions—that are transferred to an Annex I country).

This paper focuses on Clean Development Mechanism projects, since this mechanism applies to non-Annex I countries (developing countries).

Within the Kyoto Protocol, Clean Development Mechanism projects are classified in different "sector scopes." Two scopes include water-related activities:

- Scope 13: Waste handling and disposal
- Scope 15: Agriculture

These two scopes include projects where methane is the main greenhouse gas targeted.

The recognition of the magnitude and importance of nitrous oxide (N_2O) emissions from wastewater treatment plants has grown steadily over the past years. Nitrous oxide emissions are particularly relevant in nitrifying systems.[1] Nevertheless, to date, most CDM projects and methodologies focus on methane gas.

As of Jan. 26, 2009, 1,696 projects have been registered as CDM projects under the Kyoto Protocol framework, of which 405 projects are included within those two scopes (24% over total registered).[2]

---

1 Rosso, D. and Stenstrom, M. (2008), Comparative Energy and Carbon Footprints for Activated Sludge Layouts, WEF Specialty Conference: Sustainability 2008—Green Practices for the Water Environment, National Harbor, Maryland, June 22–25, 2008

2 http://cdm.unfccc.int/Statistics/Registration/RegisteredProjByScopePieChart.html

CLEAN DEVELOPMENT MECHANISM PROJECT DEVELOPMENT

Project activities must comprise a series of characteristics to be able to qualify as a CDM project. First, it has to be demonstrated that it contributes to GHGs emission reduction beyond the baseline situation (business-as-usual). This condition, generally referred to as additionality, involves proving that the implementation of the project could not have been possible (technically and financially) without the issuance of certified emission reductions, generally referred to as CERs. Other conditions are that the project must not be forced by local regulations and it should contribute to sustainable development.

In many countries, methane emissions from wastewater treatment activities are not regulated. Thereby, utilities tend not to invest in methane avoidance or methane capture projects unless there is a potential for carbon credits.

Trading CERs depends on the type of project involved and how it is developed. In many cases, to help finance capital expenditures, carbon credits are sold before project development. In other cases, the developer may opt to sell the credits upon issuance, i.e., once the reduction is effectively achieved, in which case a better price per ton of CERs is likely. Unfortunately, delays in carbon credits issuance may have a financial impact on those projects waiting to get CERs issued.

A typical CDM project cycle comprises of several stages (see Figure 14-1).

CDM projects start with a conceptual design and an assessment of its feasibility. Then, a Project Design Document (PDD) is prepared. The PDD contains all information pertaining the mitigation project as well as the calculations required to estimate emission reductions and a monitoring program to be implemented once the infrastructure needed is in place. Both the calculations of emission reductions and the monitoring program must follow an Approved Methodology, which are methodologies previously presented to and approved by the CDM Panel within the UNFCCC. If there was no Approved Methodology available for the project in question, a new methodology must be developed and presented by the project proponents.

Once the PDD is ready, it is presented to the Designated National Authority (usually the Ministry of the Environment, or its equivalent, of the hosting country). To carry on with the approval cycle, the Designated National Authority must issue a note indicating that the proposed project contributes to the sustainable development of the host country (*sine-qua-non* condition). Then the project goes through validation and registration (generally done by a consulting firm or technical group registered at the UNFCCC and acting as a Designated Operational Entity).

Once this part of the cycle is completed, the project can be built and operated. During its operation, the project will be monitored according to the methodologies presented in the PDD. A second Designated Operational Entity will be in charge of verifying and certifying that the emission reductions are actually taking place. Then, the CERs are issued by the UNFCCC, registered in the Annex I country that purchased the CERs.

Projects in countries that have not ratified the Kyoto Protocol may follow a shorter path, as shown in the figure.

Recently, many projects related to climate change mitigation in water utilities faced problems during the monitoring and the verification stages as the CERs they effectively generated were not close to those predicted in the Project Design Document.

Experience shows that project developers need to be conservative when estimating the Project Design Document emission reductions from a potential project and focus on the development of a good monitoring plan to reduce the gap between estimated CERs and actual (issued) CERs, thereby, increasing project reliability and improving trading opportunities.

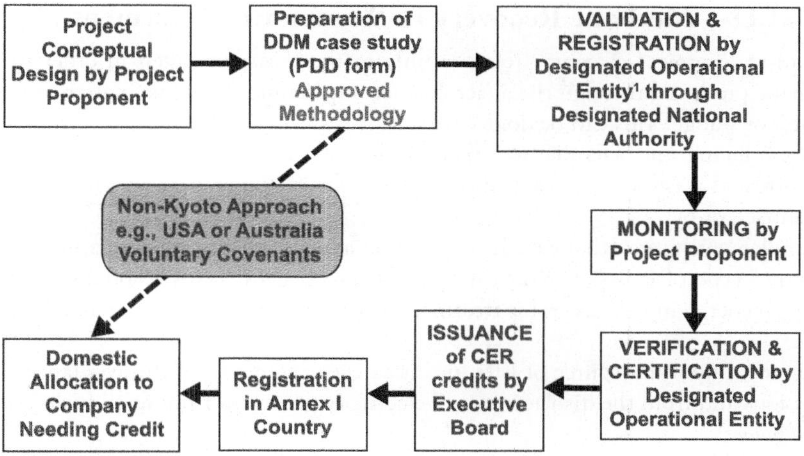

Figure 14-1. Typical CDM project cycle

Brazil and Mexico are leaders in CERs issuance in the wastewater sector in Latin America, with all CERs to-date coming from animal waste management systems. Seventeen registered projects from Brazil have already issued CERs, while Mexico has 16 projects. Other countries that issued CERs in the water sector in Latin America are Chile and Ecuador.

APPROVED METHODOLOGIES AND CASE STUDIES IN THE WATER SECTOR

The United Nations Framework Convention on Climate Change (UNFCCC) has approved a series of methodologies which serve as tools to estimate *ex-ante* emission reductions (ERs) and monitor actual ERs achieved by a wide variety of project activities under the CDM frame.

In the water sector, GHG reductions have focused on methane gas produced through the anaerobic degradation of organic matter contained in wastewater and with indirect emissions from energy consumption during wastewater treatment.

So far, UNFCCC has approved methodologies covering the following emission reduction aspects:

- Capture and flaring of generated methane
- Energy generation with captured methane
- Replacement of anaerobic lagoons with aerobic systems
- Methane ERs through aerobic co-composting with organic solid waste
- Switching to less energy consuming treatments (e.g., from activated sludge to upflow anaerobic sludge blanket reactors, or UASB)
- ERs in sludge management (treatment and disposal)

What follows is the description of the approved methodologies for small and large scale projects in the water sector, together with case studies that illustrate respective emission reductions.

AMS III.H—Methane Recovery in Wastewater Treatment

This small-scale project category[3] refers mainly to the measures taken in order to recover the methane produced through the anaerobic decomposition of organic matter contained in wastewater or sludge. This can be done either by:

- Introducing an anaerobic treatment system with methane recovery to a currently untreated wastewater stream (discharged to an anoxic or near-anoxic natural watercourse), or by
- Implementing a methane capture system in an existing anaerobic treatment.[4]

Another type of emission reduction project included in this methodology consists in changing an energy intensive aerobic treatment system for an anaerobic system with methane recovery.

The most important source of ERs in this category comes from the avoidance of methane emissions and from the displacement of electricity generated from fossil fuels *with biogas generated power.*

AMS III.H Case Study 1

AgriVerde Ltd. registered in 2007 a CDM project for recovering methane from the anaerobic treatment system of its PKS Milano Pinang Awan palm oil mill in Sumatera Utara, Indonesia.[5]

This palm oil mill processes almost 300,000 ton/year of fresh fruit bunches, generating approximately 194,000 m^3/year of wastewater with high organic content (50,000 mgCOD/L).

The palm oil mill effluent (POME) is treated through a pond system consisting of a cooling pond, a sludge pond, a neutral pond, two anaerobic lagoons, facultative lagoons, an aerobic lagoon, and sedimentation ponds. The depth of each anaerobic lagoon is 4.5 m and the average temperature in the region is 27.8°C, both factors favorable to creating anaerobic conditions.

The project consists in the installation of sealed high-density polyethylene (HDPE) geomembranes covering the anaerobic lagoons, a routing system, and enclosed flares to combust the produced methane.

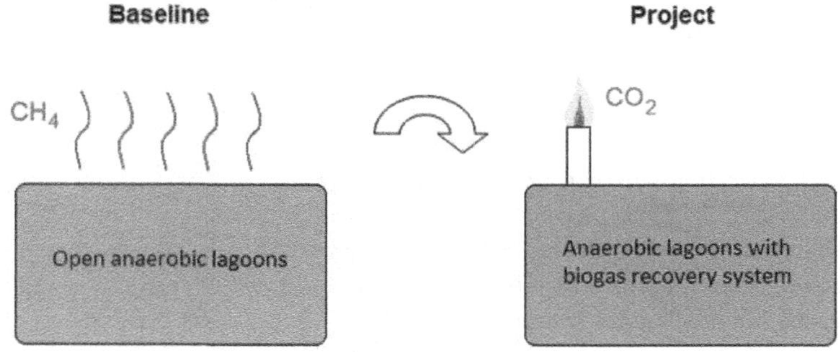

3 UNFCCC classifies projects as small scale when the expected emission reduction is less than 60,000 tCO_2-eq./year. Approved methodologies (AMS) for this type of projects are considerably simpler than other approved methodologies (AM).

4 Approved methodology AMS III.H—*Methane Recovery in Wastewater Treatment,* Version 05.

5 UNFCCC CDM Simplified Project Design Document for *Methane Recovery in Wastewater Treatment Project AIN07-W-01, Sumatera Utara, Indonesia,* Version 01.

In the baseline scenario, anaerobic lagoons emit 36,000 tCO_2-eq./year. Emissions from other lagoons are not taken into account since they will remain approximately the same. After project implementation, methane emissions from anaerobic lagoons are cut by 90%, with a 10% remaining emissions from the leakage inherent to the capture-and-flaring system.

Also, the project contributes to local sustainable development by improving air quality through the reduction of odor, and by including the installation of an extensive monitoring system to calculate actual ERs and assure compliance with local environmental regulations.

Summary

- Processes 300,000 ton/year of fresh fruit bunches
- Generates 194,000 m^3/year of high COD wastewater
- Average temperature in the region: 27.8°C
- Lagoons depth: 4.5 m
- Baseline: open anaerobic lagoons
- Project: installation of a biogas recovery and flaring system
- Emission reduction: 32,000 tCO_2-eq./year.

AMS III.H Case Study 1

Another case study where AMS III.H has been applied is in the municipal wastewater treatment plants of the city of Chennai, India.

In 1999, the government launched The Chennai City River Conservation Project, with the purpose of implementing a sewage treatment system and stopping untreated sewage flows into the city's waterways.

The program included the construction of additional wastewater treatment plants (WWTPs) at four locations (Kodungaiyur, Koyambedu, Nesapakkam, and Perungudi), all consisting in activated sludge process with anaerobic sludge digesters (with no biogas recovery system), totaling a treatment capacity of 264,000 m^3/day.[6]

The project developers learned about the potential for energy generation, which could significantly reduce electricity purchases from the grid. Then, they decided to install biogas capture and power generation systems in all four locations.

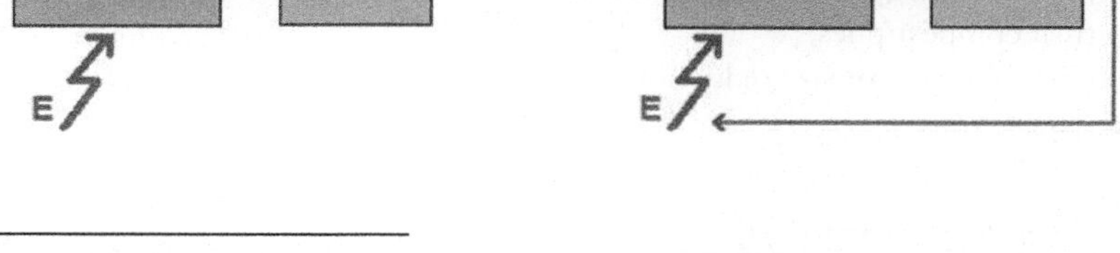

6 UNFCCC CDM Simplified Project Design Document for *Methane Capture and Power Generation at Sewage Treatment Plants, Chennai, India*, Version 01.

The physical component is composed of:

- The gas holder receiving biogas from the digester,
- The alkali scrubbing unit for removal of hydrogen sulfide, and
- The gas based power generation unit.

The total installed capacity for electricity generation is 3,160 kW, supplied at 415 V to the internal power distribution system.

The total amount of GHG avoided from release to the atmosphere by the project activity is estimated at 57,000 tCO_2-eq./year. Out of these emission reductions, 18% comes from displaced electricity from the grid, and the rest from captured and burned methane.

Furthermore, the project contributes to enhance to the local socio-environmental situation by employing around 30 local workers for each plant and improving the electricity scenario in the region.

Summary

- Chennai City WWTPs: activated sludge process + sludge digester
- Sludge generated by 264,000 m^3/day of wastewater
- Baseline: anaerobic sludge digester without biogas recovery system
- Project: installation of biogas recovery system and gas-based power generation unit
- ERs: 57 000 tCO_2-eq./year, 18% from displaced electricity, 82% from captured methane.

AMS III.F—Avoidance of Methane Emissions through Controlled Biological Treatment of Biomass

The approved methodology for small-scale AMS III.F relates to projects that avoid methane production and consequent release to the atmosphere. Methane gas is the result of anaerobic degradation of organic matter (including sludge), occurring at a solid waste disposal site lacking a biogas capture and flaring system.[7]

The proposed project activity avoids methane emissions through the replacement of the current disposal method with a controlled biological treatment. The proposed treatment can be either:

- Aerobic treatment by composting and proper soil application of the compost, or
- Anaerobic digestion in closed reactors equipped with biogas recovery and combustion/flaring system.

This methodology is also applicable to co-treatment of solid biomass waste with wastewater. In the absence of this project, wastewater would have been treated anaerobically, without biogas recovery. Hence, co-treatment of wastewater reduces those emissions, while using wastewater as a source of moisture and/or nutrients.

Project emissions consist of:

- Emissions generated by additional transport of residues and compost (and wastewater when applicable),
- Emissions generated by consumption of electricity from the grid or fossil fuels:
  - To produce energy used on-site for aeration,
  - To turn compost piles,
  - To chop biomass for size reduction,
  - To screen waste,

7 Approved methodology AMS III.F—*Avoidance of Methane Emissions Through Controlled Biological Treatment of Biomass*, Version 04.

  - To dry the final compost product, and
  - To treat wastewater runoff.
- Methane emissions from physical leakage of the anaerobic digester,
- Methane emissions during composting process,
- Methane emissions from improper storage of waste or final products.

AMS III.F Case Study

Gianyar Regency, in Bali, Indonesia

Until 2008, most of the collected waste of Gianyar Regency (Bali, Indonesia) was dumped on a landfill without methane recovery.

Waste characteristics are approximately as follows:

- 85% of organic matter, which could be composted,
- 5% of inorganic material (mostly plastic) that could be recovered and sold to recyclers, and
- 10% that can be safely disposed in the landfill.

A CDM project to expand the capacity of an existing waste recovery and composting facility from 4 tons/day to 50 tons/day was proposed to cope with all waste in Gianyar Regency (500,000 habitants).[8]

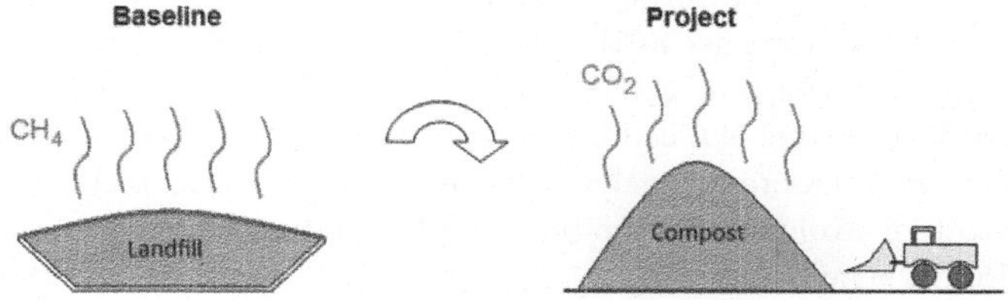

The existing 400 m^2 building would accommodate two additional shredders, blowers to aerate the compost piles, and the existing baler. A new covered 5,000 m^2 area under light steel construction would house:

- Waste sorting on 700 m^2,
- Composting in table piles on 3,200 m^2,
- Compost curing and compost storage on 1,100 m^2, as well as an additional compost sieve and compost mixer.

For managing and handling of compost and waste, on-site additional vehicles were required.

The project activity will require an investment of US$380,000. The waste processing capacity of the existing facility will be expanded to almost 15,000 ton of waste per year. An average annual emission of 8,000 tCO_2-eq., generated by decomposition of organic matter disposed of in a landfill will be avoided. After project implementation, GHG emissions were estimated to descend to 300 tCO_2-eq./year, comprising electricity use (12%), additional transport of waste and compost (8%), and on-site fossil fuel use (80%).

8 UNFCCC CDM Simplified Project Design Document for *Gianyar Waste Recovery Project, Bali, Indonesia*, Version 03.

Summary

- Gianyar Regency, population: 500,000
- 85% of waste is organic
- Baseline: 15,000 t/year of waste to landfill
- Project: expand capacity of composting facility by 15,000 t/year
- Investment: US$380,000
- ERs: 8,000 tCO_2-eq./year

AMS III.I—Avoidance of Methane Production in Wastewater Treatment through Replacement of Anaerobic Systems by Aerobic Systems

This small-scale methodology estimates and monitors GHG emission reductions obtained when replacing an anaerobic treatment system with no biogas capture by an aerobic system. Aerobic systems do not generate methane. However, being more energy intensive, they may generate additional GHG emissions at the electrical generation stations.[9]

To guarantee anaerobic conditions, the lagoons (baseline scenario) have to meet the following criteria:

- At least 2 m deep,
- Monthly average temperature on the project site should not be less than 15°C (at least during part of the year, since methane emissions are only calculated for those months when average temperature is above 15°C), and
- Volumetric loading rate of Chemical Oxygen Demand > 0.1 kg COD*m^{-3}*day^{-1}.

The most important fact to be analyzed in this kind of projects is the balance between the methane generation avoidance by aeration, and the indirect CO_2-eq. emissions from energy generated for the aeration system.

AMS III.I Case Study

An example of this project category is the Granja Tres Arroyos Methane Avoidance in Slaughterhouse Effluent Project, Buenos Aires, Argentina.

This facility processes an average of 150,000 heads/day, generating 4,400 m^3/day of high COD effluent (7,300 mg/L). The treatment system consists of an anaerobic lagoon system.[10]

The project activity is based on the installation of a primary treatment system consisting of:

- Two parallel sieves,
- A mixing tank,
- A dissolved air flotation system (to separate oil and fat which are then centrifuged and the oil is sent to be used as fuel in a boiler), and
- A secondary treatment system based on three superficially aerated lagoons placed in series.

Sludge generation is estimated to be insignificant, needing to be extracted only once every five years.

---

9 Approved methodology AMS III.I—*Avoidance of Methane Production in Wastewater Treatment through Replacement of Anaerobic Lagoons by Aerobic Systems,* Version 05.

10 UNFCCC CDM Simplified Project Design Document for *Granja Tres Arroyos Methane Avoidance in Slaughterhouse Effluents Project, Buenos Aires, Argentina,* Version 01.

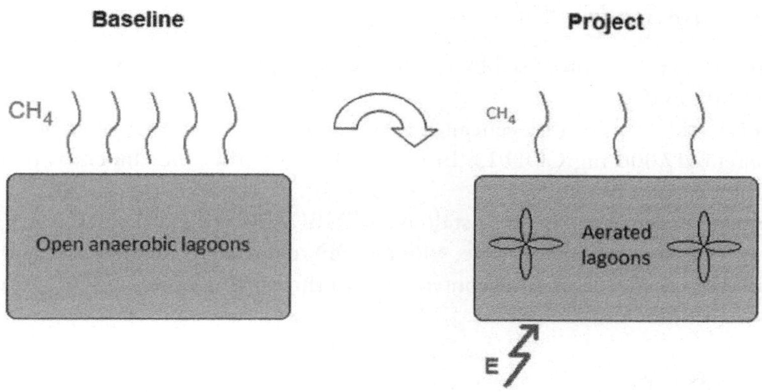

An *ex-ante* calculation of the potential CERs to be achieved estimated a reduction of 12,200 tCO_2-eq./year, during a crediting period of seven years. Of the 4,200 tCO_2-eq./year emitted by the project activities, 16% correspond to electricity use and the other 84% to methane emissions from the anaerobic zones of the aerobic lagoons.

In addition, the project contributes to achieve sustainable development goals in the host country by:

- Increasing job opportunities for qualified workers during the construction, installation, and operation of the facility,
- Improving air quality through odor reduction, and
- Contributing to a cleaner energy matrix by partially replacing natural gas in the boiler with recovered oil from the treatment system.

Summary

- Processes 150,000 heads/day
- Generates 4,400 m^3/day of effluent with 7,300 mg COD/L
- Baseline: anaerobic lagoons
- Project: addition of superficial aeration system to existing lagoons
- ERs: 12,200 tCO_2-eq./year

ACM0014—Mitigation of Greenhouse Gas Emissions from Treatment of Industrial Wastewater

This consolidated methodology is applied in projects where anaerobic reactors, with methane recovery, are installed upstream of an existing lagoon system.

As in AMS III.I, the baseline lagoons must establish anaerobic behavior guidelines.[11]

A significant portion of the chemical oxygen demand (COD) is degraded in the new reactor. Methane generated is captured and routed to a flare or energy device.

Reactor effluent is then sent to aerobic treatment or to the existing lagoon system. Since a large portion of the organic content is degraded in the new anaerobic reactor, less CO_2-eq. is released compared to the baseline scenario.

11 Approved consolidated methodology ACM0014—*Mitigation of Greenhouse Gas Emissions from Treatment of Industrial Wastewater,* Version 02.1.

ACM0014 Case Study

A project of this type was proposed by an Indonesian tapioca starch mill with a production capacity of 200 ton/day.

The operation of this facility generates approximately 2,700 m^3/day of effluent with high organic content (17,000 mg COD/L). In the baseline scenario, the effluent is treated in an open anaerobic lagoon system.[12]

The project developer plans to install an ANUBIX-B™ reactor, which is of the UASB (upflow anaerobic sludge blanket) type, with a COD removal efficiency of at least 90%. The effluent with the remaining organic content goes to the existing open lagoon, emitting a significantly smaller amount of methane.

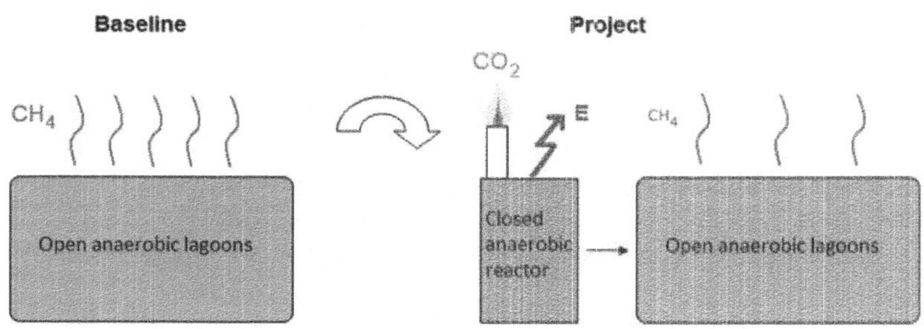

The collected biogas will be used for energy generation for on-site use, displacing carbon intensive electricity currently used, and generating this way, an additional GHGs emission reduction.

The yearly ERs are expected to reach up to 72,000 tCO_2-eq., during a 10-year crediting period.

Summary

- Produces 200 ton/day of tapioca starch
- Generates 2,700 m^3/day of effluent with 17,000 mg COD/L
- Baseline: open anaerobic lagoons
- Project: Installation of a UASB reactor previous to the lagoon system (including energy generation with collected biogas)
- ERs: 72,000 tCO_2-eq./year

ACM0010—Consolidated Baseline Methodology for GHG Emission Reductions from Manure Management Systems

This approved methodology involves manure management projects where the existing anaerobic treatment system is replaced by one or a combination of various animal waste management systems (AWMSs), resulting in less methane and nitrous oxide emissions.[13]

There are certain conditions that the baseline system must meet for this methodology to be applicable:

---

12 UNFCCC CDM Project Design Document for *Methane Avoidance through Utilization of Anaerobic Reactor for Wastewater Treatment at Tapioca Starch Factories*, Ketapang, Indonesia, Version 02.

13 Approved consolidated methodology ACM0010—*Consolidated Baseline Methodology for GHG Emission Reductions from Manure Management Systems*, Version 05.

- Livestock population of cattle, buffalo, swine, sheep, goat, or poultry must be managed under confined conditions,
- Annual average temperature in the project site should be higher than 5°C, and
- The retention time on the anaerobic treatment system must be at least one month.

ACM0010 Case Study

The Methane Capture and Combustion from Swine Manure Treatment for Peralillo study is a good example of this project category. Agrosuper, the project developer, is Chile's largest pork production company with over 1 million pigs.[14]

In the baseline scenario, manure and other organic waste from approximately 100,000 pigs is treated in 4.5-m deep open anaerobic lagoons.

The project activity involves covering the existing lagoons with a high density polyethylene geomembrane to capture methane gas. Methane gas is recovered and used in a boiler, where sludge from the digester is heated. Sludge is then re-introduced in the lagoon, thereby improving biogas production and anaerobic treatment efficiency. Excess methane is burned in an enclosed flare system.

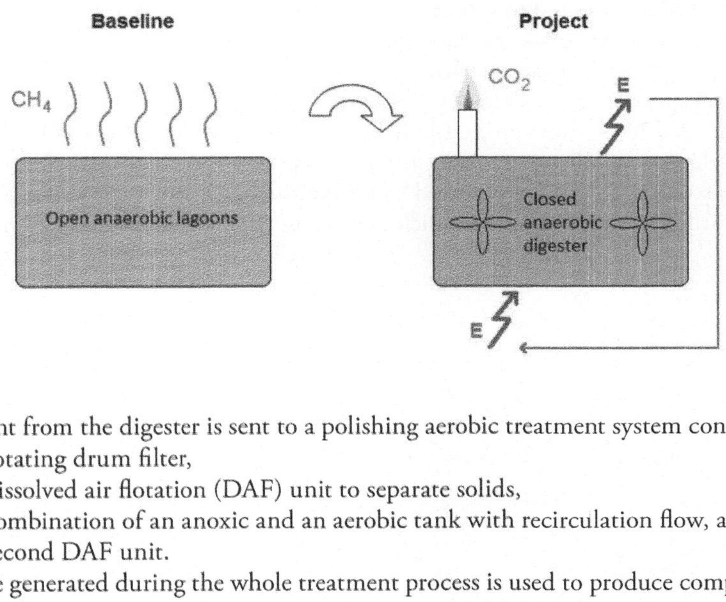

Effluent from the digester is sent to a polishing aerobic treatment system consisting of:

- A rotating drum filter,
- A dissolved air flotation (DAF) unit to separate solids,
- A combination of an anoxic and an aerobic tank with recirculation flow, and
- A second DAF unit.

Sludge generated during the whole treatment process is used to produce compost for soil application.

The emission of 80,000 tCO_2-eq./year in the baseline scenario (all coming from anaerobic decomposition in open lagoons) is reduced by the project activity to 25,000 tCO_2-eq./year (still emitted from the lagoons due to flaring inefficiencies), thereby generating 55,000 CO_2-eq./year, during a seven-year crediting period.

Summary

- Peralillo farm handles 100,000 pigs
- Baseline: 4.5-m deep open anaerobic lagoons

14 UNFCCC CDM Project Design Document for *Methane Capture and Combustion from Swine Manure Treatment for Peralillo, VI Region, Chile*, Version 03.

- Project: installation of biogas recovery system, flare, and heating system
- ERs: 55,000 tCO_2-eq./year

CONCLUSIONS

In Latin America, and in other regions in the developing world, climate change mitigation activities are mainly driven by the Kyoto Protocol.

In the water sector, several methodologies to certify GHG emission reductions have been approved by the UNFCCC. Most approved methodologies are related to methane emission reductions in wastewater and sludge treatment.

An evaluation of the methodologies approved for calculation of emission reductions, reveals a need for further research. Parameters used in these methodologies are not always standard to water industry, mathematical models are not always correct, and most of the calculations focus mostly on methane gas emissions. For instance, nitrous oxide (N_2O), a powerful greenhouse gas, is generally not considered in these methodologies. Recent research suggests that N_2O emissions in wastewater treatment plants can be of the same order of magnitude as CH_4 emissions.

New methodologies need to be developed for the water sector. Large-scale cleanup programs, where effluent being discharged untreated to anoxic water bodies will be treated at new wastewater facilities present an excellent opportunity to certify emission reductions. However, so far, there are no methodologies approved to estimate and monitor these reductions.

As of Jan. 26, 2009, 1,696 projects have been registered under the Kyoto Protocol framework in developing countries, of which 405 projects (24% over total registered) are related to the water sector. The activity observed so far indicates that the water sector is active and moving toward better standards and policies related to climate change mitigation.

INTRODUCTION

The climate is already changing largely because of human activities and even with aggressive efforts to control future emissions of greenhouse gases, climate will continue to change and the rate of change may accelerate. Thus, utilities have to adapt to the consequences of the climate change. These papers are organized into two sections: planning for climate change and the strategies some utilities are already using to adapt to climate change.

Planning for Climate Change

While many government agencies have made steps to address mitigation, few have moved toward adaptation. This section provides practical examples of how to build adaptation to climate change into existing planning processes or activities.

Understanding risk is a critical aspect of planning for climate change. Freas, Bailey, Munevar, and Butler look at several approaches that utilities can take to understand their risks including the threshold approach. This approach begins with defining thresholds based on knowledge of the system, and the scenario approach begins with outputs from climate models. The total water management approach considers the planning and operation of an entire water resources system.

Bakker and van Schaik discuss the use of a portfolio approach to climate change adaptation that builds on the principles of integrated water management. Using case studies from around the world, they present an overview of demand and supply measures being implemented.

"Half Full or Half Empty? Either Way it is Time to Plan" is the aptly entitled article by Standish-Lee, Loboschefsky, and Lecina. They review strategies utilities can take to increase long-term water supplies, and they discuss approaches utilities can take to incorporate climate change into long-term planning. They also discuss practices by several water utilities in the western United States.

Adaptations can be divided into *no regrets* and *low regrets*. Smith explores these concepts as he reviews the risks water managers will likely face from climate change and some strategies that can be used to identify adaptations.

Paper 15. Incorporating Climate Change in Water Planning

Paper 16. Climate Change and Water Utilities

Paper 17. Half Full or Half Empty? Either Way It's Time to Plan

Paper 18. Climate Change Is Real: How Can Utilities Cope With Potential Risks?

Case Studies and Practical Actions

The final section of the book presents case studies from around the world. The studies demonstrate how different utilities and government agencies are taking practical actions to ensure resilience to climate change.

Deslauriers, Holmes, McDonald, and Loiacono lay out principles wastewater managers in San Francisco can follow for adapting to climate change. They focus on the San Francisco Public Utilities Commission (SFPUC) 30-year Sewer System Master Plan and the consequences of change in precipitation intensity duration frequency (IDF) curves.

The San Francisco area perspective on adaptation and mitigation is given by Wallis, Ambrose, and Chan through their analysis of the planning activities of the East Bay Municipal Utility District (EBMUD) in Oakland, Calif. They describe how many potential climate change impacts are integrated into EBMUD's long-range planning, water supply planning, and strategic planning and budgeting. Wallis et al. also address EBMUD's plans on reducing its greenhouse gas emissions.

In the Netherlands, water shortage is not generally considered an issue, whereas flooding is a constant threat. Climate change may change this situation with low river flows and groundwater levels coupled with higher pollutant loads causing new challenges. Berg, Frijns, and Zwolsman discuss demand and supply actions to cope with these changes, including the idea of flexible sources. Mitigation measures to reduce the Dutch water utilities climate footprints and the inclusion of energy use reduction strategies into adaptation are also discussed.

In Singapore, water security is the over-riding issue. Wui discusses the diversified strategy that the Singapore Public Utilities Board has adopted. The four "National Taps" approach includes the use of NEWater (reclaimed water for drinking and industrial use) and desalination. Technology and innovation figure highly in Singapore's strategy with a strong research and development program supporting their initiatives. The paper also discusses how Singapore is addressing the projected increase in sea-level rise and flooding, as well as their holistic response to climate change through their National Climate Change Strategy.

Southern California has had water scarcity and water rights challenges spanning back over a century. Climate change is expected to exacerbate this situation. Laugier, Means, Daw, and Hurley discuss the potential risks that climate change may have on water quantity and quality from changing snow pack, sea-level rise, and other issues and the impact to water users including water utilities, agriculture, and hydro-electricity generation. The authors theorize that in Southern California, there will be an increased emphasis on demand management, and they conclude with general recommendations for water utility adaptation strategies.

Australia has faced a severe and on-going drought in most of its major cities. Howe and Rhodes describe how Melbourne assessed its risk from climate change and has put in place a flexible adaptation strategy to cope with potential futures. The strategy takes an intensive look at the water supply system but also characterizes the risk and potential adaptations for areas of infrastructure often overlooked by the wastewater and drainage systems.

In the Scandanavian countries of Denmark, Sweden, and Norway, climate change is expected to increase the frequency and intensity of storm events, resulting in increased sewer overflows and urban flooding. Mark, Svensson, König, and Linde discuss how each of the Scandanavian countries is progressing with actions to combat climate change, including mitigation guidelines for urban sewer systems and guidelines for analysis of urban drainage impacts. The paper is supported by case studies of four areas where the changes in urban flooding under different climate scenarios are presented.

Paper 19. Planning Strategy in a Changing Climate

Paper 20. Climate Change: Charting a Water Course in an Uncertain Future

Paper 21. Implementation of Climate Adaptation and Mitigation Strategies for Drinking Water Production in the Netherlands

Paper 22. Meeting the Challenges of Climate Change: Singapore

Paper 23. **Climate Change and Adaptation in Southern California**

Paper 24. **Melbourne Water Climate Change Study: Impact and Adaptation of Melbourne's Water Supply, Wastewater, and Stormwater Systems**

Paper 25. **Climate Change Impacts on Urban Drainage Systems in Scandinavia**

– 15 –

Incorporating Climate Change in Water Planning

Kathy Freas, Bob Bailey, Armin Munévar, and Susan Butler

The global water crisis and the related influences of climate change on water supply are universally recognized. The Pew Center report on global climate change (Frederick and Gleick 1999) states, "The socioeconomic implications of both climate and nonclimate impacts on water supply and demand will depend in large part on the ability to adapt to change and on whether water managers and planners take action." At the January 2008 World Economic Forum (2008) meeting in Davos, Switzerland, attendees made water supply a top priority; discussions of the effects of population growth and climate change on water management dominated the program. In February 2008, eight major US water agencies united to form the Water Utility Climate Alliance, acknowledging that plans for future investment in water infrastructure must accommodate climate change projections (SFPUC 2008). Clearly, the specter of climate change is recognized by the water community as a significant challenge to sustainable and reliable water supplies.

Among the most important effects of climate change are those related to the influence of increasing temperatures on seasonal patterns of precipitation and evapotranspiration and the effects these changes have on water availability and the overall water cycle. Generally it is expected that these climate changes will result in increased precipitation at higher latitudes and reduced precipitation in the subtropics and the southwestern United States, more intense but less frequent rainfall events, reduced soil moisture and increased evapotranspiration, longer drought periods, increased precipitation as rain and reduced snowpack storage, and increased frequency of extreme weather events such as hurricanes and flooding.

Current and projected changes in climate can reduce the flexibility and reliability of existing water management systems and challenge the ability of these systems to deliver sustainable water supplies, provide adequate flood management, and ensure adequate flows for environmental purposes. In response to these challenges, water managers and other water community leaders are looking for new ways to assess the vulnerability of their systems to the impacts of climate change and to plan and develop water management systems for a future that will be strongly influenced by climate change.

"By recognizing the imminent water crises around the globe, we have the opportunity, or perhaps the responsibility, to rethink how we consider and manage a resource so fundamental to our long-term survival and prosperity," said Ralph Peterson, chair and CEO of CH2M HILL. "A diversified water supply portfolio is one of the best approaches to minimize the future risks of population growth, increasing demand, and climate change. Such a diversified portfolio is most easily achievable if we adopt a new attitude about water management that is based on an integrated understanding and management of the water cycle."

CLIMATE CHANGE RISK ASSESSMENT AND TOTAL WATER MANAGEMENT FOR SUSTAINABLE WATER SUPPLIES

Two powerful tools—climate change risk assessment and total water management (TWM)—are used to determine the vulnerability of water management systems to climate change. These tools are also used to drive the development of integrated water management strategies and infrastructure to provide more resilient water supplies in the face of climate change.

Climate Change Risk Assessment

Climate change risk assessment is used to systematically evaluate the vulnerability of water systems to climate effects. Over the past decade, dramatic improvements have occurred in global climate modeling and the scientific understanding of climate processes. However, uncertainty remains in the projections of climate change effects and therefore in the risks associated with these effects. To adequately assess and respond to potential climate change risks to water supply, a systematic approach that examines a range of climate change projections as well as the sensitivity of water system components to those projected changes is necessary.

One approach for assessing climate change risks to water systems and for guiding subsequent decision processes for adapting to climate change impacts uses a dual analytical framework—the threshold–scenario risk assessment framework (Figure 15-1). The framework comprises two parts: the more qualitative threshold risk assessment approach and the quantitative scenario risk assessment approach. The two parts can be used separately or sequentially, with the quantitative scenario approach typically following use of the qualitative threshold approach when results of the qualitative approach suggest that greater detail is needed to determine adaptations for water systems.

Threshold Risk Assessment Approach

The threshold approach relies on the experience and judgment of professionals who are familiar with the function of the built and natural components of the water system being considered. In guided discussions designed to identify the components and functions of water systems that are most susceptible to significant climate change impacts, these professionals can assess the linkages among the functions of water system components and climate and the cumulative effect these changes have on the overall water system. In some cases, this qualitative threshold approach is sufficient to advise decisionmaking regarding adaptations or other actions necessary to address anticipated effects of climate change.

The threshold approach involves the following steps:

1. Define the thresholds or performance criteria of the water system components. This step establishes the range of normal function for each water system component outside of climate change considerations. Conditions outside of this range are assumed to disrupt the function of the component (e.g., a diversion structure can capture flows only in the range of 5,000–7,000 cfs or current estuarine ecological function is maintained by salinity ranging from 5,000 to 10,000 mg/L).
2. Establish the climate variables of importance (e.g., precipitation, temperature, wind) and explore the sensitivity of various water system components to these climate variables. This step establishes the changes that could result in violation of threshold or performance criteria, thus identifying the sensitivity of each system component to potential climate changes.
3. Determine water system responses as a result of potential changes in climate variables and determine the range of anticipated changes in climate variables based on readily available global climate models or a literature review. Once thresholds,

sensitivities of water system components, and the range of anticipated changes in climate variables are established, the components that are especially susceptible to disruption of function from climate changes can be identified. The results of this step put focus on those system components for which modifications or adaptations for climate change may need to be considered.

4. Develop adaptation strategies, i.e., actions, operations, or infrastructure changes, that provide the system with increased ability to adapt and therefore reduce or eliminate the adverse effects of climate change on water system function. These adaptations can be adjusted and repeatedly tested through the following steps to ensure their effectiveness:
   a. Assess vulnerability of the system to climate change impacts and evaluate the ability of the system components to adapt to possible impacts that violate thresholds in light of proposed adaptation strategies. Some water systems may be sufficiently adaptable to provide sustainable water supplies regardless of limited changes in the function of individual components caused by climate change.

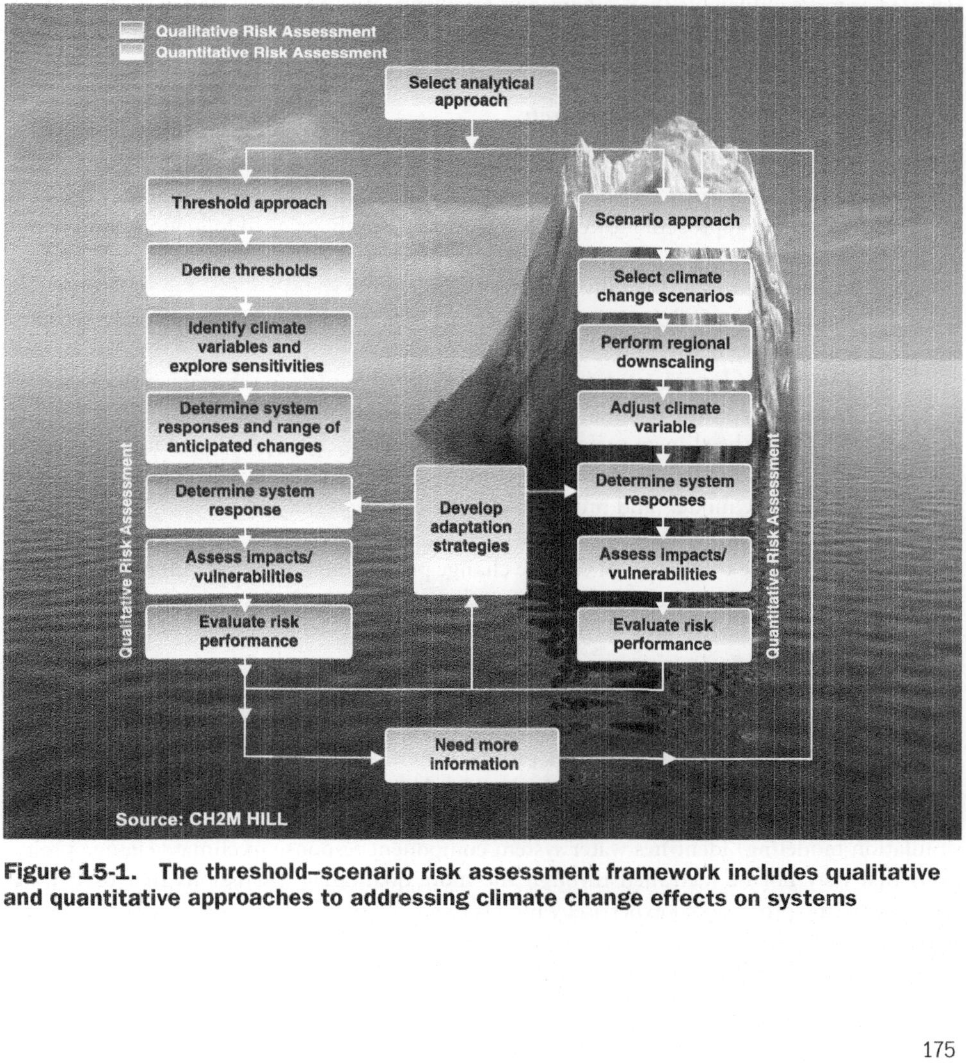

Figure 15-1. The threshold–scenario risk assessment framework includes qualitative and quantitative approaches to addressing climate change effects on systems

b. Assess the system's risk performance by evaluating the degree of uncertainty of climate change projections (robustness) driving the vulnerability of system components. Do the projections fully reflect the array of applicable climate change scenarios? If the projections have a high degree of uncertainty, the risk of a system's vulnerability may be reconsidered. For example, temperature increases and sea level rise are among the more robust of climate model projections, but precipitation projections involve significant geographic uncertainty.

c. Evaluate overall system risk and identify areas needing greater study, if any. Risk should be based on the vulnerability of system components and the robustness of the projection. A cumulative assessment of the effects of climate change on water system components is necessary to determine potential effects on the overall function of the water system. This cumulative assessment further informs the selection of adaptations to attenuate climate change effects.

This threshold approach identifies system components that are dependent on the status of climate variables (e.g., precipitation, temperature) and the overall system risk to climate change, resulting in a preliminary risk assessment based on the professional judgment of experts who know the system and the planning area. Specific system components can be revealed as more vulnerable than others, potentially warranting more detailed study to assess the quantitative nature of the risk. This involves the second portion of the dual framework: the scenario risk assessment approach.

Scenario Risk Assessment Approach

This more quantitative approach can provide a more detailed evaluation of the range of potential climate impacts on a water system. The scenario approach can be used separately from the qualitative threshold approach. However, it is more commonly invoked following the outcome of the threshold approach.

The scenario approach generally includes the following steps:

1. Select a range of climate change scenarios that are based on commonly accepted emission scenarios and global climate models.
2. Downscale (translate) the projections of global models to the scale of the system of interest (typically regional) using accepted techniques.
3. Establish the climate variables of importance (e.g., precipitation, temperature).
4. Determine system responses to climate change effects by downscaling historical hydrologic models and projected climate change scenarios.
5. Develop and refine adaptation strategies.
6. Determine robustness of climate change projections and assess vulnerability of the system to climate change impacts in light of proposed adaptation strategies.
7. Evaluate the overall system performance.

The quantitative scenario approach incorporates many of the same steps included in the qualitative threshold approach. However, this approach elicits a range of water system component responses by employing simulations of physically based linkages among climate, watershed, and system function (i.e., the process would use hydrologic models for changes in runoff or integrated hydrodynamic/habitat models for estuarine ecological function).

Inclusion of the physically based responses of water system components (usually through simulation modeling) identifies water system component response to climate change projections that may not be identified through the more qualitative discussions that used water system–specific professional expertise, which is the basis for the threshold approach. The scenario approach can also incorporate the quantitative analysis of many climate scenarios and global climate models to capture the full range of future uncertainty and yield statistically defensible results—a level of complexity beyond that achievable with the qualitative threshold approach.

Advantages of a Dual Analytical Framework

Both the threshold and scenario approaches produce assessments of system vulnerability and risk from climate change, but at different levels of detail. The threshold approach can be used to prescreen climate change impacts, followed by the more quantitative scenario approach, which provides a more detailed understanding of the system response to climate change.

This dual framework provides a customizable, transparent, defensible, and flexible process for identifying climate change impacts and developing adaptations, as necessary. It does not require excessive levels of detailed analysis but allows more detailed analyses when necessary. The framework also ensures that the context and direction for technical analyses are consistent throughout the process; support the integration of the analyses; and formally describe and capture impacts, uncertainties, and risks to support decisionmaking processes.

Total Water Management

Once system risk is characterized, the threshold–scenario risk assessment framework can be integrated with general elements of the TWM approach. This approach can be customized for developing and implementing appropriate responses to climate change impacts.

TWM integrates the functions of the components of the built and natural water cycle in such a way that the entire water system can be planned and operated more holistically to provide more sustainable water supply solutions. The major components of the built and natural water cycle on which TWM approaches generally are based are shown in Figure 15-2. A more detailed array of water cycle components frequently considered when developing TWM plans includes various combinations of watershed dynamics: groundwater, surface water, storage, reservoir operations, and conveyance options; conjunctive use and aquifer storage and retrieval; demand variables including municipal and industrial demands (Figure 15-2), agricultural, conservation, and instream and other ecological requirements; treatment options including desalination and reuse; and stormwater and flood management.

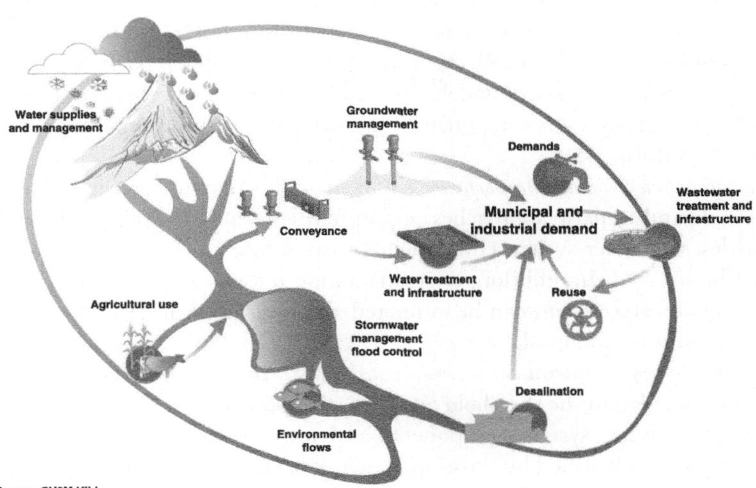

Source: CH2M HILL

Figure 15-2. Total water management planning often involves incorporation of the major components of the built and natural water cycle (or water system) or subsets of these major components

A TWM approach facilitates long-term planning, promotes consistency and efficiency, optimizes use of the water system, encourages and facilitates regional planning, provides flexible solutions, and enhances communication and community support for an understanding of the projected effects of climate change on water systems and the necessary modifications to address those effects.

Plan Elements of TWM

Previous studies identified six fundamental elements of the general TWM planning process, all of which are useful when developing sustainable water systems in response to projected climate change impacts (Patwardhan et al. 2007). TWM can be used to guide water system modifications in response to the outcome of the threshold or scenario climate change risk assessment.

- *Define goals and alternatives.* For climate change applications, development of goals is an iterative process that is an integrated part of the threshold and scenario approaches (Figure 15-1). Overall goals are related to developing a water system that is resilient to climate change. Alternatives for achieving this goal arise from an understanding of system vulnerability and how vulnerability is minimized as identified through the threshold–scenario risk assessment process and the adaptations developed as a result.
- *Collect and analyze data.* Necessary and sufficient data to conduct a climate change risk assessment should be identified based on the decision to use either the threshold or scenario risk assessment approach. For the threshold approach, sufficient data or information on the function of system components are required for a thorough qualitative discussion of potential effects of projected climate change on water system components and the system as a whole. For the scenario approach, appropriately broad climate scenarios must be selected along with the appropriate downscaling methods and physical system models needed to quantitatively assess water system component responses to projected climate change effects. Processes similar to those used for the threshold approach may be used to develop and refine adaptations intended to reduce water system vulnerability.
- *Evaluate and select alternatives.* The process for assessing the ability of proposed adaptations to attenuate the projected effects of climate change on the water system involves iterative discussions of water system component thresholds and the sensitivity of system components to climate changes using the threshold approach. With the scenario approach, physically based models are used to determine the effects of proposed adaptations on water system component responses and to assess remaining vulnerabilities and risk performance.
- *Develop or select models to perform TWM analyses.* As noted in the discussion of the threshold and scenario approaches, appropriate conceptual or physically-based models that describe water system component function and function of the overall system must be selected. In addition, once adaptation options are selected, the function of the overall water system can be evaluated and optimized, if desired, by using any of several optimization models.
- *Conduct an impact assessment including stakeholder involvement.* The impact assessment process included in the threshold and scenario approaches determines climate change impacts on water system components and whole systems based on vulnerabilities and responses identified by those approaches. Stakeholder involvement in the impact assessment can include the expert technical community, the water management and user communities, environmental organizations, and regulatory agencies, depending on the technical, geographic, and political scope of the water system in question.

- *Develop an implementation plan.* As appropriate adaptations are developed to address the projected impacts of climate change on the water system, implementation plans that are flexible and that prioritize actions to address the most significant of anticipated impacts are important. Informing the public and other stakeholders regarding the assumptions and processes for developing and implementing water system adaptations to climate change as well as visible milestones indicating progress can promote acceptance and active support for changes and the financial challenges that can be associated with them. Development of effective implementation plans can be facilitated by using an optimization platform that integrates across water system components and that includes financial modules that can guide the structure and other financing necessary to develop the system. Funding of major items (capital and operations and maintenance) as well as institutional funding are critical elements of the implementation plan. These efforts often involve long-range plans and require careful consideration to ensure that key items are funded.

CASE STUDY: INTEGRATING CLIMATE CHANGE IN WATER MANAGEMENT PLANNING

The Lower Colorado River Authority—San Antonio Water System Water Project

Planning for long-term water supply reliability in the face of increasing population and climate change effects is under way in Texas. Through a climate change risk assessment approach, combined with innovative TWM considerations, the Lower Colorado River Authority (LCRA) and San Antonio Water System (SAWS) are developing a robust plan for water supply infrastructure in central Texas—one of the fastest growing regions in the United States.

Water planning for the LCRA and SAWS regions (Figure 15-3) suggests that water supplies will be insufficient to meet the increasing municipal and industrial demands and that supplies for irrigated agriculture—a cultural and economic mainstay of the region—will be reduced significantly in Colorado, Wharton, and Matagorda counties as their populations continue to grow. Although San Antonio has an aggressive, highly effective water conservation program, it won't be sufficient to meet future water needs. In addition to the limited supplies that exist today, climate projections indicate that portions of the watershed may experience greater drying and reduced runoff, while a rise in sea level could change the Matagorda Bay salinity and ecosystem function.

LCRA and SAWS have entered into a long-term agreement to study and potentially implement a plan that would increase available water supplies for agriculture in the lower Colorado River basin and develop an interbasin transfer of up to 150,000 acreft of surface water from the lower Colorado River basin to the San Antonio area. The plan provides the basis for a TWM approach that includes options among sources, storage, operations, and other infrastructure improvements as well as conservation to enhance system reliability. The associated studies are rigorously evaluating the options to provide long-term water supply sustainability. LCRA and SAWS currently are in the fifth year of technical studies to determine the feasibility of the proposed interbasin project.

Currently, the Highland Lakes reservoir system, which is managed by LCRA at the upper end of the system, and run-of-the-river rights throughout the basin are used to meet water demands for municipal, industrial, and agricultural interests throughout the LCRA basin. The plan currently being studied, which would reduce stored water releases from

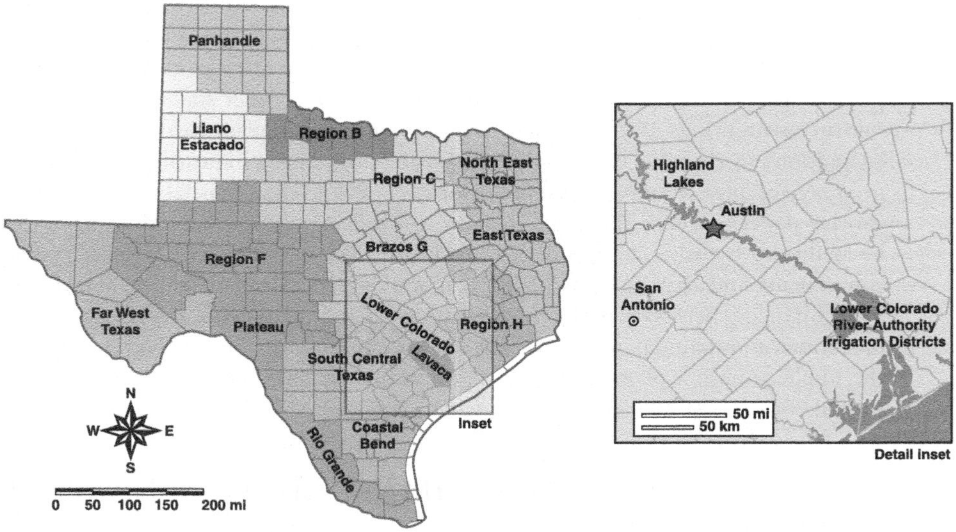

Figure 15-3. Central Texas and Lower Colorado River Authority (LCRA) regions (A) and location of the LCRA project (B)

Highland Lakes to maintain higher lake levels for recreation and other purposes, would also increase total supplies in the basin through agricultural conservation, additional storage of surface water in a downstream off-channel reservoir, and development of groundwater supplies to supplement surface water supplies for in-basin agricultural needs and environmental flows (Figure 15-4). The schematic shows how the project would operate. The green boxes indicate the sources of water for the project. The gray ovals indicate how the water will be used. Stored water releases from the Highland Lakes and run-of-the-river rights would be used to meet inbasin agricultural demands and environmental needs as well as be diverted to off-channel storage to provide up to 150,000 acre-ft of reliable supply for San Antonio. Groundwater and conservation strategies would be developed to meet agricultural needs.

Studies to determine the feasibility of the plan include evaluation of the effects of various options on the function of the lower Colorado River basin through time. The studies include environmental conditions (water quality, Matagorda Bay health evaluation, aquatic habitat, and wildlife); water supply (surface water availability, groundwater for agriculture, and agricultural conservation); siting of engineering facilities; and socioeconomic studies.

On the basis of feedback from stakeholder groups and the project's science review panel, the project team is also evaluating the relationship between climate and water system performance. An understanding of this relationship will provide a foundation for building in flexibility and adaptability to the 80-year water supply plan. The study results will help decisionmakers and stakeholders assess potential impacts, evaluate tradeoffs, and develop adaptation strategies as climate changes are observed. These results will also instill greater confidence about water supply investments.

The project evaluation is applying the tiered threshold–scenario risk assessment framework to assess climate change risks to sustainable water supplies for the LCRA–SAWS water

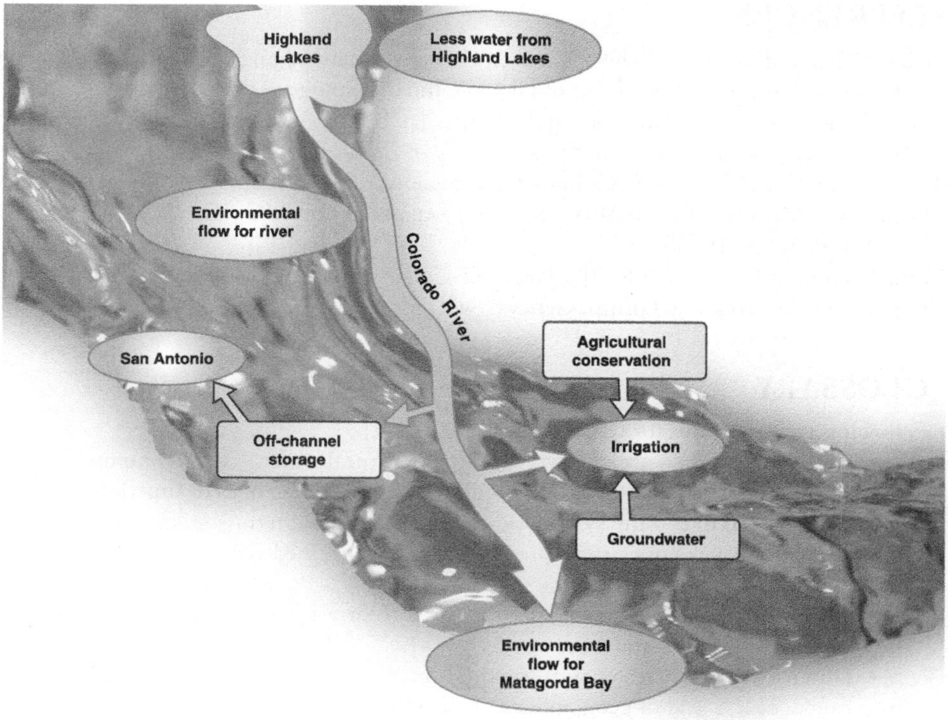

Figure 15-4. The Lower Colorado River Authority—San Antonio Water System Water Project

project. Currently, the project team is evaluating system component and overall system vulnerabilities to potential climate change impacts using the threshold approach. Results are expected to identify vulnerabilities to water system function that can be addressed through integrated water system modifications.

Because this project represents a significant investment for both agencies, developing ways to anticipate and adapt to possible changes over time will ensure that the project maintains its value as a key piece of the water supply portfolio in the two river basins. Considering the possible effects of changing climate parameters using the tiered assessment framework during the planning phase will put LCRA and SAWS in a better position to develop structural, operational, and policy strategies that reflect TWM elements and can lead to a water system that is more resilient to climate change.

ACKNOWLEDGMENT

The authors thank LCRA project manager Leah Manning; SAWS project manager Gary Guy; and the entire LCRA–SAWS Water Project team in the development of this article.

REFERENCES

Frederick, K., and Gleick, P. 1999. Water & Global Climate Change: Potential Impacts on U.S. Water Resources. Pew Center of Global Climate Change, Arlington, Va.

Patwardhan, A.S., Baughman, A.T., and Thorpe, J. 2007. Developing and Implementing a TWM Strategy—Approaches and Examples. *Jour. AWWA*, 99:2:64.

SFPUC (San Francisco Public Utilities Commission). 2008. Major U.S. Water Agencies Form New National Climate Alliance. sfwater.org/detail.cfm/MC_ID/ 20/MSC_ID/357/ MTO_ID/552/ C_ID/3867.

World Economic Forum. 2008. The Power of Collaborative Innovation. Annual Meeting Report. World Economic Forum, Geneva, Switzerland.

GLOSSARY

Following are definitions for terms used in this article:

- **Ability to adapt**—capability of built and natural water system components to accommodate changes in climate without significant changes in function
- **Climate change**—any systematic change in the long-term statistics of climatic elements (e.g., temperature, pressure, winds) sustained over several decades or longer
- **Climate variable**—physical quantity used to measure climate such as temperature, precipitation, and wind
- **Downscaling**—mathematical method to transform climate change projections from a coarse grid or geographic scale to a finer scale
- **Robustness**—the degree of certainty of climate-change projections
- **Scenario approach**—use of multiple courses of action or events based on a plausible range of operating conditions derived from historical information and/or model results; refers to quantitative climate change risk assessment process that is part of the overall tiered threshold–scenario risk assessment framework
- **Sensitivity**—inclination of the function of water system components to be compromised by climate changes that result in exceeding the threshold limits of those components
- **System risk**—as part of the threshold–scenario risk assessment approach described in this article, the measure of the consequences to a water system where thresholds of normal function are exceeded as a result of climate change
- **Threshold**—water system component limits within which the component functions normally
- **Vulnerability**—the susceptibility of a water system to adverse effects of climate change; vulnerability is a function of a system component's sensitivity to climate change and the ability to adapt to climate change
- **Water system or system response**—the integrated and cumulative result of the responses of water system components, built and natural, applied across all components of a water system

Originally printed in Journal AWWA *June 2008.*

– 16 –

Climate Change and Water Utilities

Marloes H.N. Bakker and Henk van Schaik

INTRODUCTION

Climate Change Increases Water Stress

Impacts of climate change on water resources will be felt in virtually every aspect of natural resources management. All decisions about long-term water planning, the design and construction of new water supply infrastructure, the type and acreage of crops to be grown, urban water allocations and rate structures, reservoir operation, and water supply and demand management depend on climatic conditions and what humans do to respond and adapt to those conditions. Many uncertainties remain about the timing, direction, and extent of these climatic changes, as well as about their societal implications. Indeed, the most important effect of climatic change for water systems will be increasingly the overall uncertainty managers face. These uncertainties greatly complicate rational water resource planning and have contributed to the ongoing debate over how to respond.

Globally, the negative impacts of future climate change on freshwater systems are expected to outweigh the benefits. By the 2050s, the area of land subject to increasing water stress due to climate change is projected to be more than double that with decreasing water stress. Areas in which runoff is projected to decline face a clear reduction in the value of the services provided by water resources. Increased annual runoff in some areas is projected to lead to increased total water supply. However, in many regions this benefit is likely to be counterbalanced by the negative effects of increased precipitation variability and seasonal runoff shifts on water supply, water quality, and flood risks.

MANAGING WATER STRESS ON SUPPLY SIDE AND DEMAND SIDE

Managing Water-Related Stress: A Case of Supply and Demand Management

A characteristic of water demand is its inexorable rise over many years; continued growth is projected over coming decades. Meeting this increasing demand from existing resources is self-evidently an uphill struggle, particularly in water stressed/scarce regions in the developed and developing world alike. There are typically three potential responses that can be takenin two different management areas:

- Supply-side: meeting demand with new resources.
- Demand-side: managing consumptive demand itself to postpone or avoid the need to develop new resources.
- A combination of the two (see Table 16-1).

Table 16-1. Supply- and demand-side strategies for different sectors. Table adapted from IPCC (2001).

SUPPLY SIDE	DEMAND SIDE
MUNICIPAL WATER SUPPLY	
Increase reservoir capacity Extract more from rivers or groundwater Alter systems operating rules Interbasin transfers Desalinisation	Incentives to use less (pricing) Legally enforceable water use standards Increase use of grey water Reduce leakage Development of nonwater–based sanitation systems Seasonal forecasting
INDUSTRIAL AND POWER STATION COOLING	
Increase source capacity Use low-grade water	Increased water-use efficiency and water recycling
HYDROPOWER GENERATION	
Increase reservoir capacity	Increasing efficiency of turbines: encourage energy efficiency
NAVIGATION	
Build weirs and locks	Alter ship size and frequency
POLLUTION CONTROL	
Enhance treatment works	Reduce volume of effluents to treat Catchment management to reduce polluting runoff
FLOOD CONTROL	
Increase flood protection Catchment source control to reduce peak discharges	Improved flood warning and dissemination Curb floodplain development
IRRIGATION	
Increase irrigation source capacity	Increase irrigation-use efficiency Increase drought-tolerant varieties Change crop patterns

There is considerable pressure from the general public, regulatory agencies, and some governments to minimize the impacts of new supply projects (e.g., building new reservoirs or inter-regional transfer schemes), implying the emphasis should be shifted toward managing water demand by best utilizing the water that is already available.

Water Supply Management

Water supply can be seen as the process of self-provision or provision by third parties of water of various qualities to different users. As a fundamental driver of the hydrological cycle, climate determines how much water is available (supply) in the short and long term.

A few examples that illustrate how countries deal with a decrease in water by implementing measures on the supply side are presented here.

Ethiopia: Construction of Dams

Ethiopia is a landlocked country in the eastern horn of Africa. The need for water in Ethiopia is severe; their access to water supply is one of the lowest in the world. Along with limited food supply, during times of drought, water-related diseases are rampant. Surface water sources such as springs and ponds dry up. What limited water sources remain become heavily contaminated by environmental waste, such as human and animal excreta which is washed in when the limited rains do come. The stagnant water serves as a breeding place for mosquitoes. Projected future climate trends are decreases in rainfall amount exacerbated by higher evaporation rates associated with the increasing temperatures. Projections of precipitation are more uncertain than projections on temperature and considerable regional variations exist. According to the country's First National Communications to the UNFCCC, precipitation is expected to decrease in the northern regions, while southern areas could see an increase of as much as 20%. Solutions for Ethiopia lie on the supply side via dam

Umgeni Water Supply—A "Total Approach"

An interesting example of a combined or total approach, entailing measures on both the demand and the supply side, is that of Umgeni Water. Umgeni Water is one of Africa's most successful organisations involved in water management and the largest bulk water supplier in the province of KwaZulu-Natal, South Africa. It is one of the largest catchment-based water utilities in Southern Africa, supplying safe drinking water to almost 4.8 million people annually. Climate change has been identified as an inherent risk and is therefore incorporated into the Corporate Risk Register.[1] It is recognised that investigations are required to establish a better understanding of the potential impacts of climate change at a local scale, such that appropriate control measures can be instituted to bring the residual risk to a level that the organisation is willing to accept. Based on the outcomes of this project, Umgeni Water will need to develop adaptive sustainable strategies to mitigate against the risks associated with these potential impacts. Water resource development plans, system operating rules and disaster risk management plans will contain the means of implementing these strategies. As such, Umgeni Water is systematically including climate change risk on both the supply and demand side of the water supply equation.

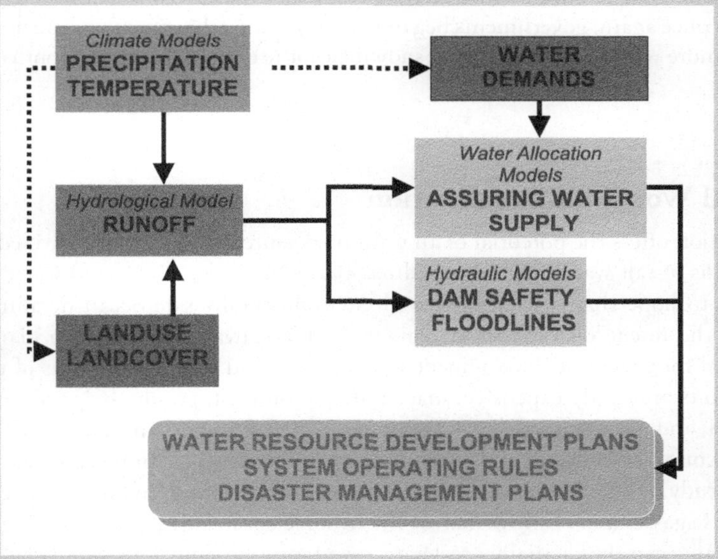

1 Umgeni Annual Report 2007–2008.

Figure 16-1. The Umgeni water strategy: catchment-based impact scenario assessment process. Source: Steve Gilham, Umgeni Water, South Africa.

constructions, but it should be noted that investments to reach water security require 144 years at 5% of GDP.

Perth, Australia: Desalination

The city of Perth in Australia is mainly going for supply-side management by introducing desalination technologies. The city of Perth is facing a serious water shortage. Over the past 10 years or so, the city has seen a 21% decline in rainfall, but the stream flow into dams—the actual amount running into storage — dropped about 65%. The Water Corp. turned to the nearby Indian Ocean to help solve the water shortage problem. The Kwinana Desalination Plant south of the city opened in November of 2006. The facility, the first of its kind in Australia, covers just a few acres in an industrial park next to the ocean. It is located just south of Perth, and turns water from the Indian Ocean into about 150 million litres of drinking water per day, roughly 20% of Perth's daily consumption. That makes the plant the single largest source of water for the city. Some predict that desalination will account for at least half of Perth's water in the next 30 years.

United States of America: Rethinking Big Dam Policies Because of Climate Change

The era of massive dam construction in the West—which tamed rivers, swallowed towns and created irrigated agriculture, cheap hydropower, and persistent environmental problems—effectively ended in 1966 with the completion of Glen Canyon Dam in Arizona. But the population of the Western states grew nearly 20% in the 1990s, to more than 64 million, and continues to swell. Demand for water from growing cities, industry, agriculture, and struggling fish runs is already high. Increasing the pressure are fears that climate change will cause rain instead of snow to fall in winter, reducing the snowpack that provides water in summer months. So once again, governments began studying dams, this time to create huge reservoirs to capture more winter rain and spring snowmelt for use in dry summer months.

Critical Words on Desalination

Desalination offers the potential of an unlimited source of fresh water purified from the vast oceans of salt water that surround us. The public, politicians, and water managers continue to hope that cost-effective and environmentally safe ocean desalination will come to the rescue of water-short regions. While seawater desalination plants are already vital for economic development in many arid and water-short areas of the world, many plants are overly expensive, inaccurately promoted, poorly designed, inappropriately sited, and ultimately useless.[1] The potential benefits of ocean desalination are great, but the economic, cultural, and environmental costs of wide commercialization remain high. A study by the WWF found that desalination uses large amounts of energy, emits greenhouse gasses and destroys marine life in some coastal areas. The building of desalination plants can thus actually accelerate greenhouse gas emissions and increase climate change.

In many parts of the world, alternatives can provide the same freshwater benefits of ocean desalination at far lower economic and environmental costs. These alternatives include treating low-quality local water sources, encouraging regional water transfers, improving conservation and efficiency, accelerating wastewater recycling and reuse, and implementing smart land-use planning.

Source: Cooley et al. (2006)

1 At present, the only significant seawater desalination capacity is in the Persian Gulf, on islands with limited local supplies, and at selected other locations where water options are limited and the public is willing to pay high prices.

Water Demand Management

Water demand is interpreted as demand for services, including drinking water, irrigation, hydro, navigation, and even protection in case of storms. Essentially, the concept in water management is water security. Countries or basins are considered highly stressed if the per capita water availability is less than 1,000 m^3 per year based on actual runoff (Falkenmark 1989). This amount includes water for drinking, food production, industry, hydropower, etc. Estimates of the number of people living in such severely stressed basins range from 1.4 billion to 2.1 billion (Kundzewicz et al. 2007).

Water demand management can be defined as any method—whether technical, economic, institutional, financial or social—that will accomplish one (or more) of the following five tasks:

- Reduce the quantity or quality of water required to accomplish a specific task.
- Adjust the nature of the task or the way it is undertaken so that it can be accomplished with less water or with lower quality water.
- Reduce the loss in quantity or quality of water as it flows from source through use to disposal.
- Shift the timing of use from peak to off-peak period.
- Increase the ability of the water system to continue to serve society during times when water is in short supply.

Water demand management strategies and tools therefore enable water-use efficient, equitable, and sustainable practices and policies. Essentially, it requires a change in behaviours and practices in the way in which water is used, particularly in the agricultural sector where most of the water is consumed.

Water demand management is a strategic and effective adaptive no-regret strategy to the current challenge of water scarcity and will become more so as climatic variability and climate change impacts intensify. It increases social resilience and contributes to preparedness policies. There are several difficulties to consider in this process: first, the impacts of climate change on the water sector are complicated and to a large extent unpredictable. Second, many impacts may be nonlinear and chaotic, characterized by surprises and unusual events. These difficulties complicate the development of measures.

Examples of demand management measures include:

- Campaigns for water economy at household and industrial level (Atlanta). For instance, cards in hotel rooms asking "Our State is experiencing severe water shortage. Please join us in our efforts to conserve this precious resource;"
- Economy in water use (economic shower heads, water free toilets, water-economic washing machines);
- Pricing;
- Restrictions on users, for instance, lawn watering only at night (Australia, UK, US);
- Restrictions on allocations, for instance, contracts with utilities, farmers (Philippines and North East Brazil, elaborated in the two examples below that demonstrate how tailored climate information can help water managers allocate water); and
- Regulations on building codes and settlements (in flood-prone areas) connected with insurance (US).

Below, there are more detailed examples of where implemented demand management measures can be found.

Metro Manilla, Philippines—Allocation Managing Demand Using Climate Forecasts

The reservoir that supplies water to Metro Manila is increasingly vulnerable to hydrologic variability, both drought and flood. A large irrigation area also relies on water from the reservoir in this shared water system. In times of water scarcity, questions of where the water should be allocated become critical and often contentious. Research institutions work with the urban water supply service, irrigators, and national-level agencies to create anticipatory strategies for managing water crises.[1] Together, they are exploring economic mechanisms such as option contracts, index insurance, and building capacity in the use of climate information and forecasts in order to allocate water efficiently and equitably.

Ceará, Brazil—Allocation Water Through Demand Management

Drought is a major challenge for those who rely on water in Ceará, Northeast Brazil, such as farmers and irrigators, as well as the Fortaleza metropolitan area. Scientists have developed a long-range forecast (up to 18 months in advance) of the inflows to the major reservoir system. Simulations have shown the potential for improved reliability of water deliveries for all users when forecasts are integrated with reservoir releases. Continuing work focuses on capacity building with local stakeholders in order to operationalize forecast use.

Sydney, Australia—Portfolio of Water Demand Management Options Implemented

While Sydney, on Australia's south-ast coast, responded to the on-going drought issues by moving toward partial desalination,[2] it is worthwhile to note that this city also has one of the largest demand management programs in the world—a program that includes a household retrofit program, active leakage reduction, water regulatory requirements on new and remodelled buildings that require a rainwater tank or grey water system, extensive recycling, and on-going restrictions.

Water and Adaptation: A Global Overview

Water managers all around the world are implementing all three measures. However, the impacts of climate change will be felt differently in different geographical regions, hence the responses taken all around the globe differ, although decision makers and managers always need to address the challenges in an integrated fashion.

In Latin America, for instance, some adaptation practices have been put in place, including the use of managing transbasin diversions and the optimization of water use. In Africa, local communities and farmers have developed adaptation schemes to forecast rainfall using accumulated experience. Farmers in the Sahel also use traditional water harvesting systems to supplement irrigation practices (IPCC 2007).

Water managers in a few countries, including the Netherlands, Australia, the UK, Germany, the US, and Bangladesh, have begun to address directly the implications of climate change as part of their standard flood and water supply management practices. These adapta-

1 http://portal.iri.columbia.edu/portal/server.pt/gateway/PTARGS_0_5280_2009_0_0_18/water.pdf.

2 The city gave concept approval for the desalination project and project approval for the desalination plant at Kurnell, an even larger plant than Perth's.

tions have generally taken the form of alterations to methods and procedures, such as design standards and the calculation of climate change allowances. For example, such adaptations have been implemented for flood preparedness in the UK and the Netherlands (Klijn et al. 2001, Richardson 2002), for water supply in the UK (Arnell and Delaney 2006); and for water planning in general in Bangladesh.

Below, an overview, which cannot be exhaustive since many individual measures can be used in many locations, gives some other examples of the adaptation measures taken by water utilities, both planned and autonomous,[3] currently in use across the world, as presented in the latest regional chapters in the IPCC Working Group reports (IPCC 2007). Note that examples of "concrete" actions in the water sector to adapt specifically and solely to a changing climate are very rare. This is partly because climate change may be only one of many drivers affecting strategies and investment plans (and it may not be the most important one over the short-term planning horizon), and partly due to uncertainty in projections of future hydrological changes.

Africa

- Enhancing resilience to future periods of drought stress by improvements in present rain-fed farming systems through improvements in the physical infrastructure including: water harvesting systems, dam building, water conservation and agricultural practices, drip irrigation, and development of drought-resistant and early-maturing crop varieties and alternative crop and hybrid varieties

Asia

Improvement to agricultural infrastructure including:

- Pasture water supply
- Irrigation systems and their efficiency
- Use/storage of rain and snow water
- Recycling and reuse of municipal wastewater, e.G., Singapore
- Reduction of water wastage and leakage and use of market-oriented approaches to reduce wasteful water use

Australia and New Zealand

- Treatment plant to supply recycled water
- Reduce channel seepage and conservation measures
- Pipelines to replace open irrigation channels
- Improve water-use efficiency and quality
- Drought preparedness, new water pricing
- Installation of rainwater tanks
- Seawater desalination

3 A distinction is frequently made between autonomous and planned adaptations. Autonomous adaptations are those that do not constitute a conscious response to climate stimuli, but result from changes to meet altered demands, objectives, and expectations which, whilst not deliberately designed to cope with climate change, may lessen the consequences of that change. Such adaptations are widespread in the water sector, although with varying degrees of effectiveness in coping with climate change. Planned adaptations are the result of deliberate policy decisions and specifically take climate change and variability into account, and have so far been implemented infrequently (IPCC 2007).

Europe

- Demand-side strategies such as household, industrial, and agricultural water conservation, repairing leaky municipal and irrigation water reservoirs in highland areas, and dykes in lowland areas
- Supply-side measures such as impounding rivers to form instream reservoirs, wastewater reuse and desalination systems, and water pricing

Latin America

- Rainwater catchments and storage systems
- "Self-organisation" programmes for improving water supply systems in very poor communities
- Water conservation practices, reuse of water, and water recycling by modification of industrial processes and optimization of water use

North America

- Improved water conservation and conservation tillage
- Investments in water conservation systems and new water supply and distribution facilities

Polar Regions

- Flow regulation for hydro-electric production, harvesting strategies, and methods of drinking water access

Small Islands

- Desalination plants
- Large storage reservoirs and improved water harvesting
- Protection of groundwater, increasing rainwater harvesting and storage capacity, use of solar distillation, management of stormwater and allocation of groundwater recharge areas in the islands

CONCLUSIONS

The vulnerability of the water industry in terms of water security and water quality to the impacts of climate change is firmly established. Water professionals and utilities acknowledge that climate change is unequivocal, and shares an increasing concern that it needs to take action now to mitigate and adapt. Many water utilities have therefore begun to respond to climate change through "adaptation" measures to modify plans and operations to minimize impacts. These adaptation efforts fall into two broad categories. The first consists of vulnerability analyses that are intended to identify the most near-term priorities in places where impacts could be felt the soonest. The second is long-term planning, or more formally, integrated resource planning (IRP) that adopts the broadest possible strategic view of how a utility can plan to cope with such systemic changes over the longer term. In addition, utilities are responding through adoption of measures to help mitigate the onset of climate change by reducing energy consumption that contributes to the production of greenhouse gases.

Portfolio Approach

The long-term response that is most prevalent in discussions of climate change is fortunately one that is familiar to many water utilities—IRP, often called integrated resources planning, or integrated water resources management (IWRM), although it is sometimes undertaken without being explicitly labelled as such. A hallmark of this approach to long-term planning is the adoption of a very broad view of the problem that "integrates" all facets of it—environmental, socioeconomic, and engineering—as a basic strategy for keeping a wide range of options open and providing a maximum of flexibility. All of the advice on adaptation to climate change begins with the same message: employ a portfolio approach—maintaining a maximum degree of flexibility and resiliency. The continued improvement and refinement of best practice in IRP will be of significant value to water utilities in coping with climate change.

An essential part of maintaining a broadly "integrated" approach is the continuous involvement of all stakeholders. IRP is most appropriate to problems of sweeping proportions that involve complex trade-offs between multiple objectives and multiple constraints. The best solutions are made possible in such situations because IRP recognizes that stakeholders have the capacity to redefine some of the objectives and constraints when necessary to avoid an impasse. But this only works if stakeholders are fully involved.

The bottom line in water supply planning has always been a matter of coping with variability. With the coming changes in climate, there will be a heightened need to respond to increased variability. The net effect of the direct impacts of global warming will be to change the variability of key parameters affecting the quantity and quality of water that would normally be expected to be available at any specific time and place. In addition, the capability to store water in various forms and the demand for water will be changed.

Given the sweeping extent of these changes, the IRP approach of taking the broadest possible view of the problem is indeed an ideal adaptation strategy. Working within broadly established system boundaries, it is possible to approach optimization in a manner that first derives maximum advantages from flexible operating strategies to expand the number of ways in which supplies can be managed to meet demands. This can have the benefit of deferring irreversible capital projects that may present more risk in a changing environment. Water utilities already employ sophisticated modelling tools (RIVERWARE, BASINS, OASIS, WEAP) to design such operating strategies, and it seems likely that applying such modelling tools to the design of adaptation strategies will be as, or more, important a priority as the climate models previously discussed in the examples of Manilla and Ceará.

Significantly, the IRP approach also features comprehensive assessment of strategies that can be applied to manage the water demand side of the equation. Warming processes will lead to altered demand patterns as a result of seasonal shifts in precipitation, more evaporation, more frequent heat waves, and more extensive droughts. Conservation programs offer a bonus in reducing both water supply needs and energy use. Bolstering conservation incentives (and disincentives to outdoor water use) may become more essential if warming processes otherwise increase water demands, especially during coincident peak demand periods when both water supply and electric power capacities are stretched to their limits.

REFERENCES

Cooley, H., Gleick, P.H., and Wolf, G., 2006. Desalination, with a grain of salt. Pacific Institute, June 2006. Available at http://www.pacinst.org/reports/desalination/desalination_report.pdf.

Cromwell, J.E., Smith, J.B., and Raucher, R.S. 2007. Implications of Climate Change for Urban Water Utilities. Prepared for the Association of Metropolitan Water Agencies. Washington, D.C. and Boulder, Colo. 2007. Available at: http://www.amwa.net/galleries/climate-change/AMWA_Climate_Change_Paper_12.13.07.pdf.

Dietz, T., Rosa, E.A., and York, R. 2007. "Driving the human ecological footprint." Frontiers in Ecology and the Environment. February 2007. Available at http://www.stirpat.org/frontiers_hi_res.pdf.

El Pais. June 2005. See online archive at http://www.elpais.com.

Falkenmark, M. 1989. The Massive Water Scarcity Now Threatening Africa—Why Isn't It Being Addressed? *Ambio*, 18, 112–118.

Fraile, I. 2006. Water Management in Spain. Background paper to *Making the Most of Scarcity: Accountability for Better Water Management Results in the Middle East and North Africa.* World Bank, Washington, D.C.

IPCC. 2001. *Climate Change 2001: Synthesis Report. A Contribution of Working Groups I, II, and III to the Third Assessment Report of the Intergovernmental Panel on Climate Change* Watson, R.T. and the Core Writing Team (eds.). Cambridge University Press, Cambridge, UK, and New York. 398 pp.

IPCC. 2007. *Climate Change 2007: Synthesis Report. Contribution of Working Groups I, II and III to the Fourth Assessment Report of the Intergovernmental Panel on Climate Change.* Core Writing Team, Pachauri, R.K., and Reisinger, A. (eds.). Geneva, Switzerland. 104 pp.

Kemper, K., Dinar, A., and Blomquist, W. 2005. *Institutional and Policy Analysis of River Basin Management Decentralization.* World Bank, Washington, D.C.

Kundzewicz, Z.W., Mata, L.J., Arnell, N.W., Döll, P., Kabat, P., Jiménez, B., Miller, K.A., Oki, T., Sen, Z., and Shiklomanov, I.A. 2007: Freshwater resources and their management. *Climate Change 2007: Impacts, Adaptation and Vulnerability. Contribution of Working Group II to the Fourth Assessment Report of the Intergovernmental Panel on Climate Change,* M.L. Parry, O.F. Canziani, J.P. Palutikof, P.J. van der Linden, and C.E. Hanson, eds. Cambridge University Press, Cambridge, UK. pp. 173–210.

Umgeni Water. 2007. Umgeni Annual Report 2007–2008: Growth, Development and Sustainability. Available at http://www.umgeni.co.za/Uploads/annualreport2007/pdfs/Annual%20Report.pdf.

– 17 –

Half Full or Half Empty? Either Way It's Time to Plan

Perri Standish-Lee, Erik Loboschefsky, and Kristina Lecina

William Shakespeare wrote in Henry VI, "smooth runs the water where the brook is deep." Unfortunately, life is not so smooth where water supply is scarce, and utilities that are responsible for sustaining water resources in arid and semi-arid regions face turbulent times. Effective management of both water quality and quantity are especially critical for communities furnished with the short end of the divining rod.

WESTERN UNITED STATES FACES DROUGHT CONCERNS

Arid and semi-arid regions comprise approximately 40% of the earth's surface area and sustain more than one billion people but experience only 2% of the global runoff (UNEP 2000). In the United States, several consecutive years of abnormally high temperatures, below-average precipitation, and increased water demands left the Colorado River at an all-time low in 2004 (USBR 2004). Water surface levels in reservoirs throughout the western United States, such as Lake Powell and Lake Mead, are currently at near-record lows. Additional precipitation during 2005 led to hopes of alleviating this drought; however, above-average temperatures and below-normal precipitation in recent months have refuelled drought concerns.

Aggravating the situation is the decline in water quality that results from diminished flows. In the Colorado River basin and in much of the West, salinity concentrations in rivers and streams are directly influenced by flow volume. Figure 17-1 illustrates how salinity concentrations in the Colorado River dropped in the mid-1980s when river flows were at historically high levels and, conversely, in the early 1990s when flows were low and salinity concentrations were high (Colorado River Basin Salinity Control Forum 2005).

The Sierra Nevada are an important source of water supply for much of the western United States. The snowpack from these mountains can be visualized as a thermal storage reservoir, retaining snow that eventually melts and provides water during the spring and summer months. However, even in Northern California where water resources are relatively more abundant than in Southern California, climate changes could have a profound effect on water availability. Although total annual precipitation values are predicted to remain relatively stable, most climate change models predict diminished snowpack levels in the Sierra Nevada (Cayan et al. 2001). Figure 17-2 compares recent snowpack conditions with projected conditions for the year 2060.

With a reduced snowpack, meeting water supply demands during dry seasons would become more difficult. Some models predict higher flows during the rainy season and lower flows during the dry season. In the dry season, problems of reduced supply are

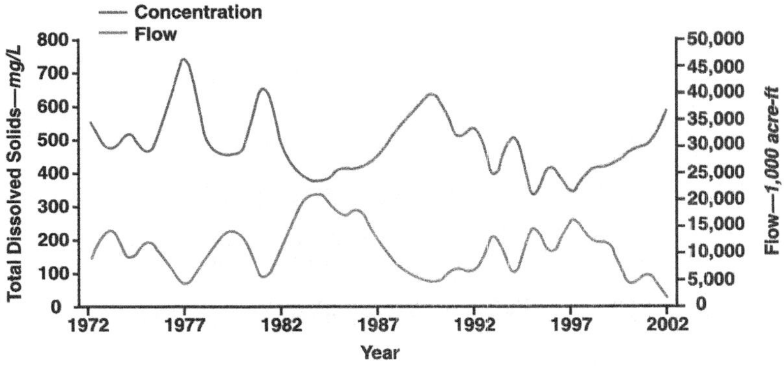

Figure 17-1. Colorado River inflow to Lake Powell and total dissolved solids concentrations

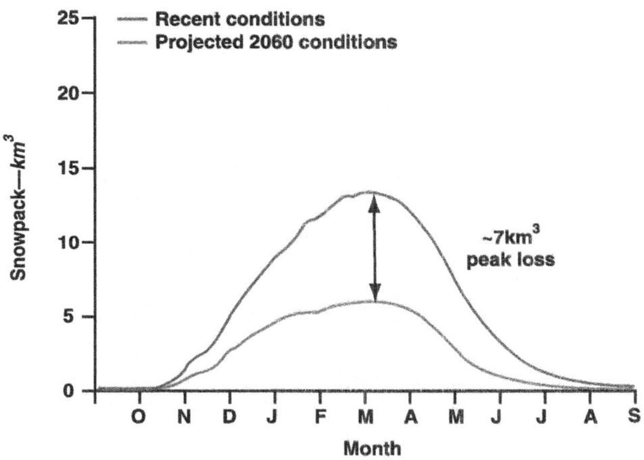

Figure 17-2. Natural storage in the Watershed—recent and projected snowpack storage

likely to be exacerbated by declining water quality resulting from diminished flows. In the San Francisco Bay estuary, for example, such changes would yield much higher dry-season salinities and elevated bromide concentrations (Figure 17-3).

STRATEGIES VITAL TO LONG-TERM WATER SUPPLY

Existing water resources must be sustained to meet water demands of the urban, industrial, and agricultural sectors, which compete for limited supplies. Sustainability in all three sectors requires effective utilization of water supplies, including implementation of water-saving practices and development of emergency water shortage/drought plans. It requires communities and water agencies to work together to optimize existing water sources without affecting future generations' abilities to do the same. Examples of some entities cooperating to achieve sustainability are:

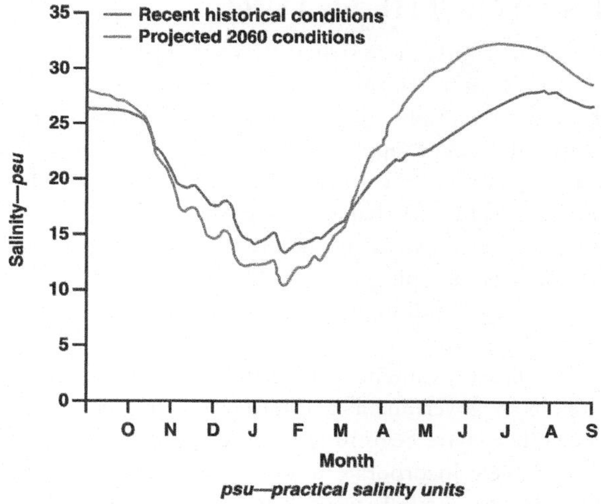

Figure 17-3. Mean annual cycle of salanity change in San Pablo Bay

- In the 1970s, the seven states of the Colorado River basin implemented a salinity control program that is believed to be an important factor in having reduced the overall Colorado River salinity by as much as 100 mg/L (Colorado River Basin Salinity Control Forum 2005).
- The Contra Costa Water District's construction of Los Vaqueros Reservoir allows high-quality water to be pumped into the reservoir in spring months when the Sierra Nevada snowpack melts and flows into the delta. This water acts as a reserve for late summer months when the salinity levels in the delta are high because of low river flows.

Strategies for helping achieve water sustainability in arid regions include:

- technology transfer;
- public outreach;
- conservation and other activities that increase water use efficiency;
- quantification of effective as well as ineffective water management practices;
- assessment of potential future supplies; and
- incentives for water transfers, socioeconomic development, and efficient water use.

Identification of practical and acceptable water supply options has become a fundamental element of water strategy development. Technological progress and regulatory changes throughout the past decade have opened numerous water supply options that were once considered infeasible because of exorbitant costs or health concerns. Supply options that should be considered in the development of successful sustainability strategies include:

- water conservation;
- reuse of treated wastewater and stormwater;
- desalination of ocean and brackish waters;
- stormwater detention;
- water transfers, water banking, or aquifer storage and recovery; and
- out-of-basin importation of water supplies to basins in need.

The ultimate choice of any supply option should be driven by the option's ability to provide water of suitable quality and to function as a long-term source of supply.

UTILITIES PLAN FOR THE FUTURE

Emerging regulations are becoming more stringent. Water utilities with superior water quality are able to meet the regulations with minimal changes to their treatment systems. However, utilities that rely on degraded water sources are finding it necessary to upgrade treatment systems and/or adopt new sources of supply, both of which can be quite costly.

New policies and practices that encourage conservation offer hope for parched communities, especially those in arid climates that are hardest hit by drought. Some water demand–management measures—such as restricting the planting of high water–demand crops and encouraging water friendly landscaping in both the public and private sectors—generate substantial water savings. Incorporation of water-saving devices in households yields smaller but equally important savings.

The introduction of measures intended to reduce water demand often arises from ad hoc analysis and decisions by governments at times of extreme water shortage. Water utilities in the West and Southwest are beginning to develop water strategies and master plans that effectively and proactively incorporate water demand–management measures, as well as technologies and practices to address water quality issues. The Southern Nevada Water Authority (SNWA) in Las Vegas, Nev., and the Zone 7 Water Agency in Livermore, Calif., offer two examples.

SNWA

The SNWA implemented plans to manage, protect, and augment the local water supply, which primarily depends on Colorado River water delivered from Lake Mead. First, in 1996, the SNWA implemented a comprehensive water resource plan that outlined a portfolio of existing and future resource options to help meet future demands (SNWA 2005). The original plan, which has been reviewed annually, emphasized water conservation, environmental protection, and the need for flexibility in the timing and implementation of additional resources and associated treatment and delivery infrastructure. That plan launched a series of regional water conservation programs to achieve 25% regional conservation by the year 2010.

When surface water levels in Lake Mead began to drop in early 2000 as a result of the severe drought in the Colorado River basin (Figure 17-4), the SNWA also adopted a comprehensive regional drought plan to accelerate local water conservation and mitigate potential water shortages. Through conservation initiatives and subsequent drought plan measures,

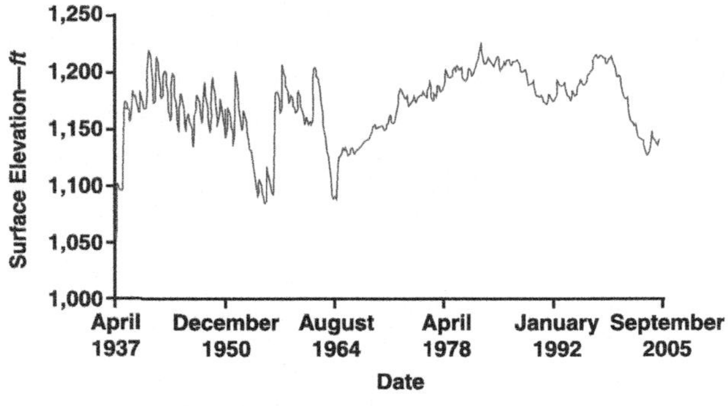

Figure 17-4. Historical surface water levels in Lake Mead

Southern Nevada's annual water consumption has decreased during the past three years by approximately 20 bil gal, despite the addition of nearly 250,000 new residents and 38 million annual visitors.

Additionally, in 2004 the SNWA implemented two plans to accelerate the development of three water resource projects to provide additional groundwater and surface water resources to Southern Nevada. This work included the initiation of an integrated water planning process—involving all relevant parties—to assess funding and develop alternatives for proposed projects (SNWA 2005).

To augment its existing surface water rights to Lake Mead, the SNWA has begun to develop two other smaller, local surface water sources. In addition, the utility is expanding its water rights in other areas, establishing water banking programs within the Las Vegas Valley and in Arizona and California, reusing water supplies for nonpotable purposes, and returning highly treated wastewater flows to the Colorado River for return-flow credits. Throughout these planning efforts, the SNWA has worked closely with local and regional stakeholders to identify solutions that maximize the benefits to all stakeholders.

Zone 7 Water Agency

The Zone 7 Water Agency, which is developing strategies to address water quality issues associated with reduced freshwater flows in the San Francisco Bay estuary, serves as another example of a major utility planning for the future of its water supply. Zone 7's salt-management strategy, developed to address increasing levels of total dissolved solids in the groundwater basin, supports increased conjunctive use with shallow groundwater demineralization to offset current and future salt-loading to the main groundwater basin. In addition, Zone 7 is adding a new surface water treatment plant to address long-term water quality concerns and respond to short-term concerns such as taste and odor, total organic carbon, and high bromide concentrations from the South Delta that occur during droughts. The Altamont Water Treatment Plant will incorporate coagulation, membrane filtration, ozonation, and biologically activated contact to reduce activated organic carbon. During periods of high bromide concentrations, an application of acid, ammonia, or chlorine will be used before ozonation to reduce bromate formation.

COLLABORATION AND COMMITMENT ARE KEY

As population growth, climate change, and water quality degradation further affect the limited water resources of the western United States, local governments, water utilities, and the public must continue their efforts to sustain supplies. The human population can contend with water scarcity only through responsible development, effective water resource management, and the realization that the world shares the same finite water supply.

REFERENCES

Cayan, D.M. et al. 2001. Investigation of Climate Change Impacts on Water Resources in the California Region: Department of Energy Accelerated Climate Prediction Initiative (ACPI) Progress Report. Scripps Institution of Oceanography, San Diego.

Colorado River Basin Salinity Control Forum. 2005. *2005 Review: Water Quality Standards for Salinity, Colorado River System.* Colorado River Basin Salinity Control Forum, Bountiful, Utah.

SNWA (Southern Nevada Water Authority). 2005. SNWA Resource Plan. www.snwa.com/html/wr_resource_ plan.html (accessed April 2006).

UNEP (United Nations Environment Programme). 2000. The Watershed: Water From the Mountains Into the Sea. www.unep.or.jp/ietc/Publications/Short_Series/LakeReservoirs-2/index.asp (accessed April 2006).
USBR (US Bureau of Reclamation). 2004. USBR Lower Colorado Region. www.usbr.gov/lc (accessed April 2006).

Originally printed in Journal AWWA *June 2006.*

– 18 –

Climate Change Is Real: How Can Utilities Cope With Potential Risks?

Joel B. Smith

The climate is changing, and the rate of change is likely to accelerate. A utility's water quality, supplies, and demands may have changed already. How will water resources and water utilities be affected, and what strategies can be implemented to prepare for future change?

Greenhouse gases (GHGs) exist naturally in the atmosphere, trapping heat and making life on Earth possible. Human activities—fossil fuel burning, deforestation, and agricultural practices—increase GHG concentrations. Carbon dioxide is the most important GHG emitted by humans, and its concentration has increased by one-third since the 19th century. Figure 18-1, displays CO^2 trends during the last 50 years. Human activities have also increased other GHG concentrations, such as methane, nitrous oxides, and chlorofluorocarbons. These increased GHG concentrations are trapping more energy in the atmosphere. The Intergovernmental Panel on Climate Change (IPCC)—the authoritative source on the subject—concluded that there is a 90 percent chance that warming, particularly in the last 50 years, is caused mostly by human activities.

Since 1900, average global temperature has increased by 1.3°F, as shown in Figure 18-2. In addition, the temperature rise has accelerated in recent years. Sea level is also rising, while snowpack and glaciers around the world are generally decreasing.

Figure 18-3 displays US temperature and precipitation trends from 1931–2005. Most of the country warmed during this period, with the western United States warming the most. Interestingly, some areas in the Southeast did not warm.

Precipitation changes are more complex. On average, US precipitation increased 7 percent from 1931 to 2005, but the change was not uniform. The eastern half of the country generally got wetter and cooler. In the West, some areas got wetter, some drier, and some saw little change. Although average precipitation increased, the largest precipitation events have seen the largest increases, which may have caused more flooding.

FUTURE CHANGES

Climate will continue to change because

- even if GHGs do not increase in the future, global temperatures would still rise about 1°F because it takes decades for the Earth's climate to catch up with increased GHGs.
- GHG emissions and concentrations are likely to continue increasing because humans are emitting more GHGs with each passing year. Some of the increase can be attributed to population growth, economic growth, and lifestyle choices, including size of homes, cars, and types of appliances.

Because uncertainty surrounds these and other factors, the IPCC developed a set of scenarios of future GHG emissions and how CO_2 concentrations could rise from the present-day 380 ppm by volume to 500–1,000 ppm by 2100. This means that temperatures could rise faster than they did in the 20th century. The IPCC projects global average temperatures will

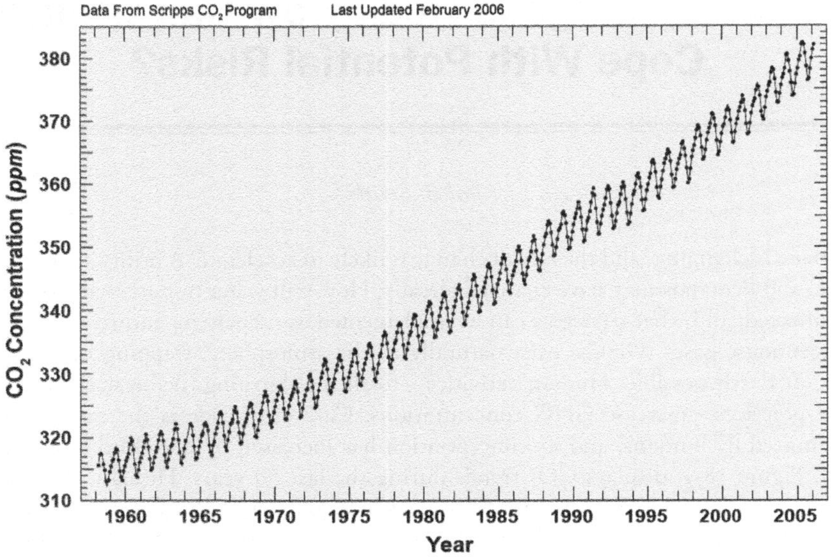

Figure 18-1. Carbon dioxide concentrations 1958 to the present. CO_2 levels have increased significantly in the last 50 years.

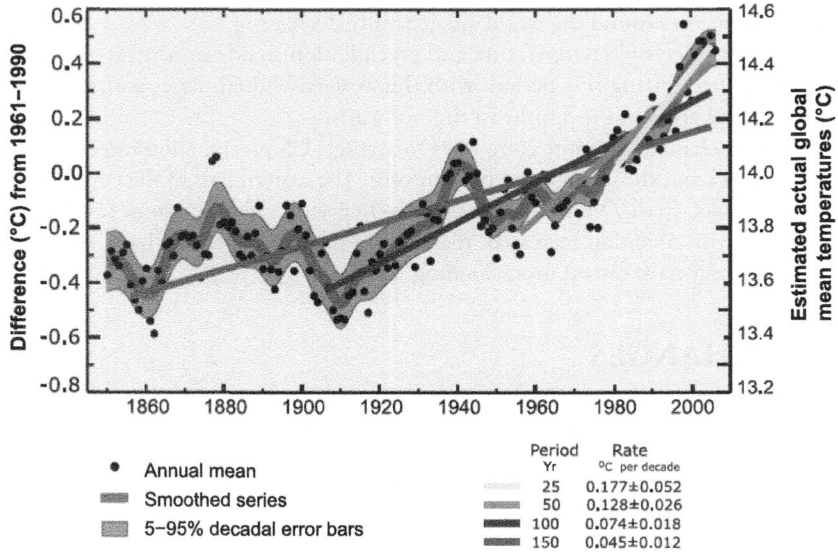

Figure 18-2. Global average temperature. Temperature rise has accelerated in recent years.

rise by at least 2°F—and perhaps as much as 11°F—by 2100. Sea level is projected to rise 1–3 feet or more. However, due to uncertainty about how much and how quickly major ice sheets could melt, it's unclear how much sea level could rise this century.

HOW WILL US CLIMATE CHANGE?

On average, US climate will be warmer and wetter, but not all areas will warm the same, and some areas are projected to become drier. Land areas will warm more than oceans. As snow and ice melt, more land in high altitudes will be exposed to sunlight, resulting in more warming.

This means that higher latitude and inland areas generally will warm more than coastal areas. For example, one would expect Minnesota to warm more than Mississippi. On average, the lower 48 states are projected to warm one-third more than global average temperatures. In other words, if average global temperatures increase by 3°F, US temperatures can be expected to increase by 4°F.

Changes in precipitation are more complex and vary by season. The IPCC concluded that there is a two-thirds chance that the northeastern US will see an increase in average annual precipitation, while the Southeast will see a decrease in average annual precipitation.

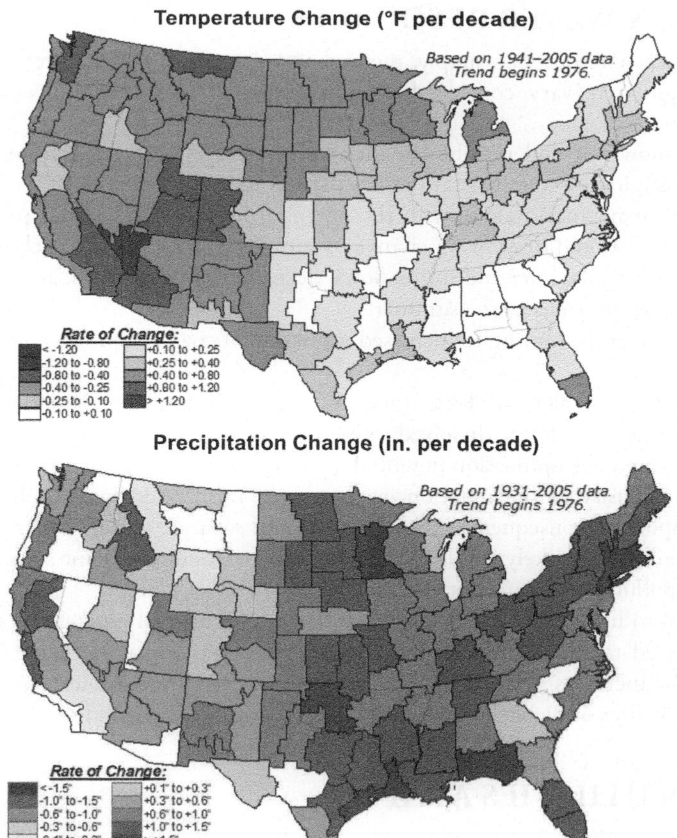

Figure 18-3. Temperature and precipitation changes for the United States. Most of the country has warmed in recent decades, and precipitation has increased.

For many water suppliers, changes in seasonal precipitation can be more important than annual averages. Winter precipitation is projected to increase in most of the northern US but decrease in the Southwest. Most of the country is projected to have no change in summer precipitation, but parts of the Northwest and Southeast are projected to have decreased summer precipitation. Note that even if precipitation remains unchanged, higher temperatures can make conditions drier by reducing soil moisture, streamflow, and lake levels through increased evaporation.

The following changes are expected:

- A longer growing season
- More warm nights
- More heat waves
- More days without precipitation
- More intense precipitation events and fewer days with snow

In addition, hurricanes and tropical storms would become more intense. Scientists aren't sure if the number of hurricanes and tropical storms will increase, but those that do occur will likely have stronger winds, more rain, and higher storm surges, which will be elevated already because of higher sea levels.

EFFECTS ON WATER RESOURCES

All aspects of water resources—supply, quality, and demand—will be affected by climate change. Changes may vary considerably from basin to basin, but expect to see these occurrences:

- Earlier snowmelt and possibly smaller snowpacks. Higher temperatures will cause more precipitation to fall as rain rather than as snow.
- Change in snowpack, along with the potential for increased winter precipitation in northern areas and drier or no change in summer precipitation, which generally will lead to increased winter and spring flows, decreased summer flows, and increased pollutant concentrations in the summer.
- Higher sea levels, leading to increased salination of coastal aquifers and higher storm surges.
- Warmer water, which will lead to increased water demand, particularly for irrigation and outdoor uses; lower dissolved oxygen levels; increased disinfection by-products; and increased eutrophication potential.

Increased intensity of precipitation events would raise peak flood probabilities. This would have important consequences for combined and separate sewer systems. In addition, more intense rain would likely result in more intense pulses of nonpoint source pollutants, increasing the pollutant and turbidity concentrations in water bodies.

In addition to hydrologic cycle and demand changes, natural vegetation could change. Fires are projected to increase, particularly in the West. Following fires, precipitation can cause substantial increases in turbidity. Pest infestations could increase in many areas, which can lead to forest fires or diebacks, further increasing water body turbidity.

HOW CAN UTILITIES ADAPT?

Water planners already project for changes in population, income, tastes, and other factors. Climate change should be added to the list. As in the accompanying case studies, utilities are already coping with climate variability and preparing for low-probability events.

The first step is to understand how a utility may be affected by climate change. There are two approaches. One is to use climate models to estimate how climate in the region can change and then analyze the consequences of climate change for runoff, water quality, and demand. The other approach, called a "threshold approach," involves examining the system to assess where it may be close to thresholds that are sensitive to potential climate changes. For example, seasonal supply pattern changes could necessitate use restrictions, or an increase in peak flows could overwhelm storage or treatment systems. When such vulnerabilities are identified, global climate models can be consulted to determine the likelihood of exceeding the thresholds.

The second thing to consider is adaptations. There are two types of adaptations: no regrets and low regrets.

No Regrets

These adaptations make sense even without considering climate change but make more sense in addressing climate change. Such adaptations include:

- Use water more efficiently through water conservation.
- Implement flexible mechanisms for allocating water supplies, such as price-based systems or sharing agreements among users—agreements that allow municipalities to temporarily purchase water from farmers during droughts.
- Diversify supplies by increasing water reuse, desalination, and interbasin transfers.
- Reduce climate variability risks, such as exposure to floods and droughts. Note that development patterns can increase the exposure of people and property to flood and drought even if the climate doesn't change. As more people move into flood- and drought-prone areas, the effects of climate change intensify.
- Improve forest management to reduce fire and pest infestation risks.

Low Regrets

These are adaptations that are justified only by climate change. Because these adaptations may take many years to yield benefits and because of uncertainties about regional climate change, these adaptations shouldn't involve significant increases in resources. Such adaptations often involve modifying long-term decisions to incorporate risks of climate change. So, if infrastructure is being built anyway, it may make sense to change the size to accommodate climate change. The cost of such adjustments may be low.

Adopting a risk management paradigm makes sense in light of climate change. The effects of climate change are uncertain, but some low-probability effects may have high consequences. Although challenging, preparing for such events makes sense. This means addressing the most likely outcomes, as well as low-probability and high-consequence outcomes.

Most important, recognize that the climate is changing, and the change is likely to accelerate. It is prudent that water managers understand the potential risks and prepare for the future accordingly.

Originally printed in Opflow *February 2008.*

– 19 –

Planning Strategy in a Changing Climate

Sarah Deslauriers, Lydia Holmes, P.E., Steve McDonald, P.E., Jon Loiacono, P.E.

INTRODUCTION

According to the Intergovernmental Panel on Climate Change (IPCC) Fourth Assessment Report (AR4) released in February 2007, human actions are "very likely" (90% or greater probability) the cause of global warming. In addition to nature's contribution of GHGs (e.g., water vapor, carbon dioxide, methane, nitrous oxide, and other gases), human-generated emissions of those GHGs as well as a suite of synthetic GHGs (e.g., chlorofluorocarbons, sulfur hexa-fluoride, and other gases) are accumulating in the atmosphere and are "very likely" responsible for enhancing the earth's greenhouse effect and resulting in global climate change.

This paper provides details of the projected climate change impacts on water resources management, water and wastewater facilities, and stormwater infrastructure. Recommendations are provided to water and wastewater agencies for planning for climate change impacts, with special focus on applying projected changes in precipitation patterns to intensity–duration–frequency (IDF) curves for planning purposes. A case study of the analysis and recommendations made for the San Francisco Sewer System Master Plan (SSMP) related to the impacts of global climate change over the next 30 years is included. This paper concludes with a list of organizations in the field of climate change whose research can be applied to water resource planning, the design of water and wastewater treatment facilities, and the design of stormwater infrastructure.

IMPACTS OF GLOBAL CLIMATE CHANGE ON THE WATER AND WASTEWATER INDUSTRY

Global climate change results in direct and indirect impacts. Direct impacts are those related to the increasing atmospheric temperatures, such as an increase in source water temperature, thermal expansion of the marine mixed layer of the ocean, and melting land ice and snow, to name a few. Indirect impacts include decreased capacity for dissolved oxygen in source waters having higher temperatures, changes in weather (more specifically precipitation patterns), and rising sea level caused by the thermal expansion of the marine mixed layer of the ocean and melting land ice.

The key climate variable of concern to both water resources and stormwater management is precipitation, especially at a daily or finer time interval. Since about 1970, total annual precipitation across the U.S. has remained slightly above the 20th-century mean and averaged about 5 percent more than the previous 70 years.

There is no consensus on forecasts for total annual precipitation; however, it is projected that the probability of 24-hour extreme precipitation events will increase by a factor of 2 or more by the end of the 21st century (relative to those that occurred during the year 2000)

(Kharin and Zwiers 2005). Reservoir and storm water collection and distribution system designs are traditionally based on historical weather conditions; this planning strategy is no longer sufficient.

More than 50 percent of the nation's population lives on 12,000 miles of coastline (including Hawaii and Alaska). In addition to planning for the projected changes in design storm, coastal facilities and infrastructure must consider a projected rise in sea level. Currently, the designs of coastal facilities and infrastructure are based on historical tidal levels and weather patterns. These designs should be re-examined for projected conditions.

In addition to the physical impacts of climate change, there are now regulatory impacts. For example, in September 2006, the California legislature passed Assembly Bill No. 32 (AB 32), the Global Warming Solutions Act, which became effective Jan. 1, 2007. This Act is the first regulatory program in the U.S. that will require public and private agencies, including publicly owned treatment works (POTWs), statewide to reduce GHG emissions to 1990 levels by the year 2020 and 80 percent below 1990 levels by 2050. The GHGs to be regulated are carbon dioxide, methane, nitrous oxide, hydrofluorocarbons, perfluorocarbons, and sulfur hexafluoride. In response to AB 32, a statewide POTW steering committee was established. The committee identified three potential impacts of AB 32 on California POTWs: GHG emissions reporting, GHG emissions reductions, and implementation of a cap-and-trade program. As a result of AB 32, other states are beginning to propose regulations for reporting and capping GHG emissions, and the USEPA is currently drafting a mandatory reporting rule to become effective in 2009.

PLANNING FOR A CHANGING CLIMATE

Long-term planning of water supply and sewer infrastructure typically involves the projection of future needs based on historical demand and weather patterns. However, over the past several decades, as evidence of climate change has become increasingly accepted, water quality managers are beginning to seek integrated strategies for adaptation and mitigation.

Adaptation

An adaptive planning strategy considers future climate changes as factors in long-term designs for infrastructure and water management decisions. Such factors include

- Rising snowline levels and earlier spring run-off from snowmelt requiring re-examination of water storage management decisions.
- Changing precipitation patterns challenging the timing of scheduled releases and refill of reservoirs. Without adequate knowledge of the changing precipitation patterns, there is an increased risk of flooding upstream of reservoirs.
- Rising sea levels enabling saltwater to intrude farther inland, encroaching on fresh water aquifers and estuaries.

To help water managers and planners handle these issues, the American Water Works Association published the following set of recommendations (American Water Works Association 1997):

- Water agencies should consider re-examining engineering design assumptions, operating rules, system optimization, and contingency planning for existing and planned water-management systems under a wider range of climatic conditions than traditionally used.
- Water agencies and providers should explore the vulnerability of both structural and nonstructural reservoir operation and water supply management water systems to plausible future climate changes, not just past climatic variability.

- Governments at all levels should re-evaluate legal, technical, and economic approaches for managing water resources in the light of possible climate changes.
- Water agencies should cooperate with leading scientific organizations to facilitate the exchange of information on state-of-the-art thinking about climate change and impacts on water resources.

Wastewater facilities and sewer infrastructure are also sensitive to changing precipitation patterns (i.e. an increase in extreme precipitation events). In addition, coastal wastewater facilities and sewer infrastructure are vulnerable to inundation caused by both rising sea levels and changing precipitation patterns.

Sea Level Rise

Sea level rise is already occurring. The global average rate of sea level rise is estimated between 1.8 to 2.8 mm/yr (0.07 to 0.11 in./yr). A study published in 2005, argues that there is no significant difference in the rate of coastal and global average sea level rise between 1950 and 2000 as predicted by 20th century climate model simulations, and that the best estimate of global and coastal average sea level rise is 1.8 mm/yr (0.07 in./yr) (White et al. 2005). However, altimetry data obtained between 1993 and 2003 show nonuniform geographical distribution of sea level change, with some regions exhibiting trends about 10 times the global mean (Cazenave and Nerem 2004). Some uncertainty exists, because of uncertainty about the emissions of greenhouse gases, population growth rates, government policies to address emissions, and the actual dynamics of the oceans and ice sheets.

The IPCC released the Fourth Assessment Report (4AR) in early 2007, and projects sea levels to rise between an average of 0.18 and 0.59 m (7 and 23 in.) by 2100 (Meehl et al. 2007). In comparison with IPCC's 2001 projections of 0.09 and 0.88 m (3.5 and 35 in.) for 2100, the range of projected sea level rise has narrowed. These projections are based on physical models that do not reproduce the current rate at which the polar ice caps are melting. Since the release of the 4AR, there is extensive, new published research that shows major advances in the science of sea level rise that have increased the uncertainty of IPCC projections. Future rates of sea level rise are likely to accelerate.

Peer-reviewed studies now project a rise in mean sea level with a full range of 0.51 to 1.40 ms (20 to 55 in.) by 2100. The Independent Science Board suggests the IPCC conservative estimates be viewed as minima for planning purposes (Mount 2007).

Changes in Precipitation Patterns

Precipitation (i.e., extreme storm events) is the key climate variable of concern to stormwater management, especially when it must be projected at a daily or finer time interval (Watt et al. 2003). Although projections for annual mean precipitation vary, most regional climate model results suggest that the extreme daily precipitation rate will increase. Global simulations show that the percentage increase in extreme rainfall is greater than the percentage increase in mean rainfall. The likelihood of extreme 24-hour precipitation events is projected to double by the end of the 21st century (relative to those events that occurred during the year 2000) (Kharin and Zwiers 2005).

Mitigation

The IPCC Assessment Reports consider scenarios that use the rate of change in atmospheric concentration of carbon dioxide to evaluate the impact this concentration will have on global climate change. There are scenarios that provide hope for mitigating global climate change through the reduction of GHG emissions. A mitigative planning strategy considers measures to prevent future climate change (through the reduction of GHG emissions per million gallons of pumped and/or treated water or wastewater) as factors in long-term designs for infrastructure and water management decisions. Such factors include

- Increasing energy efficiency of facility pumps and treatment processes.
- Reducing truck transport of chemicals. This can be accomplished by investigating alternatives to reduce the chemicals consumed in treatment processes and by seeking out chemical manufacturers that are located closer to the facility site.
- Reducing truck transport of solids. This can be accomplished by investigating alternatives to reduce the quantity of solids hauled long distances, and by seeking out disposal or land application sites that are located closer to the facility site.
- Adding or increasing energy production by water and wastewater agencies of renewable sources (i.e., cogeneration, in-conduit hydropower, and biogas).
- Adding renewable energy sources, such as wind and solar power generation.

Nationally, water and wastewater utilities account for four percent of electricity use, which includes 56 percent of municipal energy use (Electric Power Research Institute 2002). Eighty percent of the energy used by water and wastewater utilities is attributed to water pumping, while 20 percent is caused by treatment (Electric Power Research Institute 2002). However, in California, water-related energy use consumes 19 percent of the state's electricity as shown in Figure 19-1 (California Energy Commission 2005), 30 percent of its natural gas, and 88 billion gallons of diesel fuel every year. While the energy consumption numbers presented in this figure do not include energy consumption related to chemical production or for the transport of those chemicals and hauling of solids to a disposal/land application site, this type of analysis provides insight as to where the focus of energy efficiency improvement projects should be made.

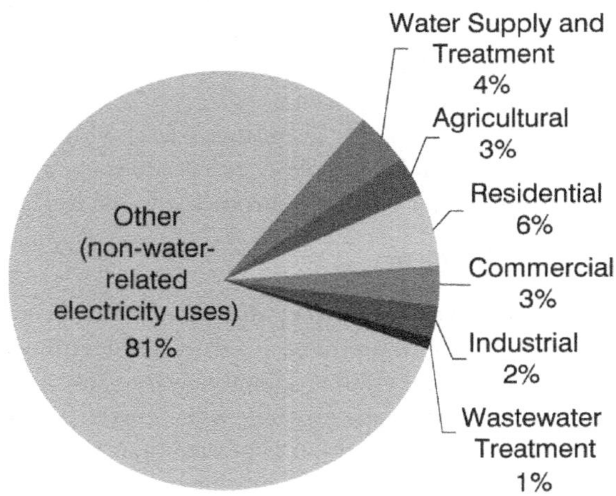

(California Energy Commission 2005)

Figure 19-1. Electricity use in the state of California for the year 2001

CASE STUDY: SAN FRANCISCO PUBLIC UTILITIES COMMISSION 30-YEAR SEWER SYSTEM MASTER PLAN

This case study provides details of an analysis completed for the San Francisco Public Utilities Commission (SFPUC) 30-year Sewer System Master Plan (SSMP). The case study focuses specifically on how to address sea level rise and changes in the intensity, duration, and frequency of storm events.

In 2006, Carollo performed an extensive literature review of the effects of global warming on future climate change, with an emphasis on sea level rise and changes in precipitation patterns (i.e., design storm). The rate of sea level rise at San Francisco for the past decade has been 3.2 mm/yr (0.13 in./yr), while the global rate has an estimated range of 1.8 to 2.8 mm/yr (0.07 to 0.11 in./yr). In 2001, the IPCC released the Third Assessment Report and projected that the mean sea level may rise between 0.09–0.88 ms (3.5–35 in.) by the year 2100 (relative to 1990 levels). However, some uncertainty exists, because of unpredictability of GHG emissions, population growth rates, government policies addressing emissions, and the actual dynamics of the oceans and ice sheets.

San Francisco's sewer system is a combined sewer system that includes weirs along the city's perimeter for managing overflows. Some of these weirs are situated along the city's bayside below the City Datum. The City Datum is 8.61 ft above the National Geodetic Vertical Datum of 1929. Figure 19-2 shows the historical and projected monthly mean higher-high water (MHHW) levels (assuming the distribution of monthly MHHW levels remains the same) and existing levels of the city's combined sewer overflow (CSO) weirs. Figure 19-2 also shows the high estimated projections for MHHW levels using 2001 IPCC sea level rise projections for 2050 and 2100. Backflow into the weirs will occur in some areas as a result of the distribution of the high tides, long before the 2001 IPCC high projections exceed the weir levels. This is shown on the figure by the dotted lines. Because of this analysis, the city is researching duckbill valves for backflow protection in the short-term and investigating long-term solutions for potential sea level rise.

In the design phase of a sewer system, municipalities, including the city and county of San Francisco, should estimate average annual overflow frequency and volume. Metcalf & Eddy developed Figure 19-3 using data from the National Weather Service San Francisco rain gauge. Figure 19-3 divides data of average rainfall in San Francisco into phases of the Pacific Decadal Oscillation (PDO), including positive (warm) and negative (cool) periods. The PDO is known to influence the natural variability of regional California precipitation. The figure shows lower than average rainfall during the cold period and higher than average rainfall during the warm period. The figure also shows that from the mid-1980s to present day the running five-year average rainfall for San Francisco has been mostly above the long-term average, showing an increase in average rainfall.

Historically, the design of the city of San Francisco's combined sewer system has been based on an intensity–duration–frequency (IDF) curve for the 1941 five-year design storm. IDF curves were developed during the master planning process using NOAA's National Weather Service (NWS) San Francisco rain gauge data from 1975 to 2004 of rainfall instead of the overall period of record (1907–2005) to include factors such as global warming.

It is projected that the probability of extreme (24-hour) events will increase by a factor of two or more by the year 2100 (relative to those that occurred during the year 2000) because of climate change. This means that 24-hour events with return periods of 10-, 20-, 50- and 100-years will occur twice as often (or be reduced by a factor of two or more) by the year 2100 (Kharin and Zwiers 2005). In other words, the current 24-hour event with a 10-year return period could have a five-year return period by the year 2100.

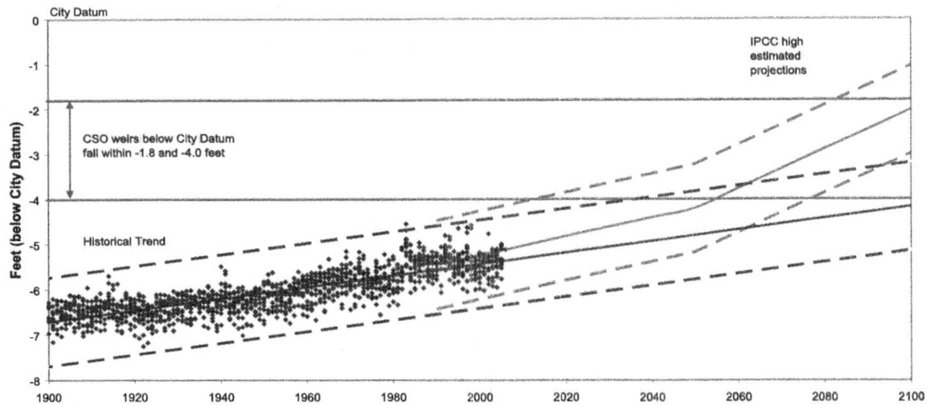

Figure 19-2. Historical monthly MHHW levels relative to the San Francisco City Datum and 2001 IPCC high estimated projections, in addition to the levels at which a portion of the city's CSO weirs exist. The dotted lines represent two standard deviations from the mean.

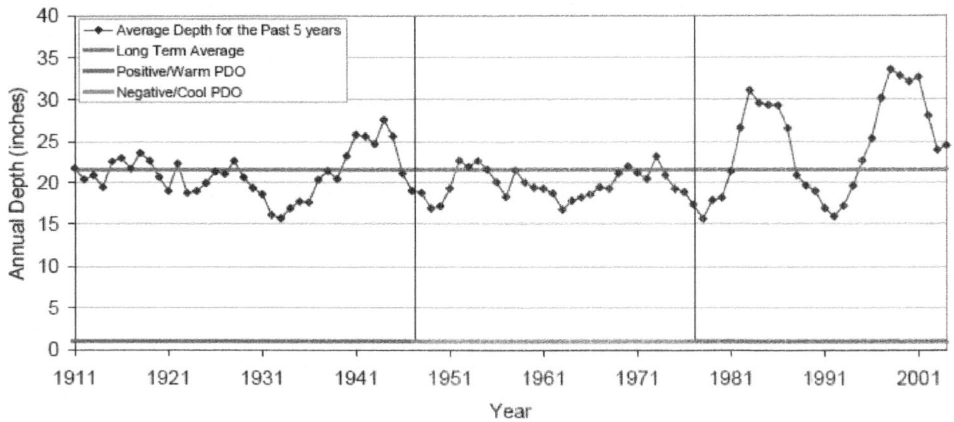

Figure 19-3. Running five-year average rainfall at NWS San Francisco rain gauge

The analysis of projected rainfall events for the San Francisco Bay area resulted in the following general conclusions to be considered in the master planning of future wastewater treatment and collection facilities for the city. These findings are summarized as follows

- By 2060, there will be an average of 8 percent increase in the magnitude (total rainfall) of 24-hour rainfall events (having 10-, 20-, 50-, and 100-year return periods) (Kharin and Zwiers 2000).
- By 2100, there will be an average of 15 percent increase in the magnitude of 24-hour rainfall events (having 10-, 20-, 50- and 100-year return periods) (Kharin and Zwiers 2000).

The general conclusions are that the frequency of rainfall events of extreme intensity (and for periods of up to 24 hours) are going to increase in the long-term, but the total average annual precipitation will either remain the same or will not increase significantly. Consequently, the *design storm* for master planning sewer system improvements for the next 30 years should consider these long-term trends. Carollo recommended conducting a sensitivity analysis to assess the risk of current infrastructure designs being compromised or overwhelmed by future climate changes.

Given the previous general conclusions, an analysis of San Francisco rainfall for the past 30 years was conducted to

- Develop a typical five-year rainfall period for evaluating combined sewer overflow (CSO) control alternatives for San Francisco.
- Develop design storms for evaluation of the system during discrete storm events.

Analyses were conducted using the hourly rainfall recorded at the NWS rain gauge located in San Francisco. Rainfall statistics were developed for each storm event during the available period of record (1907–2005) including total storm depth, peak intensity, and duration. Annual rainfall statistics including total annual rainfall depth and number of storms were also determined for the years 1907–2004 (complete data for 2005 were not available at the time of this analysis). Storm depth is important for CSO control measures that involve storage, while peak intensity controls the occurrence of individual CSOs.

The procedure to develop the typical rainfall period involved selecting a five-year span that closely resembles the long-term average in terms of the number and distribution of storm events. Individual storm events were then added and subtracted so that the resulting five-year period matched average trends.

The first step in this process was defining the long-term statistics that the typical period should match. The selected five-year period was tailored to match rainfall averages for the past 30 years instead of the total period of record to take into account factors such as climate change and El Niños. The period selected is 1977 through 1981.

For the design storms, an analysis was conducted based on the past 30 years of rainfall recorded at the NWS rain gauge. IDF curves were developed based on information provided by the state climatologist's office, and an extrapolation was used to show the three-month, six-month, and one-year return period events. The resulting individual design storms were compared to the 1941 IDF curve for the five-year design storm currently used in the city of San Francisco for sewer system design. The current 1941 IDF curve has higher peak intensity for very short times (5–10 minutes) and shows lower overall volume compared to the new IDF curve developed as part of this analysis.

Additional analyses were conducted to assess the spatial variation in rainfall throughout the city and to ascertain if the design storms required adjustment for use in different parts of the city. Annual rainfall, storm depth, and storm intensity at the various gauges from the San Francisco Hydraulic–Hydrologic Data Acquisition Rain (SFHHDAR) network were compared to the rainfall recorded at the NWS rain gauge. It was found that the NWS rain gauge consistently showed higher numbers than those recorded at other city rain gauges. To remain conservative, it is proposed that no factor be applied to adjust for spatial variation.

For drainage design work, IDF curves were developed and compared to those currently in use by the city (1941 IDF curve for five-year design storm) and to those developed for the 1973 Wastewater Master Plan. Figure 19-4 and Table 19-1 show the comparison of the 1941 IDF curve and the new IDF curve for the five-year design storm, while Figures 19-5 and 19-6 show comparisons of the 1973 Wastewater Master Plan IDF curves for the 10-year and 25-year design storm, respectively, with the new IDF curves. For greater durations, the recent analysis yields comparable or larger intensities than the 1941 and 1973 hyetographs but shows lower total precipitation for those events.

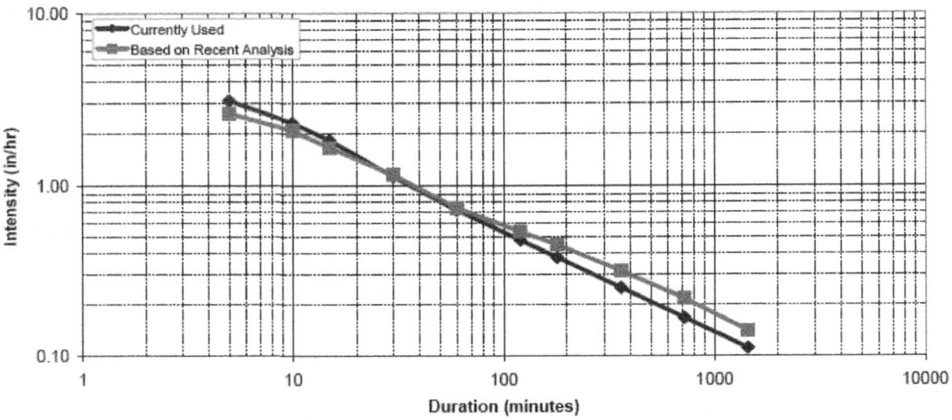

Figure 19-4. Comparison of 1941 and new IDF curves for the five-year return period

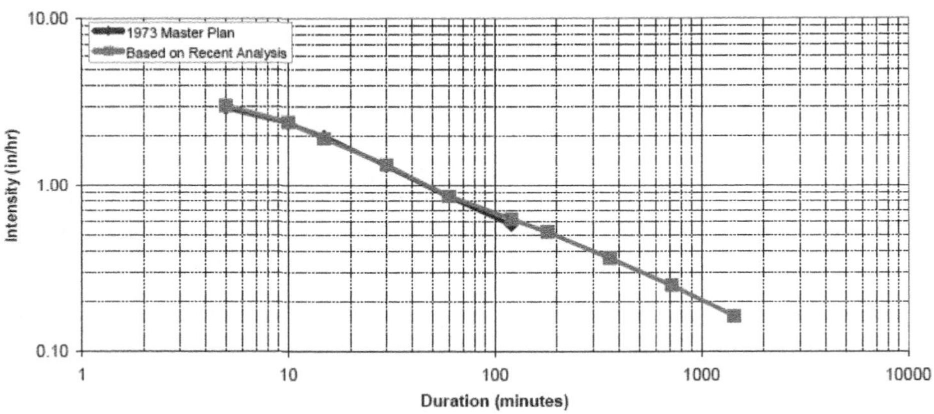

Figure 19-5. Comparison of 1973 and new IDF curves for the 10-year return period

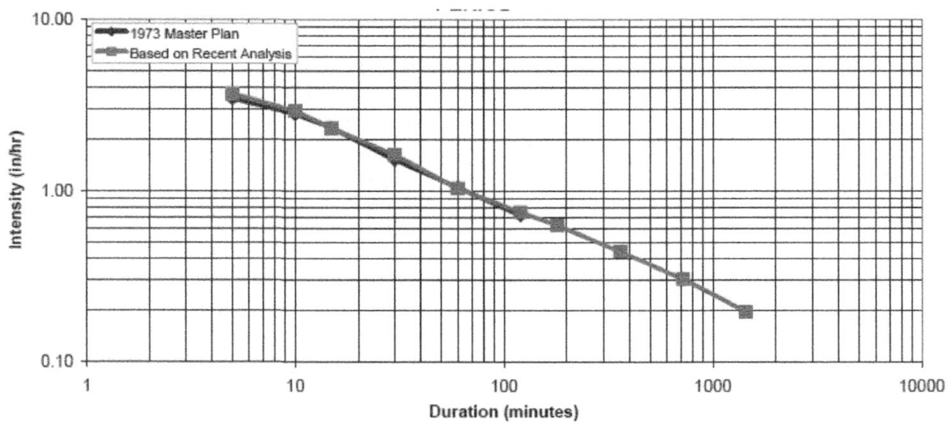

Figure 19-6. Comparison of 1973 and new IDF curves for the 25-year return period

Table 19-1. Comparison between 1941 and new five-year design storm

Duration	5Y Intensity (in/hr)	
	1941	New
5 Min	3.13	2.64
10 Min	2.32	2.10
15 Min	1.84	1.67
30 Min	1.15	1.16
1 Hr	0.73	0.74
2 Hr	0.48	0.54
3 Hr	0.38	0.45
6 Hr	0.25	0.32
12 Hr	0.17	0.22
1 Day	0.11	0.14

The five-year design storm from the new analysis could be considered conservative; however there are significant implications in using it instead of the current design storm (1941 five-year IDF curve). For example, if the five-year design storm from the new analysis is used, it potentially changes the "level of service" provided, which has system-wide cost, design, and regulatory implications. Therefore, until more accurate projections are generated from regional climate models for the San Francisco area, it is recommended that the current design storm continue to be used for sizing and design of collection system infrastructure.

It is widely recognized that there are global (and regional) climate changes caused by human and natural occurring activities and that these changes will continue at a rate faster than experienced in human history. Therefore, the potential impacts of climate change on San Francisco's sewer system should be investigated and updated regularly (every five years) and the findings incorporated in the planning and design of new infrastructure.

CONCLUSION

Global climate change is already occurring. Sea levels are rising, precipitation patterns are changing, snowlines are rising, and spring runoff from snowmelt is occurring earlier. Scientists agree that the impacts are likely to accelerate over the next couple decades; however, there are opportunities to mitigate climate change by reducing GHG emissions. Water quality agencies need to adopt more integrated adaptive and mitigative management strategies starting with the American Water Works Association recommendations.

RESOURCES FOR WATER AND WASTEWATER INDUSTRY

The following list, thought not comprehensive, provides names and locations of institutions that have active groups studying climate change and the projected impacts.

- Lawrence Berkeley National Laboratory, Berkeley, Calif.
- Lawrence Livermore National Laboratory, Livermore, Calif.
- National Center for Atmospheric Research, Boulder, Colo.
- National Oceanic and Atmospheric Administration's National Climatic Data Center, Asheville, N.C.
- Pacific Institute for Studies in Development, Environment, and Security, Oakland, Calif.
- University of California, San Diego—Scripps Institute of Oceanography, San Diego, Calif.

REFERENCES

American Water Works Association. 1997. "Climate Change and Water Resources." Committee Report of the Public Advisory Forum. *Journal AWWA*, Vol. 89, Issue 11, pp. 107–110. Denver, Colo.

California Energy Commission. 2005. "California's Water–Energy Relationship" Final Staff Report. Sacramento, Calif.

Cazenave, A., and R.S. Nerem. 2004. "Present-day sea level change: Observations and causes." *Reviews of Geophysics*, Vol. 42, RG3001. Washington, D.C.: American Geophysical Union.

Delmotte, and J. Jouzel. 2005. "Stable Carbon Cycle–Climate Relationship During the Late Pleistocene." *Science*. 310. 1313–1317. Washington, D.C.

Electric Power Research Institute (EPRI). 2002. "Water and Sustainability: U.S. Electricity Consumption for Water Supply and Treatment—The Next Half Century" (Volume 4). Palo Alto, Calif.

Mount, J. 2007. "Sea Level Rise and Delta Planning." Memorandum to Blue Ribbon Task Force, Independent Science Board. San Francisco, Calif. http://www.deltavision.ca.gov/BlueRibbonTaskForce/April2008/Item2_Attachment1.pdf.

Kharin, V.V., and F.W. Zwiers. 2000. "Changes in the Extremes in an Ensemble of Transient Climate Simulations with a Coupled Atmosphere–Ocean GCM." *Journal of Climate*, 13, 3760–3788. Victoria, B.C., Canada.

Kharin, V.V., and F.W. Zwiers. 2005. "Estimating extremes in transient climate change simulations." *Journal of Climate*, 18, 1156–1173. Victoria, B.C., Canada.

Kharin, Viatcheslav V., Francis W. Zwiers, X. Zhang, and G.C. Hegerl. 2007. "Changes in temperature and precipitation extremes in the IPCC ensemble of global coupled model simulations." *Journal of Climate*, 20, 1419–1444. Victoria, B.C., Canada.

Meehl, G.A., T.F. Stocker, W.D. Collins, P. Friedlingstein, A.T. Gaye, J.M. Gregory, A. Kitoh, R. Knutti, J.M. Murphy, A. Noda, S.C.B. Raper, I.G. Watterson, A.J. Weaver, and Z.-C. Zhao. 2007. "Global Climate Projections." *Climate Change 2007: The Physical Science Basis*. Contribution of Working Group I to the Fourth Assessment Report of the Intergovernmental Panel on Climate Change. S. Solomon, D. Qin, M. Manning, Z. Chen, M. Marquis, K.B. Averyt, M. Tignor, and H.L. Miller (eds.), Cambridge, U.K., and New York: Cambridge University Press.

Siegenthaler, U., T.F. Stocker, E. Monnin, D. Lüthi, J. Schwander, B. Stauffer, D. Raynaud, J.-M. Barnola, H. Fischer, V. Masson-Delmotte, and J. Jouzel. 2005. "Stable Carbon Cycle–Climate Relationship During the Late Pleistocene." *Science*. 310. 1313–1317. Washington, D.C.

Watt, W.E., D. Waters, and R. McLean. 2003. "Climate Variability and Urban Storm Water Infrastructure in Canada: Context and Case Studies." Toronto–Niagara Region Study Report and Working Paper Series, Report 2003-1. Waterloo, Ont.: Meteorological Service of Canada.

White, N.J., J.A. Church, and J.M. Gregory. 2005. "Coastal and global averaged sea level rise for 1950 to 2000." Geophysical Research Letters, 32, L01601. Washington, D.C.: American Geophysical Union.

Originally a 2008 AWWA Utility Management Conference proceeding.

– 20 –

Climate Change: Charting a Water Course in an Uncertain Future

Michael J. Wallis, Michael R. Ambrose, and Clifford C. Chan

Climate change is a growing threat to our entire planet. It threatens communities that have enjoyed plentiful water supplies for decades and that have planned their water systems on the basis of historical water supply records. Climate change science is more convincing than ever, and although the full effect of climate change has not been felt, the warning signs of change are clear. Hot spells are lengthening, sea levels are rising, less snow blankets western mountains, and rising river temperatures pose a threat to our natural resources. The path our predecessors followed to build our water supply and infrastructure, the path we learned to follow in turn, may not be the way forward. For today's Western water managers—who inherited elegantly engineered systems built with confidence that winter snows bring spring runoff to fill reservoirs—difficult decisions loom about how to best provide for the needs of future generations. Water managers can no longer rely on stochastic hydrology models using historical records and must adapt to meet a different, undefined future.

Managing water resources in a warming climate demands water utilities' attention today, even though climate change forecasts and probabilities continue to evolve in the scientific community. The increasing certainty of a warmer planet creates little of the specificity water utilities need to plan for the future. It may be years before reliable predictive models of regional environmental impacts are available, but waiting for those models is impractical. Water utilities must judge their local vulnerability to climate change and develop mitigation and adaptation strategies flexible enough to accommodate uncertain climate change predictions.

Water utilities can learn together how to plan for, prepare for, and mitigate climate change. The East Bay Municipal Utility District (EBMUD) of Oakland, Calif., has developed mitigation and adaptation strategies to deal with the changing climate and its effects on water resources.

THINK GLOBALLY ABOUT CLIMATE CHANGE

The recently released Fourth Assessment Report (AR4) by the Intergovernmental Panel on Climate Change asserts that the warming of the climate is unequivocal and assigns a better than 90% probability that the observed warming is human-caused (IPCC 2007). Significant findings for water utilities are:

- Two greenhouse gases (GHGs), atmospheric carbon dioxide (CO_2) and methane, exist at higher levels today than in the past 650,000 years. Eleven of the past 12 years (1995–2006) are among the warmest on record, and in the past 100 years, global average temperatures increased approximately 1°C.

Figure 20-1. EBMUD uses available capacity at its wastewater treatment plant to convert trucked-in nonhazardous waste (such as food processing waste and wastewater sludge) to methane, which is used to generate renewable energy

- For the western United States, temperatures could rise 2–7.5°C by the end of this century. Precipitation will fall less often as snow and more often as rain, and storm tracks are projected to move steadily northward, making central California drier than it is today. The snow-covered area in the mountains has decreased by approximately 5% over the past 40 years, and the spring runoff now occurs earlier in the year.
- Sea level has risen approximately 6 in. over the past century and is projected to rise another 0.6–1.9 ft by the end of the century. New data on the melting of the Antarctic and Greenland ice sheets suggest that these figures could be low.
- Heat waves, defined as at least five consecutive days with a maximum temperature higher than the average by at least 5°C, have become more frequent. Heat wave durations are projected to increase to 85 days, and the growing season to lengthen 19–28 days.

As gloomy as all that sounds, some research indicates that GHG emissions continue to outpace the worst-case scenario in AR4.

ACT LOCALLY TO UNDERSTAND CLIMATE CHANGE

EBMUD provides an average of 215 mgd of water to more than 1.3 million customers on the east side of California's San Francisco Bay. Almost all of EBMUD's water supply originates in a 577-sq-mi watershed in the Sierra Nevada. EBMUD's two primary water supply reservoirs—Pardee and Camanche—located on the Mokelumne River, provide water supply, flood protection, resource management, recreation, and hydropower. Water travels through three aqueducts that stretch 90 mi from the Sierra foothills across the Sacramento/San Joaquin River Delta) and into the San Francisco Bay area.

EBMUD relies on snowmelt for its water supply, and predictions of dwindling snowpack and reduced precipitation challenge the water supply planning and operational assumptions. EBMUD's water transmission system crosses the earthquake- and flood-prone delta aboveground for several miles. The delta's aging levees have made national news because they are as vulnerable to failure as the levees in the New Orleans, La., area that were devastated by Hurricane Katrina. The sea-level rise that is projected would exacerbate those vulnerabilities.

EBMUD's concerns about the effects of climate change are shared by its customers. In a recent customer survey, nearly 75% of respondents said climate change will be a problem for local water supplies in the next 50 years. When asked how concerned customers were about the effects of climate change on the availability of water, 46% rated it a "highest concern."

EBMUD has closely followed emerging climate science and has collected information about climate change in its region for years. EBMUD's efforts to grapple with the problem have focused on data gathering, scientific research, and peer consultations. EBMUD staff amassed a lot of information and began applying it to water supply and operational planning. Staff members began presenting key findings to the EBMUD board of directors and to a board-led business forum that brings together key business stakeholders for discussions of emerging water issues once a year.

By mid-2007, EBMUD established a districtwide management approach to define what it needed to know to make reasonable short-, intermediate-, and long-term decisions and coordinate research efforts. EBMUD formed a climate change committee to coordinate research across departments in its 2,000-employee organization and to project climate change effects as the basis for guiding future water supply and infrastructure planning and budgeting. Today this interdisciplinary team meets regularly to guide internal work, coordinate involvement with state and federal efforts, and plan public outreach activities. The team's work is focused on

- keeping current with science and assessing potential effects of climate change,
- reviewing observed data and identifying trends that would indicate the climate in the Mokelumne watershed is changing,
- determining water supply and infrastructure vulnerabilities,
- integrating climate change in strategic planning and budgeting decisions, and
- developing adaptation and mitigation strategies as part of a water supply management plan.

DETERMINE THE VULNERABILITIES OF A WATER UTILITY TO CLIMATE CHANGE

EBMUD's research indicated that a warming climate will significantly affect the water utility. Six key areas of potential vulnerability were identified: water supply, flood management, water demand, sea-level rise, power generation, and water quality. Because predictive models do not exist, EBMUD performed sensitivity analyses to understand the vulnerabilities and resiliency of existing and planned systems. Understanding the vulnerability of EBMUD's system forms the basis of planning adaptation strategies.

Water Supply

Most water utilities in the western United States benefit from a snow-fed system in which snowpack provides "storage," and the gradual runoff in the late spring and early summer replenishes reservoirs and meets peak warm-weather demands. In the past 50 years, the western United States has seen a decrease in the size of the snowpack, and during the past century, the percentage of runoff between April and July has decreased by approximately 10%. A California Department of Water Resources report estimated that with a 5°C rise in temperature, the April 1 snow-covered area in EBMUD's water supply watershed could decrease by as much as 60% (CDWR 2006).

EBMUD has evaluated the effects of a 28% shift in the amount of runoff, on the basis of a 3°C rise in average temperature by 2040, occurring in early spring rather than the historical April–July period. EBMUD determined that by changing the way it operates the system, there would be little effect on water deliveries. In a warming climate scenario with decreased late spring runoff, there would be fewer years in which flood-control releases would be required, but no significant effect on the amount of water carried over to the next water year. Storage in EBMUD's reservoirs exceeds the average annual runoff in the basin, and reoperating the system to capture large volumes of early runoff would reduce the effect of decreased late spring runoff.

Changes in precipitation, however, will have a greater effect on water supply. Storm tracks are predicted to shift northward, which would decrease the average precipitation for central California even though the average precipitation for the entire western United States could remain the same or increase slightly.

Flood Management

Water supply vulnerabilities to climate change are closely related to the effects of climate change on flood management. Most of the infrastructure and operating rules for reservoirs in the western United States were designed on the basis of historical annual snowpack data. Reservoirs that fill from snowmelt typically have well-defined and -regulated flood-control requirements. Water utilities reduce reservoir storage in late fall for flood control, relying on spring runoff to refill the reservoirs. In a changing climate, the balance between flood control and water supply will be more difficult to manage. If too much water is released in the fall and reservoirs do not refill, there may not be enough water to supply customers. However, if too little water is released, there is an increased risk of flooding in downstream communities. Dealing with these conflicting priorities will require changing the operation of the reservoirs.

With predictions of more extreme weather events, EBMUD performed studies looking at flood control effects on the basis of a 3°C rise in temperature. Using data from 1997—the wettest year in recent history as a result of El Niño—the studies showed that the peak water release from Camanche Reservoir would have been approximately 15,500 cfs, far exceeding the actual 5,000-cfs release that occurred. Under this scenario, significant flooding would have occurred in the urbanized areas downstream of the two reservoirs.

Water Demands

A third area of vulnerability is the growth of water demands, despite an aggressive conservation program and strong customer support for water conservation. With a warming climate, the growing season is expected to lengthen, soil moisture content will decrease,

Figure 20-2. Rising sea levels threaten levees in the Sacramento/San Joaquin River Delta. A singular levee breach in 2004, shown here, submerged EBMUD's aqueducts for more than four months.

evapotranspiration rates will increase, nights will be warmer, and more heat waves will occur. All of these effects are expected to increase customer use of water for outdoor irrigation and increase water consumption. As part of a routine update, EBMUD projected its water demands for 2040. It estimated that the warming climate will increase service area water demands by an average of 4% (10 mgd).

Sea-Level Rise

Rising sea levels are a significant risk for water utilities located in coastal areas (such as EBMUD)—and the anticipated effects in California reach far inland. The frequency of flooding is expected to increase dramatically, even with only a 1-ft rise in sea level. A recent report showed that a 1-ft rise changes a "1-in-100" storm-surge flood event into a "1-in-10" storm-surge flood event. In addition to the obvious vulnerabilities that would result from greater frequency of flooding in the San Francisco Bay communities served, EBMUD also identified vulnerabilities to its water transmission infrastructure. The Sacramento/San Joaquin River Delta, which provides drinking water to more than 20 million people, is at risk. Failure of the delta's complex and aging levee system will not only damage low-lying infrastructure, but will lead to a massive intrusion of poor-quality water into the water supply that serves much of California.

Although EBMUD does not divert its water supply from the delta, its three water supply aqueducts cross the delta on elevated supports. Failure of the delta levees could result in catastrophic damage, interrupting water deliveries to its service area for months.

Power Generation

Energy use and hydropower generation will also be affected by climate change. Energy is a key cost driver, and if rising demand escalates prices, there could be serious effects on EBMUD. Hydropower provides revenues that help offset the need for customer rate increases, and changes to water supplies could reduce the ability to rely on this revenue source. The

California Climate Change Center reported in 2006 that a warmer climate would not only increase the demand for energy but also increase the demand for peak energy use 4.1–19.3% by the end of the century (Cayan et al. 2006). In addition, if precipitation decreases or runoff patterns change significantly, hydropower generation may decrease between 10 and 30%. For EBMUD, this would result in a reduction of 18–54 GW·h in energy production out of a current annual average of 180 GW·h.

Water Quality

Climate change is expected to increase the severity of storms, and that, in turn, could increase turbidity levels in raw water supplies. This is an area of climate change vulnerability for EBMUD because its protected water source resulted in treatment systems designed for low-turbidity source water. Experience has shown EBMUD that severe storms can dramatically slow its ability to produce treated water while simultaneously increasing the costs of water production. In addition, increasing water temperature may affect water quality by promoting algae growth and result in increased algal by-products such as taste-and-odor compounds. Increases in water temperature may also affect the water system because of its fishery responsibilities, which include maintaining a "cold-water" pool in reservoirs to manage downstream river temperatures. EBMUD works in cooperation with the State of California Department of Fish and Game to manage a fish hatchery on the Mokelumne River. Temperature management is a vital part of the reservoir operation plans so that EBMUD can provide cold water during fish migration periods.

Integrating Climate Change into Long-Range Planning

Climate change cannot be stopped. Projections show that the climate will continue to warm even if emissions of human-caused GHGs were completely suspended today. As the climate warms, water utilities will need to develop adaptation strategies to cope with these changes. Some of these strategies will take decades to develop. However, waiting to adapt until the climate has warmed is akin to waiting to build lifeboats after a ship has started to sink.

After defining vulnerabilities, the next significant challenge was to integrate this knowledge into long-range planning and ongoing budget decisions. EBMUD continuously works on an array of infrastructure and water supply planning efforts designed to ensure effective operations and stewardship of natural resources. In turn, these studies inform the strategic plan and five-year capital plan, which are key tools in establishing a biennial budget.

Water Supply Planning

EBMUD is currently updating its integrated resource water supply management program (WSMP). The WSMP is a comprehensive evaluation of EBMUD's water supply needs through the year 2040. As part of the WSMP, EBMUD is evaluating numerous water supply portfolios, with components including conservation, water recycling, groundwater banking, interbasin transfers, desalination, and surface water storage. Among other criteria, each portfolio is being evaluated for its carbon footprint. The portfolios being evaluated include one with a low carbon footprint and another that is diversified in order to adapt to future changes, including climate change.

As climate and hydrology change, using past hydrologic data to make projections of future conditions may be invalid. To address this concern, sensitivity analyses are used to understand how various factors will affect water supply deliveries and the frequency and magnitude of water shortages. These factors, in turn, help assess needs to modify plans for future investments in infrastructure and equipment.

For example, because climate change projections show that weather systems will move northward to higher latitudes, EBMUD performed a sensitivity analysis to determine how a decrease in precipitation would affect water supplies. The analysis looked at how a 10–20% reduction in precipitation would affect the water system. The results showed that water deliveries decreased 4–12% during a three-year drought scenario and that by the end of the drought, reservoir storage could drop by as much as 80%.

In addition, a decrease in precipitation would increase the frequency of dry years. On the basis of current hydrology, EBMUD projects dry years in 3 out of 10 years. With a 10% reduction in precipitation, this would increase to 4 out of 10 years, and with a 20% reduction, it would increase to 5 out of 10 years.

Results from research and sensitivity analyses drive the development of adaptation strategies and water supply planning. They help focus attention on research and studies most relevant to the identified climate change–related vulnerabilities of the water system. To guide this effort, EBMUD has integrated climate change into its approach to planning for future programs and expenditures. It has also enhanced emphasis on diversifying its water supply.

Water supply diversification is a key element of adaptation for an agency like EBMUD, which is so reliant on the precipitation that falls in a small geographic area. EBMUD has been pursuing conservation, water recycling, and seeking supplemental water supplies for decades—to ensure a reliable water supply for its customers in the future. Recently, EBMUD partnered with Sacramento County Water Agency to build a regional water supply project that will provide water for EBMUD customers in dry years and the Sacramento region in all years. The project will have the ability to divert up to 185 mgd of water from the Sacramento River, delivering up to 100 mgd to EBMUD customers by accessing water from a separate, larger watershed.

Strategic Planning and Budgeting

In 2008, EBMUD incorporated a climate change strategy as part of its strategic plan. The climate change strategy publicly commits EBMUD to developing and implementing a climate change monitoring and response plan to inform future water supply, water quality, and infrastructure planning. In addition, the strategy mitigates GHG emissions and supports "no regrets" capital investment decisions—those investments that are cost-effective under a variety of climate change scenarios.

With the formation of EBMUD's climate change committee, preparation of a climate change monitoring and response plan and an update to its WSMP, EBMUD is moving forward in developing sound adaptation strategies for climate change and other potential future effects.

MITIGATING THE CLIMATE CHANGE EFFECTS OF A WATER UTILITY

Being "green" and "sustainable" are currently hot topics. Some utilities have been ahead of the current emphasis on sustainability and have worked toward mitigation of their environmental effects for many years. For those without a current internal focus on sustainability, external circumstances may soon force more attention. Customers who are being bombarded with information about corporate sustainability efforts may also want to know how their local water service provider measures up.

In addition, federal and state governments are beginning to require climate change mitigation efforts. California's 2006 Global Warming Solutions Act (AB32) takes steps to manage GHG emissions. AB32 requires California to reduce its greenhouse gas emissions to 1990 levels by 2020 through a combination of regulations and market mechanisms. In the next few years, the federal government will likely pass some form of legislation to slow emissions growth.

States are also entering into regional partnerships to address climate change. One regional agreement, the Western Climate Initiative, was launched in 2007 and established a regional GHG-reduction target shared by seven states. The agreement calls for an economy-wide regional cap-and-trade program. The latest US regional agreement to be negotiated—the Midwestern Regional GHG Reduction Accord—commits another six states to near- and long-term GHG reduction goals under a multisector cap-and-trade system.

Although the particulars for each cap-and-trade system are different, programs are generally designed to reduce a specific type of pollution (in this case GHGs) in a defined geographic region. Under such programs, each polluting entity is issued an emissions permit for every ton of GHG it produces. This is termed its cap. Based on the assumption that it will be cheaper and easier for some to reduce their emissions below their cap, the idea is that these entities will sell off their excess permits to other emitters that cannot easily reduce their emissions. This is termed the trade. Over time the caps become increasingly strict until an overall pollution reduction goal is met.

Whether a water utility chooses to mitigate GHG emissions because of leadership from its board, staff, or the community, or in response to regulatory requirements, it makes sense to prepare by quantifying and understanding current emissions and ways to mitigate them.

Creating an Inventory of Utility-Produced GHGs

Developing an inventory of current emissions using accepted protocols enables utilities to determine what operations cause emissions. A good inventory is the first step in establishing a strategy for mitigating GHG emissions, which often can save money in operating costs for a utility. Water and wastewater utilities are an indirect source of GHG emissions because of their use of energy to transport, treat, and distribute water. The US Environmental Protection Agency estimates 3% of national energy consumption is used for drinking water and wastewater services and that these services emit approximately 45 million tons of GHGs into the atmosphere each year. The California Energy Commission says water-related energy use, including customer uses, consumes 19% of the state's electricity, 30% of its natural gas, and 88 bil gal of diesel fuel every year. Water and wastewater utilities also directly contribute to GHG emissions from their fleets of vehicles for operations and maintenance, and transportation of chemicals and other supplies to support their operations. Wastewater treatment is also a direct source of methane and nitrous oxide.

The AR4 report is optimistic that climate change can be mitigated through a 60–80% reduction in GHG emissions from current levels by 2050. This is a challenging task given expected population growth and demand for resources. According to the AR4 report, emissions of the GHGs covered by the Kyoto Protocol increased about 70% from 1970 to 2004, with CO_2 (the largest source) having risen 80%. The largest growth in CO_2 emissions was from power generation and transportation.

Some actions that may be required to adapt to climate change could result in more energy-intensive processes than in the past. As discussed earlier, climate change could reduce water quality and therefore require more treatment. Supply reductions could require groundwater storage and pumping or pumping water from more distant watersheds. Water utilities must consider the objective of reducing GHG emissions while exploring ways to meet water supply, storage, and treatment needs in ways that best balance cost, reliability, and environmental consequences.

Figure 20-3. Water utilities are large consumers of energy, so one of the key ways to mitigate greenhouse gas emissions is through responsible energy management. EBMUD has an aggressive program to minimize energy use, including use of renewable resources.

Mitigation Opportunities for Utilities

Because water utilities are large consumers of energy, one of the key ways to mitigate GHG emissions is through energy management programs. Additionally, many utilities have resources (land, water, biogas) that can be used for mitigation. Land can be reforested for carbon sequestration or used to construct renewable energy projects (wind and solar power). Water sources can be used for hydropower generation. Conservation can be used to reduce the demand on potable water sources.

Water utilities use a significant amount of energy to move and treat water. Mechanical equipment such as pumps and valves are used in these processes. Equipment should be right-sized for the application to minimize wasted energy. It should also be properly monitored and maintained to ensure peak efficiency over its useful life. The best technique to save energy and mitigate emissions is to turn off unneeded equipment or processes to avoid using energy.

Water conservation programs can also be considered energy-saving when the energy cost of water is considered. Water conservation can be especially helpful in mitigating emissions in areas in which water is conveyed long distances.

Wastewater agencies' GHG emission mitigation programs can consider both energy conservation and using biogas as an energy source. Anaerobic digestion produces biogas (methane and CO_2) and biosolids. The biogas can be used for process heating, onsite electrical generation, and other uses. Electrical energy produced from biogas is considered renewable because the emissions from this combustion are part of the natural carbon cycle, unlike those from fossil fuels.

EBMUD'S GREENHOUSE GAS MITIGATION EFFORTS

EBMUD's focus on greenhouse gas mitigation is driven by a long-standing emphasis on environmental responsibility. Its board of directors believes strongly in environmental leadership.

GHG Inventory

EBMUD conducted its first GHG inventory for 2005. EBMUD caused 53,700 metric tons of GHG emissions that year—a 10% reduction from estimated year 2000 emissions. Most of its 2005 emissions (41,459 metric tons) were from indirect sources because of electrical energy use. EBMUD's total direct GHG emissions in 2005 were 12,242 metric tons. Mobile combustion sources (vehicles, portable equipment) produced 63% of EBMUD's direct GHG emissions. Natural gas–powered equipment produced the remaining 37% of direct GHG emissions. De minimis sources (< 5% of total) include propane forklifts (20 metric tons), stationary diesel generators (347 metric tons), and stationary propane generators (3 metric tons).

It's always helpful to be able to put numbers into context by looking at operations in comparison with other agencies. One of the early steps EBMUD took to help plan how to manage its effect on climate change was to join the California Climate Action Registry. The registry is a private, nonprofit organization that serves as a voluntary clearinghouse. It develops and promotes credible, accurate, and consistent GHG reporting standards and tools for organizations to measure and monitor their GHG emissions consistently across industry sectors and geographical borders.

EBMUD is also joining the Chicago Climate Exchange (CCX), the world's first and North America's only active voluntary, legally binding trading system to reduce emissions of all six major GHGs. The CCX is a cap-and-trade system whose members make a legally binding emission reduction commitment. Members are allocated annual emission allowances in accordance with their emissions baseline and reduction schedule. Members who reduce beyond their targets have surplus allowances to sell or bank; those who do not meet the targets comply by purchasing CCX carbon financial instrument® contracts. EBMUD has committed to minimum 2010 emission reductions of 6% (1.5% per year starting in 2007) below 2000 levels.

Energy Reduction Projects

EBMUD has an aggressive program to minimize energy use through careful management of facilities, off-peak pumping schedules, and use of renewable resources. The district adopted a policy to encourage installation of renewable energy by establishing criteria for investments that are cost-neutral over the life of the project. To date, two photovoltaic systems have been installed, one at a maintenance facility and the other at a water treatment plant. The systems offset about 10% of the facility load, equivalent to the electrical demand of more than 60 homes for one full year.

Energy Generation Projects

EBMUD's resource-recovery program uses available capacity at its wastewater treatment plant to convert trucked-in nonhazardous waste (such as food-processing waste and wastewater sludge) to methane, which is used to generate renewable energy. The trucked waste program provides an environmentally responsible and economical disposal alternative for a variety of liquids and food-processing solids.

Renewable energy production has more than doubled from 2 to 4.5 megawatts over six years. This facility generates enough energy to power the equivalent of 2,500 homes at one time and supplies nearly all power required by the plant. Designs in progress will add two 4.5-megawatt turbines to generate additional power that can be sold on the electric grid. This power is considered green, which allows EBMUD to sell renewable energy credits or carbon credits.

As noted earlier, EBMUD's 40-megawatt hydropower facilities produce an average of 180 GW·h per year of renewable energy from water supply operations in the Sierra. This exceeds EBMUD's annual energy use of 145 GW·h. Hydropower is a green energy source that results in renewable energy credits and generates revenue.

Reductions in Use of Fossil Fuels

In 2001, EBMUD started a pilot program to test daily use of gas–electric hybrids in its vehicle fleet. By 2004, 90% of EBMUD's vehicle fleet was converted to hybrids. Today, all of EBMUD's 59 sedans are hybrids. This saves 12,000 gal of gasoline and reduces CO_2 emissions by 103 metric tons each year. EBMUD also has an award-winning employee transportation program that offers incentives to encourage employees to carpool, take mass transit, or bike or walk to work. This program enables EBMUD to further reduce fossil fuel consumption.

CONCLUSION

EBMUD's work on assessing its vulnerabilities to climate change, integrating climate change considerations into the organizational structure and into long-range water supply planning and budget development programs, and developing greenhouse gas mitigation strategies illustrates how water utilities can be proactive in their efforts. Although the utility learned a lot about the science and feels confident that it has established a solid management approach, its strategy, by necessity, will continue to evolve. EBMUD is pursuing several projects to further reduce GHG emissions, including in-conduit hydroelectric power and additional photovoltaic-energy-generation projects.

Climate change is one of the greatest challenges facing the planet today. That challenge will require new approaches to planning and managing water systems and water supplies. As professionals who inherited the responsibility to deliver reliable water supplies, protect public health, and preserve natural resources, EBMUD must chart a new course for the industry. EBMUD's response to climate change will profoundly influence the public's trust in water supply management and infrastructure planning.

REFERENCES

Cayan, D., Lynd Luers, A., Hanemann, M., Franco, G., and Croes, B. 2006. Climate Change in California: An Overview. California Climate Change Center, University of California, Berkeley.

CDWR (California Department of Water Resources). 2006. Progress on Incorporating Climate Change Into Management of California's Water Resources: Technical Memorandum Report. CDWR, Sacramento, Calif.

IPCC (Intergovernmental Panel on Climate Change). 2007. Climate Change 2007: Synthesis Report. Contribution of Working Groups I, II, and III to the Fourth Assessment Report of the Intergovernmental Panel on Climate Change (R.K. Pachauri and A. Reisinger, eds.). IPCC, Geneva.

Originally printed in Journal AWWA *June 2008.*

– 21 –

Implementation of Climate Adaptation and Mitigation Strategies for Drinking Water Production in the Netherlands

G.A. (Gerard) van den Berg, J. (Jos) Frijns, and J.J.G. (Gertjan) Zwolsman

INTRODUCTION

Within the EU-funded TECHNEAU project, 10 major global trends have been recognised that will affect the water industry. Relevant trends for drinking water supply include climate change, urbanisation, and energy use and costs (Segrave et al. 2007). Concern for the impact of climate change on water systems and associated ecosystems is increasing. Climate change affects water system dynamics through temperature changes, changes in precipitation patterns (more extreme floods), an increase in summer evaporation, and a decrease in snow-fed river runoff (extended droughts). These changes will have major impacts on the management of river deltas and river catchments in general (EEA 2001). The capacity of water systems to cope with changing climate dynamics depends strongly on soil characteristics, presence of inundation areas, and runoff management.

Generally, climate change impact assessments focus on risks of flooding of heavily populated low-lying areas and hydrodynamic effects on navigation, agriculture, and energy production (EEA 2007). Moreover, the frequency of occurrence of low and very low river discharges increases, as calculated for the river Rhine (Shabalova et al. 2003). Extended droughts and associated water quality effects (e.g., Van Vliet and Zwolsman 2008) are lesser recognized so far, although they may have huge economic effects on agricultural and drinking water production.

Production of reliable drinking water is of major importance for society as drinking water is the number one requirement for life. Therefore, both the reduced availability of freshwater and the decreasing water quality are of growing concern to almost all countries of Europe. Water utility companies and governments are mainly focused on solving the problems of today, i.e., renovating existing water production facilities. However, the vulnerability of drinking water production demands a careful evaluation of future threats to our natural water resources.

Kabat et al. (2005) state that climate change should not only be seen as a threat, but could also create opportunities for large-scale innovations. In this paper, the results of a feasibility study on innovative adaptive strategies for drinking water production in the Netherlands are described.

WATER STRESS FOR DRINKING WATER PRODUCTION

Drinking water supply companies have always been challenged by increased variations in water availability, water quality, production, distribution, and water demand. Climate change makes the situation more dynamic, however. The tendency toward more extremes will result in a lowering of groundwater tables, a reduction of the capacity of water reservoirs, as well as water quality effects.

Groundwater and surface water dynamics are expected to change significantly due to climate change. In some eastern parts of the Netherlands, the groundwater table may be lowered tens of centimetres during dry periods (Cirkel et al. 2006). This may have serious consequences for the use of groundwater as a source for drinking water production given the effects on terrestrial ecosystems. The anticipated effects are even more pronounced when the increase in agricultural and drinking water demand during periods of water scarcity is taken into account.

The trend toward more frequent and extended droughts is expected to cause problems for drinking water supply companies with respect to contamination of surface water systems. This may ultimately result in additional and elongated intake stops. Low river discharges are generally accompanied by less dilution of the pollution load (Van Vliet and Zwolsman 2008), resulting in higher concentrations of contaminants. This is clearly illustrated in Figure 21-1. Methyl tert-butyl ether (MTBE), the most widely used oxygenate in gasoline, shows a seasonal pattern in the river Meuse with high concentrations at low discharge. Given the absence of official quality standards for drinking water and intake of surface water, drinking water supply companies commonly use an indicator value of 1 μg/L. When predicting the effects of more pronounced droughts on river quality, substitute products should also be taken into account. An example of this is the expected shift from MTBE to ETBE (ethyl tert-butyl ether) (Deeb et al. 2003).

Similarly, reduced dilution of pollution loads is expected for compounds that are emitted to the surface water at a constant rate, e.g., via effluents of industrial or domestic wastewater treatment plants (WWTP). Several emerging compounds, such as pharmaceutical residues and endocrine disruptive compounds, may consequently be observed in higher concentrations in surface water. As these compounds are considered unfavourable in drinking water, quality standards should be set to actively control their emissions to surface water. Other relevant parameters, like oxygen and algae levels (including the toxic cyanobacteria), are also significantly influenced by extended droughts and increased temperatures. Cyanobacteria are already considered a potential threat to drinking water production in the Netherlands (Hoogenboezem et al. 2004). Floods will temporarily be accompanied by increased suspended matter levels and associated contaminants in rivers (Doomen et al. 2007), giving rise to intake problems. During 2006, the Dutch water utility Evides already had to close down their direct intake of river Meuse surface water for 47 days due to the presence of high levels of suspended matter during floods. In addition, high rainfall periods are frequently accompanied by overflows of sewerage systems negatively affecting the quality of the receiving waters.

Groundwater in the coastal region of the Netherlands is very vulnerable to salinisation, due to maritime (related to recent North Sea intrusion and upconing of ancient seawater) and continental (related to agriculture, evaporation, bank infiltration, artificial recharge, and local pollution) causes (Stuyfzand 1996). For that reason, surface water is nowadays the main source for drinking water production in the western (low-lying) part of the Netherlands, including the cities of Amsterdam, Rotterdam, and The Hague. Different treatment techniques are applied to produce drinking water, in which dune infiltration plays a crucial role. These include abstraction from the rivers Meuse and Rhine, temporary storage in large natural and artificial drinking water reservoirs, extraction after artificial dune infiltration (AStR, Aquifer Storage transport and Recovery; see Figure 21-2) and bank filtration. Dutch drinking water utilities generally apply (a combination of) advanced treatment technologies (e.g., UV disinfection, hydrogen peroxide, ozone, nanofiltration, reverse osmosis) to continuously produce high-quality drinking water from (contaminated) surface water.

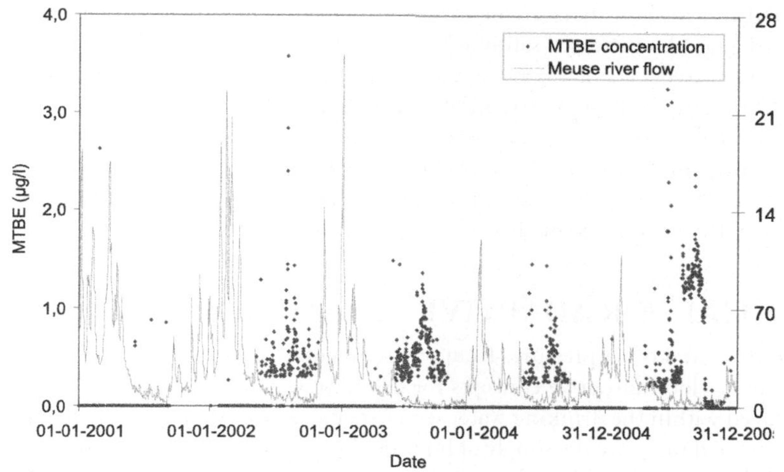

Figure 21-1. MTBE concentrations and river discharge of the river Meuse at the measuring station Eijsden during the period 2001–2005 (data from www.aqualarm.nl and www.waterbase.nl)

Figure 21-2. Infiltration pond in the dunes

Recently, changes in salinisation of the Rhine–Meuse estuary in the Netherlands within a time span of 50 years have been analysed (Doomen and Van den Berg 2005). This study shows that effects of climate change cannot be predicted reliably without a systematic knowledge of the effects of other major developments on salinisation within the delta, such as the return of natural estuarine conditions, measures to ensure freshwater supply to regional water systems (given agricultural functions), and dredging to increase the navigation depth. Future salinity gradients in the estuary are mainly determined by autonomous developments related to climate change. An increased salinisation of the Rhine–Meuse estuary is caused by a combination of increased water levels at sea, more frequent and heavier winds from sea, and lower river discharge. This will negatively influence the public (and agricultural) water supply in the area, especially during long drought spells when demand for fresh water is high. Increased salinisation will also threaten the salinity of bank filtrate abstracted close to the North Sea. Locally, the yearly average chloride concentration may increase substantially by salinisation, resulting in a decreased functionality of the abstraction.

POTENTIAL FOR ADAPTIVE STRATEGIES

Climate change and other pressures from environmental and spatial developments stress the necessity to develop innovative concepts for drinking water production. It is now generally acknowledged within the drinking water sector that adaptive strategies need to be developed and implemented to secure the supply of high-quality drinking water in the next decades. We distinguish between two different approaches, which are either demand or supply oriented. In this paper we describe four types of adaptive strategies for drinking water utitilities. These are summarised in Table 21-1 and further described in the text.

Demand strategies aim at changing the customer awareness toward the use of drinking water. Measures at the demand side ask for a sectoral approach, preferably at river-basin scale or regional scale. Such measures include education and communication to emphasise a reduction in drinking water consumption (especially to lower peak demands), introduction of progressive water pricing or the setting of specific regulations restricting or banning the use of drinking water (for e.g., car washing or lawn watering) during periods of water scarcity.

Stimulating the sustainable (and secure) use of additional water sources, such as filtered rainwater, at the household level may also increase customer awareness. Discussion and research focus mainly on the microbiological risks associated with such systems. A study by

Table 21-1. Overview of promising adaptive strategies to cope with the effects of climate change by drinking water utilities

Type	Adaptive Strategy	Examples
Demand oriented	Raising customer awareness	Education and communication
		Progressive water pricing
	Shaving off-peak demands	Restrictions or bans for certain applications during periods of water scarcity
		Use of rainwater and stormwater at household level
Supply oriented	Alternative sources concept	Use of sea water
		Use of brackish groundwater
		Use of wastewater effluents
	Multiple sources concept	Combination of individual sources
		Application of "fresh keeper" concept
	Flexible sources concept	Application of Aquifer Storage and Recovery
		Temporal use of rainfall
		Reuse of effluent from WWTPs

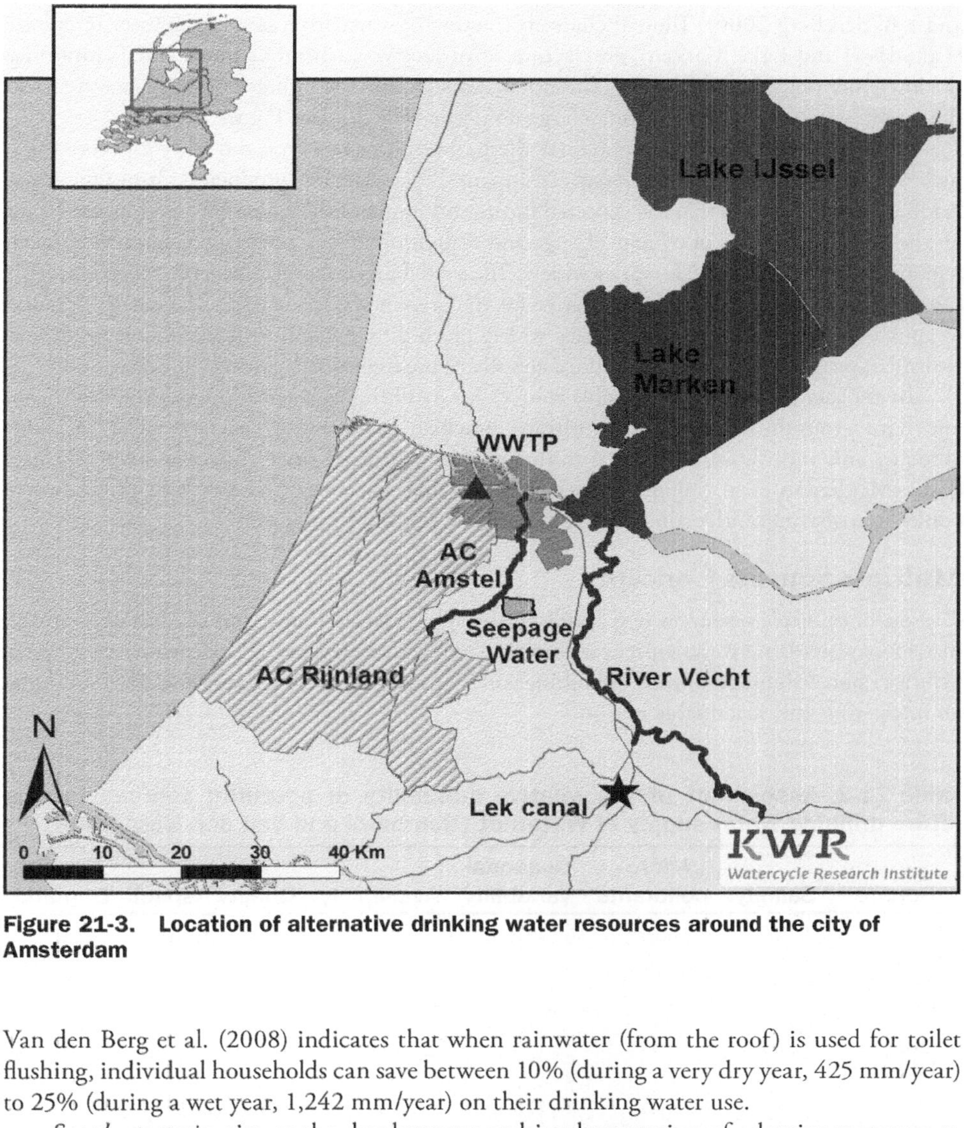

Figure 21-3. Location of alternative drinking water resources around the city of Amsterdam

Van den Berg et al. (2008) indicates that when rainwater (from the roof) is used for toilet flushing, individual households can save between 10% (during a very dry year, 425 mm/year) to 25% (during a wet year, 1,242 mm/year) on their drinking water use.

Supply strategies aim at the development and implementation of adaptive measures to manage the quantity and quality of available water sources for drinking water production.

Alternative Sources Concept

Worldwide, studies have focused on the exploitation of sea water, precipitation, brackish groundwater and domestic wastewater, using innovative treatment such as membrane technology, as alternatives to the commonly applied water resources (fresh groundwater and surface water). Some concepts are already widely applied in specific regions. Desalination, for example, is a commonly applied technique in densely populated (semi)arid areas, such as the Mediterranean countries. Drawbacks for application in other areas are the high investment costs and energy consumption in comparison with presently applied treatment technologies.

The Dutch water cycle company Waternet has investigated the potential use of alternative resources for the production of drinking water from river Rhine (Lek canal) water (Bernhardi and Van den Berg 2006). These include abstraction of water from large freshwater reservoirs (Lake IJssel and Lake Marken), abstraction of surface water from artificial canals (draining the low-lying polders) surrounding the city of Amsterdam, use of brackish seepage water, and the use of wastewater effluent from a large domestic WWTP (see Figure 21-3).

Table 21-2 shows that each potential alternative source has typical quality aspects, which may require redesign of the present treatment procedure for producing drinking water. Additionally, it is shown that the expected future continuity of these alternative resources, based on the combined effects of climate change and realisation of water management goals at local, regional, and river basin scale, varies widely. The availability of some of the proposed sources is too small to become the single alternative for the present water resource. The energy demand of applying alternative sources will vary widely given differences in infrastructural needs and treatment procedures needed to produce and distribute high-quality drinking water.

In another project, eight possible sources for drinking water production (shallow, deep, and hard groundwater, river bank filtrate, brackish groundwater, surface water, WWTP effluent, and rainwater) have been evaluated for treatment (investment and operational) costs (Meuleman et al. 2004). They concluded that rainwater and brackish groundwater could be cost-competitive alternatives for current drinking water sources.

Multiple Sources Concept

The use of multiple resources (e.g., groundwater plus surface water) and the infrastucture to distribute water from these sources within the supply area, decreases the dependency of water utilities upon a single resource. A combination of resources, as shown in Table 21-2, could be an interesting and sustainable option.

Table 21-2. Assesment of the relative suitability of potential sources for the future drinking water supply of Waternet (Bernhardi and Van den Berg 2006)

Source	Salinity	Micro-pollutants	Seasonal Variability	Availability	Quality Aspect	Energy Demand
River Rhine*	0	0	High	+	Temperature	0
Lake IJssel	0	+	High	+	Algae	—
Lake Marken	0	+	High	+	SPM	0
AC Rijnland	—	—	High	+	Chloride	—
AC Amstel	—	—	High	—	Chloride	—
River Vecht	—	—	High	—	Effluents	—
Seepage water	—	+	Low	0	Chloride	—
WWTP	0	—	Low	+	Micropollutants	—

SPM = suspended matter; AC = artificial canals; WWTP = wastewater treatment plant; — = very poor; – = poor; 0 = neutral; + = favourable.

*present source (Lek canal)

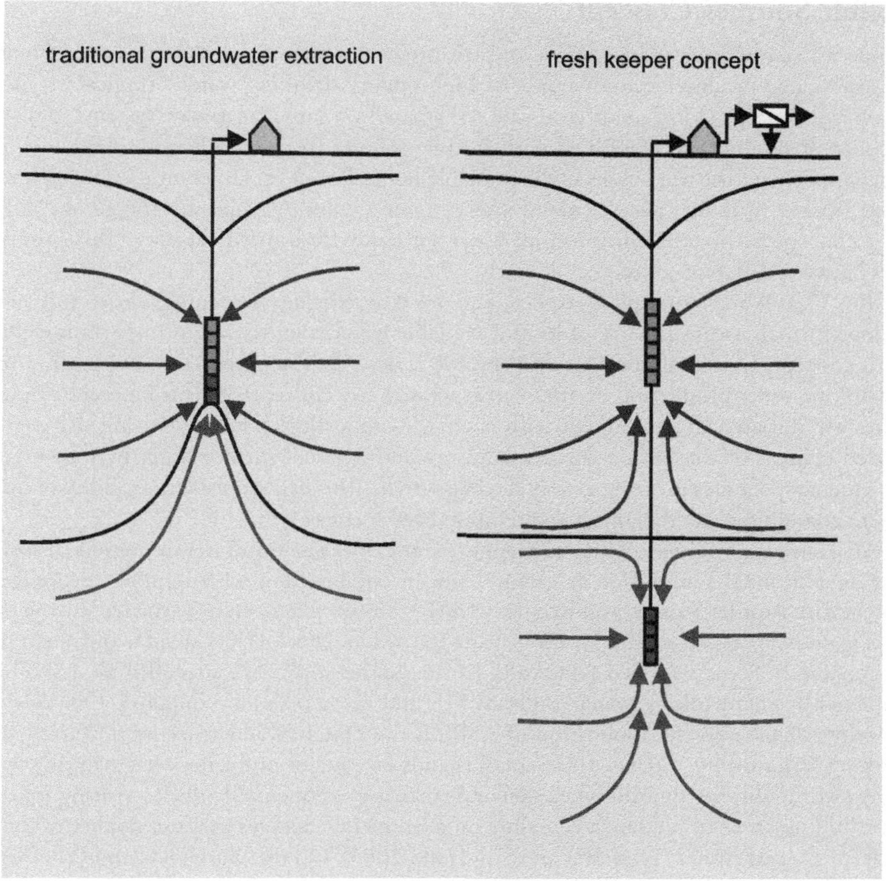

Figure 21-4. Comparison between traditional single-depth groundwater extraction (with upconing of brackish groundwater (left) and the "fresh keeper" concept (right)

A few years ago, the "fresh keeper" concept was developed in order to prevent salinisation of fresh groundwater lenses which are overlying brackish groundwater (Kooiman et al. 2004). This concept has advantages over the traditional single-depth extraction of either freshwater or brackish groundwater. Additional deep wells change the flow direction of brackish groundwater, protecting existing freshwater wells from upconing of brackish groundwater (see Figure 21-4).

The brackish groundwater extracted can be utilised for drinking water production using reversed osmosis, thus improving the capacity of well fields. The technique makes optimal use of the existing treatment and distribution infrastructure. Pilot studies are now initiated to test this concept under field conditions. The legal and hydrogeochemical aspects of membrane concentrate disposal and choice of appropriate target aquifers needs special attention (Stuyzand et al. 2007). In two Dutch pilots (by Vitens and Brabant Water water supply companies), membrane concentrates (with chlorinity up to 2,000 mg/Ll) are injected in deeper brackish aquifers.

Flexible Sources Concept

Flexible water resources management and treatment is a promising concept for continuous and sustainable production and supply of high-quality drinking water. Application of the Flex Water concept (Meuleman et al. 2004) is based on large treatment systems, designed for constant production, combined with flexible sources to deal with peak demands. This concept optimizes the use of seasonally available natural resources in combination with temporary storage to match peak demand and extreme events. A major advantage of the Flex Water concept is the possibility to link water sources with limited capacity. This promotes shifts between different sources.

The Flex Water concept is expected to be cost efficient in remote areas and newly developed urban areas. The use of locally available water resources is an important element of this concept. De Graaf and Van de Ven (2005) describe the potential use of local rainfall for drinking water production in urban areas (closed-city concept). At this moment, strategic studies are initiated to experiment with such self-supporting systems. During dry periods recycled effluent of wastewater treatment plants and regional surface water may be used as supplementary sources to create a more flexible system. This can reduce vulnerability of urban areas to changing water dynamics and temporal water stress.

An example of the Flex Water concept is the use of excess rainwater or pretreated surface water as additional sources for drinking water in combination with temporary storage in aquifers with Aquifer Storage and Recovery (ASR). Stored water can be extracted during periods of water scarcity. At several locations (e.g., in the US, Israel, UK, Canada, and Australia), this concept is already applied widescale. In the Netherlands, one succesful pilot has been completed by the drinking water utility WML and more pilots are ongoing. One of these new pilots focuses on the application of ASR on the Dutch Wadden Island of Terschelling (drinking water utility Vitens). These small islands face major problems with drinking water supply during summer months when demands increase exponentially due to visiting tourists. Succesful implementation depends mainly on minimising changes in water quality of the injected (pretreated) water (Prommer and Stuyfzand 2005). On the Wadden Islands, pretreated water will be injected in a brackish aquifer, which may induce salinisation.

POTENTIAL FOR MITIGATION MEASURES

In selecting appropriate adaptive strategies, the contribution of drinking water utilities to climate change and possible mitigation measures should also be taken into account.

Climate Footprint

The climate footprint of drinking water has been established by assessing the yearly energy consumption, direct emissions of greenhouse gasses, and indirect emissions (related to chemicals and organisation) of the production and distribution of drinking water in the Netherlands. The climate footprint is expressed in global warming potentials (GWP), i.e., CO_2-equivalents. The result is presented in Table 21-3.

In Figure 21-5, the GWP distribution of the Dutch drinking water utilities is presented. Energy consumption provides for the largest share of the GWP. The energy use is determined by the required pressure increase with pumps and the specific energy consumption of process units, such as UV lamps, saturation in flotation, and ozonation. On average, the energy use for production and distribution of drinking water is 0.47 kWh/m^3 (Vewin 2006).

Table 21-3. Total GWP of drinking water of the Netherlands, 2006

	Amount per Year	Conversion to GWP	Tonnes GWP
Water production	1,210,000,000 m^3		
Energy use			
• Electricity	568,700,000 kWh	0.59 kg CO_2/kWh	**335,530**
Direct emissions			
• CO_2 degassing	10,000,000 kg	1	**10,000**
• CH_4 degassing	1,770,000 kg	21 kg CO_2/kg CH_4	**37,170**
• N_2O ozonation	2,409 kg	310 kg CO_2/kg N_2O	**745**
Indirect emissions			
• Chemicals:			
– sodium hydroxide	8,945 tonnes	0.96	8,585
– ferric chloride	2,287 tonnes	1.15	2,630
– ferrous sulphate	6,448 tonnes	0.11	710
– hydrochloric acid	928 tonnes	0.35	325
– activated carbon	3,709 tonnes	2.8	10,385
• Materials:			
– negligible			0
• Organisation:			
– heating	2,600,000 m^3 gas	1.8 kg CO_2/m^3	**4,680**
	14,400,000 kWh	0.59 kg CO_2/kWh	**8,495**
– transport	5,600,000 L	2.45 kg CO_2/L	**13,720**
Sludge			
• Emissions	0		0
• Transport	1,500,000 L diesel	2.6 kg CO_2/L	**3,900**
Total			436,875

Source: Frijns et al. (2008)

Distribution contributes to 25% of the energy use. These figures, however, vary substantially between drinking water companies. The energy consumption depends on the scale of the production process, the source for water production, the treatment steps applied, as well as the topography and distribution distances.

Direct emissions of greenhouse gases contribute to 11% of the GWP. This is mainly caused by methane emissions from degassing groundwater (methane levels in raw groundwater used for the production of drinking water in the Netherlands range from zero to 52 mg/L CH_4, resulting in a total methane emission of approximately 2,500 tonnes methane per year (Drijver et al. 2006). Indirect emissions result from the use of chemicals and the energy use related with the organisation, i.e., drinking water company (office and transport). The energy required to produce these chemicals (e.g., sodium hydroxide and ferrous sulphate) account to the GWP of drinking water. Here also, the chemical use varies substantially between water companies.

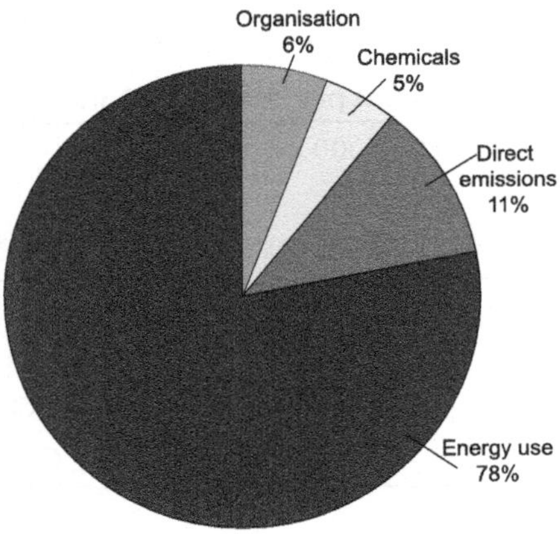

Figure 21-5. **GWP distribution of drinking water production and distribution**

The total carbon footprint of the Dutch drinking water sector is 0.44 million tonnes CO_2-eq. per year. This resembles 0.36 kg CO_2-eq./m^3 drinking water. The GWP of wastewater is about 2.5 times higher than that of drinking water (Frijns et al. 2008). Compared to the total yearly GWP of the Netherlands, 212 million tonnes CO_2-eq. (MNP 2007), the contribution of the drinking water sector is limited (0.2 %). Dutch drinking water utilities, however, are committed to reduce their carbon footprint and to implement mitigation measures.

Mitigation Measures

The energy consumption is mainly related to the applied treatment procedure. As we expect that more advanced treatment will be needed to deal with substances such as hormones and medicine residues in the future, the energy use is likely to increase. Next to the application of more advanced and energy-intensive treatment processes, the likely use of alternative sources (e.g., sea water or brackish groundwater) to deal with the effects of increasing salinisation of available water resources, may also result in higher energy consumption. The challenge thus is to improve the energy efficiency of drinking water production. Table 21-4 gives an overview of potential mitigation measures.

Energy-Efficient Technology for Drinking Water Production

Several energy-efficient technologies are being developed in the water sector. For example, the use of low-pressure mercury lamps for UV/H_2O_2 advanced oxidation is promising. To oxidize 90% organic micropollutants by UV/H_2O_2 treatment, the energy consumption of low-pressure lamps is 30–50% lower than that of medium-pressure lamps (IJpelaar et al. 2007).

The energy use of membrane filtration is relatively high, although with the introduction in the last decade of low-pressure, thin-film-condensate (TFC)-membranes, the energy use has been reduced. The recently developed ceramic membranes have a high permeability and require less energy. The new and innovative AiRO process uses air–water cleaning of vertically positioned membrane elements to control membrane fouling. As a result, only limited pretreatment is required, which reduces the energy consumption. In general, prevention of biofouling and scaling will result in membrane filtration processes requiring less energy.

Table 21-4. Overview of potential mitigation measures for drinking water production and distribution

Potential Mitigation Measure	Examples
Energy-efficient production technologies	Low-pressure UV Advanced membrane-based systems (low-pressure UV, TFC membranes, AiRO process)
Optimised distribution of drinking water	Distribution related to the area and the demand Pressure reduction and tuning
Methane recovery	Methane degassing and re-use
New water cycle concepts and household measures	Source-separated sanitation Re-use of embedded energy of water Central softening

Other developments include the efforts undertaken to improve the energy-intensive desalination processes for drinking water production from sea water. New techniques such as the membrane distillation technology Memstill are driven by minor temperature differences and are thus less energy-intensive than the conventional reverse osmosis process.

Optimised Distribution of Drinking Water

Distribution systems may be developed that take into account the distribution area and the local water demand. Related to specific distribution areas, the utilisation of pumping stations can be optimised, both for sources, distribution, and levelling down of the production (Riemersma et al. 2000). New demand models that provide stochastic residential demands per individual home and on a short time scale are now becoming available (Blokker 2006). By adjusting the production to the actual demand, a more stable production flow can be established, and thus energy can be saved.

Lowering the pressure in the distribution system will reduce the pumping energy but should not result in increased infiltration to the system and/or discomfort with the end users. The transport capacity of a pipe system can be increased by applying boosters along the pipe. Boosters can be applied to better tune the required pressure at the tap, which would have a small energy benefit.

Methane recovery

Next to energy conservation, the direct emission of methane during groundwater extraction can be reduced. It may be feasible to extract methane from large-scale groundwater wells that contain a high methane concentration. A promising technique for this is membrane degassing (Drijver et al. 2007).

New water cycle concepts and household measures

With the above-listed measures, the carbon footprint of the drinking water sector can be further reduced. However, the largest reduction can probably be achieved in new water cycle concepts and water conservation at households.

Substantial energy savings with respect to the water cycle can be achieved by introducing source-separated sanitation. Separated urine and black water collection systems (that use far less drinking water) show a positive energy balance (Roorda et al. 2008). Other favourable water cycle concepts include stormwater separation, reuse of the "embedded" energy (heat) of water, and central softening.

In recent years, drinking water utilities in the Netherlands have implemented central softening at many of their drinking water production sites. Although this has increased the energy use of the drinking water companies (by 0.02 kWh/m^3), the overall energy consumption was substantially reduced as a result of the gained energy saving at households. Soft water causes less scaling in household devices (such as boilers), which reduces the amount of energy needed to heat the water.

And it is exactly at these households where a substantial reduction of the climate footprint can be achieved. The climate footprint to produce and distribute the yearly consumption of drinking water for an average household (107 m^3) is 38.5 kg CO_2-eq. The climate footprint of a household to heat water, however, is 18 times bigger: 693 kg CO_2-eq. per year (Frijns et al. 2008). Warm water conservation at households seems to be the most effective climate mitigation measure. It is thus evident that in any climate mitigation strategy, the households need to be involved. In fact, here climate mitigation meets the demand-oriented adaptive strategy. For example, toilet flushing with rainwater not only saves about 80 litre of drinking water per day per household, but also 10 kg CO_2-eq. per year per household. Water-saving shower devices save 15 kg CO_2-eq. on water consumption and 80 kg CO_2-eq. on energy consumption per year per household.

DISCUSSION AND SYNTHESIS

Knowledge of climate change effects on drinking water supply is increasing. It is now generally accepted by drinking water supply companies in the Netherlands that climate change may affect water resources management in the near future. Measures are needed to mitigate the impacts of dynamic natural conditions on water supply, taking into account both quantity and quality aspects. In this paper, some innovative adaptive concepts to deal with the anticipated changes were presented. These include the use of new water resources, as well as multiple sources and flexible sources concepts.

An integrated approach needs to be developed that takes into account the specific functions of water systems, such as drinking water, agriculture, cooling, and ecology. A successful water management tool to improve the sustainability of resources is water allocation during periods of water shortage. In the Netherlands, a water allocation scheme such as the priority list water scarcity has recently been developed. It is anticipated that public water supply should prevail over other sectors, given the importance of clean drinking water for public health. A continuous reduction of chemical pollution of water resources by improving industrial processes and wastewater treatment, as well as introduction of cleaner agricultural practices will decrease the contamination of surface waters with pollutants that require intense treatment for drinking water production. Water managers should aim at a controlled permit system based on the maximum acceptable concentrations in the receiving water, taking into account the relevant functions of the water system. Such strategies should preferentially be developed and implemented at a river-basin scale, in accordance with the principles of the Water Framework Directive.

Adaptive strategies can not be implemented without considering the energy consumption and costs. Although the carbon footprint of production and distribution of drinking water in the Netherlands is limited compared to the total footprint, ongoing research identifies numerous options to decease the energy use by drinking water utilities and implement promising new concepts, such as a climate-neutral water cycle. We suggest that drinking water utilities should incorporate mitigation measures (e.g., energy use) in the selection of climate adaptation strategies. For example, the alternative sources strategy might result in a higher energy use for drinking water production. Demand strategies, on the other hand, might result in combined water and energy savings at households.

The development of innovative adaptive and mitigation strategies will certainly benefit from joint research on the interactive impacts of climate change and other pressures. High uncertainties associated with climate change, however, make efficient planning, prediction, and investment extremely complicated. Transition to flexible adaptive strategies will probably take decades because investment depreciation periods of assets are long. This implies that long-term strategies within an indicated timepath have to be developed. These strategies must be based on thorough knowledge of current assets to seize opportunities for change.

ACKNOWLEDGMENTS

This study is a product of the Joint Research Programme of the Dutch Water Sector (BTO). It was co-financed by the EU FP6 Integrated Project TECHNEAU (technology-enabled universal access to safe water). The results of climate mitigation measures are part of a study financed by the Dutch Ministry of Housing, Spatial Planning, and the Environment and the Foundation for Applied Water Research (Stowa).

REFERENCES

Bernhardi, L., and Van den Berg, G.A. 2006. Inventarisation opportunities for the use of alternative sources for the drinking water supply of Waternet. Kiwa report KWR06.031 (in Dutch).

Blokker, E.J.M. 2006. Modelling demand patterns: description and validation of the simulation model SIMDEUM. Kiwa report BTO2006.010 (in Dutch)

Cirkel, G., Van Grinsven, E., and Broers, E. 2006. Climate change and groundwater use. *H_2O* 2006 22, 39–42 (in Dutch).

Deeb, R.A., Chu, K.H., Shih, T., Linder, S., Suffert, I., Kavanaugh, M.C., and Alvarez-Cohen, L. 2003. MTBE and other oxygenates: Environmental sources, analysis, occurrence, and treatment. *Environ. Eng. Sci.* 20, 433–447.

De Graaf, R.E., and Van de Ven, F.H.M. 2006. The Closed City as a strategy to reduce vulnerability of urban areas for climate change. In Innovations in coping with water and climate risks: IWA conference, Amsterdam, September 25–27 2006, pp. 221–227.

Doomen A., and Van den Berg, G.A. 2005. Quick scan salinisation Rhine–Meuse estuary. Kiwa report KWR 05.074 (in Dutch).

Doomen, A.M.C., Wijma, E., Zwolsman, J.J.G., and Middelkoop, H. 2007. Predicting suspended sediment concentrations in the Meuse River using a supply-based rating curve. *Hydrol. Proc.*, 22, 1846–1856.

Drijver, B., Kappelhof, J., Polman, E., and Van Breukelen, B. 2007. Methane extraction from ground water as a sustainable alternative with economic potential. *H_2O* 2007(19), 52–55 (in Dutch).

EEA. 2001. Sustainable water use in Europe; Part 3—extreme hydrological events: floods and droughts, Environmental issue report 21, European Environment Agency.

EEA. 2007. *Climate Change and Water Adaptation Issues.* EEA Technical report No 2/2007, European Environment Agency.

Frijns, J., Mulder, M., and Roorda, J. 2008. Towards a climate neutral water cycle, Stowa report 2008-17 (in Dutch).

Hoogenboezem, W., Wagenvoort, A.J., and Blaauboer, K. 2004. The occurrence of toxic cyanobacteria in some Dutch surface waters used for the production of drinking water. RIWA report.

IJpelaar, G.F., Harmsen, D., Beerendonk., E., Metz, D., Knol, A., Fulmer, A., and Krijnen, S. 2007. Effective UV/H_2O_2 treatment of contaminated water with LP lamps. IWA 4th Leading Edge Conference, Singapore.

Kabat, P., Van Vierssen, W., Veraart, J., Vellinga, P., and Aerts, J. 2005. Climate proofing the Netherlands, *Nature*, 438 (17), 283–284.

Kooiman, J.W., Stuyfzand, P.J., Maas, C., and Kappelhof, J.W.N.M. 2004. Pumping brackish groundwater to prepare drinking water and keep salinizing wells fresh: a feasibility study. (Araguás, L., ed.) In Groundwater and saline intrusion: Proceedings 18th SWIM, Cartagena 2004, Instituto Geologico y minero de España, Madrid, pp. 625–635.

Meuleman, A.F.M., Kappelhof, J., and Ramaker, T.A.B. 2004. The flex water concept: flexible resources management of the Netherlands. In Proceedings AWWA Water Sources Conference.

MNP. 2007. Greenhouse gas emissions in the Netherlands, National Inventory Report 2007.

Prommer, H., and Stuyfzand, P.J. 2005. Identification of temperature-dependent water quality changes during a deep well injection experiment in a pyretic aquifer. *Environ. Sci. Technol.*, 39, 2200–2209.

Riemersma, M., Verberne, A., Inklaar, R., and van der Pijl, J. 2000. The energetically most optimal water distribution. *H_2O* (11), 31–33 (in Dutch).

Roorda, J.H., Koetse, E., Kunst, J., Buunen-van Bergen, A., Mels, A., Bisschops, I., Piekema, P., and Neef, R. 2008. Modern sanitation for the developed world: an example for new areas, IWA Sanitation Challenge Conference, Wageningen.

Shabalova, M.V., Van Deursen, W.P.A., and Buishand T.A. 2003. Assessing future discharge of the river Rhine using regional climate model integrations and a hydrological model. *Clim. Res.*, 23, 233–246.

Segrave, A., Pronk, W., Ramaker, T., and Zuleeg, S. 2007. Global trends affecting the water cycle; winds of change in the world of water. Techneau report.

Stuyfzand, P.J. 1996. Salinization of drinking water in the Netherlands: anamnesis, diagnosis and remediation. SGU Rapp-och Meddel. 87, Geol. Survey Sweden. In Proc. SWIM 1994, June 16–21, 1996.

Stuyfzand, P.J., Kooiman, J.W., Oosterhof, A., and Raat, K.J. 2007. Problems and solutions when recharging brackish aquifers with membrane concentrate from a less brackish groundwater source. (Fox, P., ed.) In Management of Aquifer Recharge for Sustainability. Proceedings of the 6th International Symposium on Managed Artificial Recharge of Groundwater (ISMAR6), Phoenix, Ariz., pp. 47–60.

Van den Berg, G.A., de Rijk, S., Mesman, G., Abrahamse, A., and de Graaf, R. 2008. Opportunities for decentralised drinking water supply in the Netherlands. KWR report BTO 2008.018 (in Dutch).

Van Vliet, M.T.H., and Zwolsman J.J.G. 2008. Impact of summer droughts on the water quality of the Meuse River. *J. Hydrol.* 353, 1–17.

Vewin 2006, Water in sight 2006 (in Dutch).

– 22 –

Meeting the Challenges of Climate Change: Singapore

Chang Chian Wui

INTRODUCTION

The Republic of Singapore is a small island state located at the tip of the Malayan Peninsula at the crossroads of Southeast Asia. Comprising a main island and some 63 small offshore islands, Singapore is located approximately 137 km north of the Equator. Singapore's immediate neighbour to the north is Malaysia and its immediate neighbour to the south, east, and west is Indonesia (Figure 22-1).

As a relatively low-lying, densely populated island in the tropics, Singapore is vulnerable to climate change. Much of the island is less than 15 m above sea level, with a generally flat coast. Housed within its 193 km coastline is a population of about 4.8 million, making Singapore one of the most densely populated countries in the world.

Lying at the equatorial belt, Singapore is blessed with some 2,400 mm of rainfall each year. Unfortunately, it is a small island with only 700 km^2 of land. The growing population and economy[1] exert pressure on land use, so that water catchments have to compete for land with industries, housing, commerce, roads, and schools. The lack of catchments to collect and store rainwater, high evaporative losses, and lack of natural aquifers or groundwater has led Singapore to be considered a water-scarce country by the United Nation and ranked 170th among a list of 190 countries in terms of freshwater availability.

According to the IPCC AR4, temperature rise in Singapore and the Southeast Asia region over the next century is likely to be similar to a global mean temperature rise projection of 2.8°C (1.7–4.4°C) and sea-level rise is likely to be close to global mean of 21–48 cm. Changes in annual precipitation for Southeast Asia will range from –2% to +15% with a medium of +7%. Rainfall variability will be affected by changes in ENSO but this is not well understood. The pattern is broadly one of wet season rainfall increase and dry season decrease. Extreme events are likely to become more intense in this region.

Taking these circumstances into consideration, concerns about the potential impacts of climate change on Singapore's water resources are precipitation changes and water supply resilience, flooding, and sea-level rise.[2] These potential effects are not due solely to climate change, but can be aggravated by adverse global climate change.

---

1 The per capita GDP in 2007 was US$35,329 (S$52,994). *S$1.5 ≈ US$1.*

2 Other concerns about climate change such as coastal land loss, temperature rise, energy demand, public health (heat stress and diseases), and the impacts on biodiversity are not addressed in this paper.

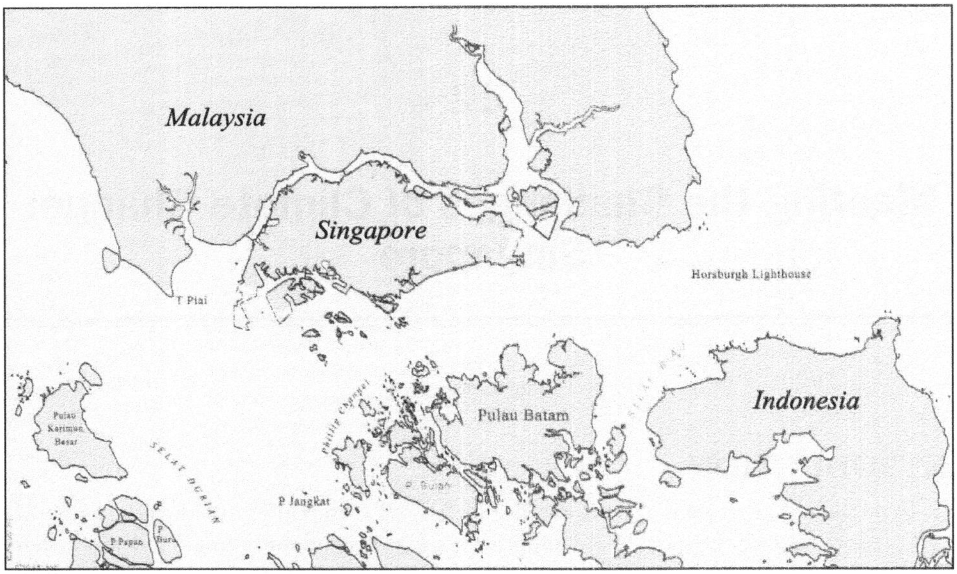

Figure 22-1. Location map of Singapore

To better understand the detailed effects and resulting impacts of climate change, the Singapore government has commissioned a study of Singapore's vulnerability to climate change. This study will project the changes in temperature, sea level, and rainfall patterns in Singapore in the next century, and the resulting impacts such as increased flooding and impacts on water resources. The study is being conducted by the local university, together with a team of foreign experts, a number of whom were involved in the drafting and review of IPCC reports. The results of the study will facilitate the identification of new adaptation measures as well as the review of existing adaptation measures. The study is expected to be completed in 2009.

INTEGRATED WATER RESOURCES MANAGEMENT

Water for All: Conserve, Value, Enjoy

PUB is Singapore's integrated national water agency, responsible for the management of the entire water loop—from the collection of rainwater to the supply of potable water, collection and treatment of used water, water reclamation, and the desalination of seawater. Singapore's integrated water resources management strategy can be summarized in six words, as in PUB's corporate tagline: Water for All: Conserve, Value, Enjoy. "Water for All" refers to the supply strategy to ensure a diversified and sustainable supply of water for Singapore. "Conserve, Value, Enjoy" describes water demand management. It reflects the approach to involve the people–public–private sectors to play their part to conserve water, keep the water catchments and waterways clean, and build a relationship with water so everyone can enjoy the water resources.

Water for All

PUB has, through a diversified water supply strategy, put in place a sustainable water supply system with four different sources, namely: local catchments, imported water, NEWater (or high-grade water reclaimed from treated used water), and desalinated water. These are known as the Four National Taps which will help meet Singapore's long-term water needs and will mitigate the impacts from climate change (Figure 22-2).

Local catchments

With a total land area of 700 km^2, Singapore has limited land for water catchment. It has, however, managed to put in place a network of drains, canals, and reservoirs to capture rainwater from half of the country's land area. The segregation of stormwater runoff from used water, judicious land use, vigilant surveillance, and strict enforcement and pro-active outreach and education programmes meant that most of Singapore could remain clean to be preserved as a water supply source. Because of this, PUB, as the national water agency, has been able to develop more water catchments, including those in housing estates and urban areas.

Singapore's newest reservoir, the Marina Reservoir, completed in 2008, is located right in the heart of Singapore's central business district. The reservoir collects runoffs from Singapore's most urbanised and largest catchment (10,000 ha), which include some of the oldest and most densely inhabited parts of Singapore. The Marina Reservoir, together with two more new reservoirs under the Punggol–Serangoon Reservoir Scheme (scheduled for completion in 2009) will bring the total number of reservoirs in Singapore to 17 and Singapore's water catchment area will increase to two-thirds of Singapore's land area.

The Marina Reservoir is not only a reservoir; it is also a tidal barrier for flood control. The seawater barrier comprises nine steel gates, each 26.8 m long. Under normal conditions, the steel gates will remain closed to separate the reservoir from the sea. During heavy rain, the steel gates will be opened to release excess stormwater into the sea when the tide is low. However, when it is not possible to do so during high tide, drainage pumps capable of pumping up to 240 m^3/s will pump out the excess stormwater into the sea. The reservoir will also offer numerous opportunities for lifestyle, recreational, and sports activities right in the heart of the city.

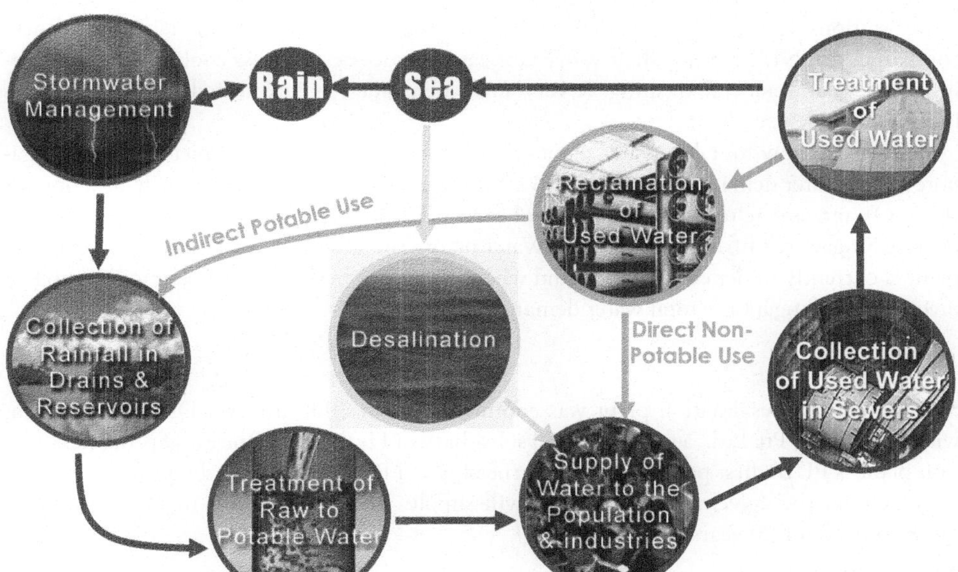

Figure 22-2. Closing the water loop

Imported water

Singapore imports water from Johor, Malaysia, through two agreements, the 1961 and the 1962 Water Agreements, for the supply of water until 2011 and 2061, respectively. Water from local catchments, together with water imported from these agreements, is sufficient to meet Singapore's needs but it has, nevertheless, supplemented them with NEWater and desalinated water. In this way, by 2011, when the 1961 Water Agreement expires, Singapore will not need to renew it. This is also to ensure that in the long term, Singapore can be self-sufficient if need be.

NEWater

In addition to water from catchments, Singapore has also included high-grade reclaimed water as part of its water inventory. NEWater is produced through a multibarrier treatment process that comprises conventional used-water treatment, micro/ultrafiltration, reverse osmosis (RO) process, and finally, ultraviolet disinfection.

To ensure the safety of NEWater, the then-Ministry of Environment (now known as the Ministry of the Environment and Water Resources) and PUB embarked on a Water Reclamation Study in 1998. Under the study, a 2.2 million gallons per day (mgd) (10,000 m^3/day) demonstration plant was set-up and operated over a two-year period to test for robustness and reliability. The quality of NEWater has also been independently verified and confirmed by an experts panel comprising international and locally renowned water experts. Overall, more than 20,000 tests were carried out for some 190 water quality parameters and the results showed that the NEWater's quality was well within the United States Environmental Protection Agency (USEPA) and World Health Organisation (WHO) drinking water standards and guidelines.

NEWater is a product of tried and tested technology that is now supplied on a large scale. NEWater is mainly supplied to industries such as wafer fabrication plants (which require very stringent and sensitive water quality requirements), petrochemical complexes and refineries, power stations, electronic companies, and commercial premises for nonpotable uses. These companies use NEWater for their wafer fabrication processes, process cooling, air-con cooling, boilers, and general washing. A very small percentage is introduced into reservoirs as indirect-potable use.

There are now four NEWater plants that can supply NEWater to meet 15% of Singapore's total water demand. In 2005, PUB awarded a 25-year design, build, own, and operate (DBOO) agreement to a local firm, Sembcorp NEWater Pte, to design, build, own, and operate Singapore's fifth and largest NEWater plant. The 50-mgd (227,300-m^3/day) capacity plant is currently under construction and when completed by 2010, NEWater will be able to meet 30% of Singapore's total water demand.

Desalination

Singapore's first desalination plant was commissioned in 2005 under a DBOO agreement with SingSpring Pte Ltd., a wholly owned subsidiary of Hyflux Ltd. The SingSpring desalination plant is PUB's first public–private partnership (PPP) project under the DBOO arrangement. Under the agreement, SingSpring will supply 30 mgd (136,000 m^3) of desalinated water to PUB for 20 years.

Meeting long-term water demand and climate change impacts

Of the four "National Taps," three—local catchments, imported, and desalinated water—are primary sources. They are, in effect, the "first drops" of water obtained from the water catchments or the sea. NEWater, on the other hand, is a secondary source created from recycling of used water from the three primary sources. Apart from freeing up a large amount of potable water for other purposes, NEWater has a strategic role in "multiplying" the water supply achieved through recycling. What this means is that if 50% of water is recycled, the water supply could theoretically be doubled. From recycling 1 drop of water, 0.5 drops of water could be obtained. This 0.5 drops can be in turn recycled to get 0.25 drops, and then 0.125 drops, and so on. Theoretically, recycling 1 drop of water results in another drop of water (0.5 + 0.25 + 0.125 + ... = 1). This is a multiplier of two. At higher recycling rate, a higher multiplier effect could be achieved. This means that instead of building new capacity through water catchment expansion, which is difficult in land-scarce Singapore, or desalination, which is costly, only half the required new capacity needs to be built and the rest made up through the multiplier effect of recycling.

Both desalinated water and NEWater have important roles to play in meeting long-term water demand and to mitigate the impacts of climate change—one as a primary source of water and the other by providing a multiplier effect through recycling. Both also have the advantage of being independent of fluctuations in rainfall, which will increase Singapore's resilience against the vagaries of the weather and uncertainties that climate change would bring. Together, NEWater and desalination has put Singapore in a strong position to meet long-term sustainability and mitigate the impacts of climate change.

Conserve, Value, Enjoy

Long-term sustainability in water cannot be achieved by boosting water supply alone. Equally important is managing water demand and encouraging water conservation. PUB recognized that encouraging the public to conserve water would not be enough and that a holistic approach would improve the chances of success. Apart from public education, pricing and voluntary efforts, mandatory requirements and technical solutions were also explored and imposed, covering maximum flow rates for fittings as well as labeling schemes to inform and educate consumers.

Unaccounted-for Water

In terms of unaccounted-for water, PUB has put in place an integrated water network management plan to ensure that leaks in our water pipe network are kept to a minimum. The plan, comprising the use of good quality network and efficient management, active leakage controls, mandatory and accurate metering policies, and strict legislation on illegal draw-offs, has enabled us to lower our unaccounted-for water from about 10% in the early 1990s to less than 5% today. In terms of leakage, the network has less than seven leaks per 100 km/year, a vast improvement from the 1980s, when they were more than 100 leaks per 100 km of pipeline per year. This has been achieved through using good pipe materials, systematically replacing old unlined cast-iron pipelines and unlined galvanized iron connections, as well as the active detection of underground leaks. By reducing water losses and keeping them in check, there is less pressure to expand water sources.

Pricing

On water pricing, a Water Conservation Tax (WCT) is imposed on every drop of potable water supplied. This is applicable to all water users, both domestic and nondomestic. The purpose of the WCT is to drive home the message that water is a scarce resource and is precious from the very first drop. In addition, the water tariff and WCT are also increased for households that consume more than 40 m^3/month to encourage water conservation.

Mandatory and Voluntary Measures

To further promote water conservation, PUB has also adopted a basket of mandatory and voluntary measures. Mandatory measures include the use of low-capacity flushing cisterns, constant flow regulators, self-closing delayed action taps, and maximum allowable flow rates. Voluntary measures include community-driven public education programmes such as the Water Efficient Homes Programme and the 10 Litre Challenge to encourage home owners and building owners to adopt water conservation measures. As part of the programme, DIY water-saving kits consisting of flow resistors, cistern water-saving bags, and leaflets on water-saving tips were supplied to residents free of charge. Results have been encouraging. Domestic water consumption was reduced from a peak of 176 L per person per day in 1994 to 157 L per person per day in 2007. With further promotional efforts and encouraging participation from the public, we expect to lower this further to 155 L per person per day by the year 2012. In 2008, PUB launched the 10% Challenge, an initiative to get the nondomestic sector, such as hotels and industries, to reduce their total water consumption by 10%.

With all these initiatives, Singapore's average annul water demand growth rate since 1995 has been kept low, at about 1.1% despite GDP growth of 5.1% per annum and population growth of 2.2% per annum.

Active Beautiful Clean Waters

Beyond water conservation, PUB has embarked on efforts to get people to value and enjoy our "blue assets." This initiative is known as the Active, Beautiful, Clean (or ABC) Waters Programme. Singapore has some 32 rivers and eventually, 17 reservoirs. Water is all around us. Thus, there is a need to ensure that this water is kept clean. In addition to source control and mitigation measures, PUB also aims to take a community approach in ensuring good water quality—by transforming the utilitarian concrete drains and canals into beautiful and vibrant community spaces for recreational activities to foster greater environmental ownership.

To garner the support of the building industry, especially developers, in implementing the ABC Waters Programme, PUB has developed a set of ABC Waters Design Guidelines. The guidelines provide professionals with technical instruction and advice on how to design water-sensitive landscaping features such as bioswales and rainwater. Such features help to soften the harsh concrete canal edges while providing treatment to the stormwater runoff before it enters the waterways through natural nutrient removal by rain gardens, aquatic plants, and retention ponds. Peak flows in the drains and waterways are also reduced as the water is retained for a longer time before reaching the waterways.

LEVERAGING INNOVATION AND TECHNOLOGIES

Innovation and the use of technologies is another important strategy for Singapore to meet long-term water demand and mitigate the impacts of climate change. A small country with no natural resources, Singapore has always used technology as a key element of its nation-building efforts. Over the years, it has invested heavily in technology to seek solutions to its water supply—the use of microfiltration (MF)/ultrafiltration (UF) membrane technology to treat the more polluted water from urban runoffs, reverse osmosis (RO) to treat seawater, and MF/RO/ultraviolet (UV) disinfection to produce NEWater.

Since the 1990s, advancements in membrane technology have significantly improved the effectiveness and reliability of membranes. With more widespread use of membranes, cost has also come down. Coupled with economies of scale due to the rapid expansion in production and productivity gains, PUB has been able to pass the savings on to customers.

NEWater prices has been reduced from S$1.30 (≈ US$0.87) per cubic metre in 2003 to S$1.15 (≈ US$0.77) per cubic metre in January 2005 and S$1.00 (≈ US$0.67) per cubic metre in April 2007. This has also encouraged nondomestic customers to make the switch from potable water to NEWater for their nonpotable uses.

Desalination of seawater is still much more energy intensive (about 3.5 kWh/m^3) compared to NEWater (about 1.0 kWh per cubic metre) or conventional water treatment and hence costly. With rising oil prices, this will be further accentuated. Future development in desalination therefore lies in reducing energy consumption. PUB is exploring various alternatives such as membrane distillation and a variable salinity plant (which is basically a two-in-one plant that can treat either freshwater or seawater). The Singapore government has also challenged the research community to come up with viable alternatives to simultaneously address the objectives of drinking water production and energy efficiency by developing methods to desalinate seawater at 1.5 kWh/m^3 or less (this is least 50% more energy efficient than current technologies). The Challenge call for proposals attracted the attention of local and overseas research institutions and companies. The proposal by Siemens Water Technologies Corp. clinched the Challenge with a novel electrically-driven desalination process. Siemens received close to S$4 million (≈ US$2.7 million) worth of R&D funds to investigate and test-bed the concept in Singapore. If successful, the concept would benefit not only Singapore but also other countries.

The focus of PUB's R&D is mainly on large-scale implementation of water solutions. These range from the development of online water-quality monitoring and management systems, to technologies to improve used water and water supply treatment systems. A number of demonstration plants have been commissioned and are up and running at our facilities. These include the Variable Salinity Plant and the Membrane Bioreactor Plant (which combines membrane separation with biological treatment to reduce plant footprint and overall energy consumption when coupled with NEWater production).

Going forward, Singapore will continue to invest heavily in R&D on water technologies. To this end, the Singapore Government in 2006 formed a National Research Foundation (NRF) and R&D funds of S$330 million (≈ US$220 million) have been set aside for the environment and water technology sector. Apart from establishing an Environment and Water Research Programme and strengthening international R&D linkages, the aim is to make Singapore a global test-bed for environmental products and solutions and to encourage the early adoption of new technologies in Singapore.

FLOODING AND SEA-LEVEL RISE

Singapore is relatively low lying and in the 1960s and 1970s, floods were common and widespread, especially in the city area. In 1978, Singapore experienced its most severe flood when a total of 512.4 mm of rain was recorded over a 24-h period (the highest in a single day and almost a quarter of Singapore's average rainfall). To compound matters, the rain coincided with a very high tide of 3.2 m (average tide is about 2.8 m). Roads were submerged under more than a metre of water. Seven people died in the 1978 flood and more than 1,000 people were evacuated from their homes by army and police boats.

Since 1973, more than S$2 billion (≈ US$1.3 billion) has been spent on the construction of new drains and canals in Singapore. These have helped to reduce flood-prone areas by more than 95% over the last few decades, even as urbanized areas have been increased over the same period. In addition, PUB has a deliberate policy to raise low-lying areas in conjunction with redevelopment proposals. The drainage infrastructure development and improvement programmes have reduced flood-prone areas from 3,200 ha in the 1970s to 124 ha today.

PUB will reduce it to less than 66 ha by 2011 through the development and improvement of drainage infrastructure in Singapore (e.g., widening and deepening of drains and canals), the completion of the Marina Barrage, as well as other flood alleviation projects. On 19 December 2006, Singapore experienced its third-highest storm on record. A total of 366 mm of rain fell over a 24-hour period. Despite the intense rain, there were no major floods and only a few isolated spots (about 15 ha) were affected.

High sea levels also pose a challenge for drainage to the sea. Higher tide levels can aggravate inland flooding during rainstorms or storm surges, where the sea is pushed to the shore by the force of the wind. In view of this, since 1991, all land reclamation projects had to be built at least 125 cm above the highest recorded tide level. With hindsight, this requirement has put Singapore in a stronger position to deal with future increases in sea levels arising from climate change as the requirement of 125 cm exceeds current IPCC's AR4 projection of the highest sea-level rise of 59 cm by the end of the 21st century.

Plans are in place to ensure that our drainage infrastructure can meet the effects of climate change. PUB is taking part in the vulnerability study to understand the impact and implications of climate change on the drainage system. The study will review drainage design criteria (intensity–duration curves), produce new flood maps taking into account climate change impacts, and review minimum platform areas for developments and land reclamation. The progressive implementation of ABC Waters projects will also help to reduce peak flows as stormwater is retained for a longer period in the catchment.

SINGAPORE'S NATIONAL CLIMATE CHANGE STRATEGY

The Singapore Government has developed the National Climate Change Strategy (NCCS), which represents Singapore's comprehensive and holistic response to climate change. The strategy documents Singapore's past and ongoing efforts on climate change and sets out future plans to address climate change in the areas of (i) vulnerability and adaptation; (ii) mitigation; (iii) competency-building; and (iv) international participation. The NCCS has been developed through a consultative, multi–stakeholder approach. The views of stakeholders and the public at large have been developed through channels such as e-consultation, dialogue sessions, and consultation forums.

A National Adaptation Taskforce, led by the Ministry of National Development, has also been formed to consider and recommend for implementation, adaptation measures relating to physical infrastructure and public health. The interagency taskforce comprises agencies responsible for land use planning, environment and water resources, housing and industrial estates, health, transport, national parks, and ports. The findings of the vulnerability study will facilitate the identification of new adaptation measures as well as the review of existing adaptation measures. Government agencies in Singapore will continue to work closely together to assess the possible impacts of climate change on Singapore, regularly review the sufficiency of Singapore's existing adaptation measures, and identify new measures and national systems to actively monitor and manage these impacts.

CONCLUSION

Singapore has taken steps to prepare for climate change. The development of the "Four National Taps" to diversify its water supply to meet its long-term water needs will also mitigate the uncertainties of climate change. Two of taps, NEWater and desalination, are not affected by the vagaries of weather. The development of NEWater is even more strategic, as it effectively doubles Singapore's water resources through recycling. Conservation and efficient management of the water network will continue to be given high priority to keep the growth of water demand low so that the need for new capacity can be deferred. Innovation and technologies has played a key role in developing water sustainability and Singapore will continue to invest heavily in R&D and new water technologies. As a result of past development planning, some of the existing measures can address to some extent some of the potential climate change impacts such as sea-level rise. A vulnerability study is currently on-going to better understand the effects of climate change in the long term.

Climate change is a long-term phenomenon and its impacts will be felt for decades. While existing measures are in place in adapt to climate change in the short to medium term, government agencies in Singapore, including PUB, will continue to review the sufficiency of existing adaptation measures and identify new measures as necessary to manage these impacts.

REFERENCES

United Nations Educational, Scientific and Cultural Organization (UNESCO), 2nd United Nations World Water Development Report (2006).

United Nations Educational, Scientific and Cultural Organization (UNESCO), 1st United Nations World Water Development Report (2002).

Singapore's National Climate Change Strategy Mar. 2008. (http://ww.mewr.gov.sg).

Tortajada, C. (2006). Case Study for the 2006 HDR—Singapore: An Exemplary Case for Urban Water Management, Human Development Report 2006.

Ng, K.H., Foo, C.S., and Chan, Y.K. (1997). Unaccounted-for Water—Singapore's experience. *Journal of Water Supply Resources and Technology – Aqua*, 46, No. 5, pp. 242–251.

– 23 –

Climate Change and Adaptation in Southern California

Maryline C. Laugier, Edwards G. Means III, Jennifer A. Daw, and Michael Hurley

BACKGROUND

The state of California continues to experience significant population growth dominated by resident birth rates and also contributed to by immigration. To meet historic and future projected water needs, California water supply heavily relies on the accumulation of winter mountain snow melting into spring and summer runoff. Much of the state's water supply is stored in an extensive system of reservoirs and aqueducts. These surface reservoirs provide storage to replenish the groundwater basins. The largest water storage and conveyance systems in California are the State Water Project and the federal Central Valley Project. Both mostly receive their water from the Sacramento Valley and serve the Central Valley. Other major water storage and conveyance systems, the All-American Canal and the Colorado River Aqueduct, divert their water from the Colorado River and serve Southern California. Another source of water supply in California is the delta in the San Francisco Bay area, which is overextended in the amount of water it supplies.

Recently observed southwestern US climate trends include changes in atmospheric temperature such as increased average temperature and increased average daily minimum temperature resulting in heat island effects; and precipitation trends, such as precipitation fluctuating over a several-year cycle and increased frequency of heavy climate events (i.e., rainfall event above the 95th percentile) or extreme climate events (i.e., precipitation totals that are exceeded no more than once per year at a given location) (IPCC 2008). Warming temperature trends in Southern California were estimated to have increased by 0.17°C (0.3°F) per decade over the past 100 years (Chung 2006). An Inland Empire Utilities Agency (IEUA) modeling study (Groves et al. 2007) predicted a summertime temperature change in Southern California between 2000 and 2030 to be between 0.1 and 2.1°C (0.18 and 3.78°F). Other climate models and simulations predict that by 2100, annual temperature will rise by 1.6 to 4.4°C (2.88 to 7.92°F) in Southern California (Cayan et al. 2007). Furthermore, increased atmospheric temperature usually results in an increase in precipitation intensity and frequency, thus leading to a greater occurrence of heavy precipitation events. This is mainly due to an increase in water vapor in the atmosphere resulting from warmer atmospheric temperatures. Additionally, a regional climate change study by Kim et al. (2001) suggested that precipitation intensity would increase as atmospheric carbon dioxide also increases. Kim et al. (2001) also described the following observed changes in precipitation events:

- A 37% increase in the average annual number of wet days in the southern Sierra Nevada. In contrast, the number of light precipitation events (i.e., less than 0.2 inch per day) remain identical or would experience a slight decrease.

- An increase in heavy precipitation from 1 to 3% of wet days.
- An increase in extreme events from 0.1 to 1% of wet days.

The study also predicted that projected changes in heavy precipitation events will most likely follow similar trends. These observed and predicted climate trends could, in turn, significantly affect water demand and supply planning.

Southern California hosts over half of the state's population, which is concentrated in and near major urban centers, but receives much less precipitation than the northern part of the state (Chung 2006). With the California population expected to increase by 25 million by 2050 and to 12 million in Southern California, water utilities will be subjected to increasing demand (absent significant additional water conservation). In addition to future population growth, other factors, such as climate change, could potentially affect future water supply reliability. Consequently, understanding the potential impacts of climate change on water is crucial to Southern California and will constitute a major challenge for water providers and water utilities.

First, this chapter will describe the impacts due to climate change and their practical effects on water utilities in terms of water quantity and water quality. Consequential vulnerabilities from this range of impacts may ultimately pose a burden on water system planning, design, and management. Thus, this chapter will also examine consequential impacts to demand and infrastructure. Faced with these additional burdens, water utilities will need to embrace these changes and work to address them. Finally, this chapter will explore utility engagement in terms of climate change mitigation, adaptation to climate change, and monitoring of regulations.

IMPACTS

The warming of atmospheric temperatures has been shown to be directly related to changes in the global hydrological cycle leading to a range of impacts that could be of concern for water utilities, if not addressed. Observed and predicted changes can be grouped in two categories, namely, water quantity and water quality. The main water quantity impacts result from changes in precipitation frequency, precipitation intensity, power generation, snowpack, and sea-level rise; while water quality impacts are typically a result of these water quantity changes. The following section briefly discusses the aforementioned impacts as well as their practical effects. This section will also describe the consequences of these impacts and practical effects on water systems, such as water demand and infrastructure.

Water Quantity

Precipitation Frequency

Precipitation in Southern California generally falls during wintertime. An IEUA study on climate change (Groves et al. 2007) indicated that the predicted range of average wintertime precipitation trends would be between a 20% decrease and a 10% increase. While issues associated with less frequent precipitation events may lead to the occurrence of drought conditions and heat waves, several issues originating from more frequent precipitation events can potentially arise. For example, greater flood risk could be of concern for water utilities. Modeling studies of flood-control effects under more-frequent-precipitation-event scenarios have shown that peak water release from reservoirs could exceed the actual release and in turn could flood downstream communities (Wallis et al. 2008). In addition, greater flow releases resulting from flooding could contribute to yield loss and diminish water availability. Greater flood risk also increases the risk for levee failure, which poses another threat to water utilities.

Florsheim and Dettinger (2007) reviewed historical levee break data for California over 155 years and inferred that 25% of breaks occurred during the 20th century and are related to fluvial processes impacted by climate change. Resizing of infrastructures such as spillways may also be required by water utilities to effectively manage greater flows as a direct consequence from future potential flooding.

Precipitation Intensity

Changes in precipitation intensity have been observed in Southern California, where extreme precipitation events increased 60 to 90% between 1948 and 2006 (Madsen and Figdor 2007). Climate models by Cayan et al. (2007) predicted an increase of daily extreme precipitation occurrences falling into the 99th percentile, from 111 to 127 over 40 years by 2100. This increased precipitation intensity may create more flooding challenges for water utilities. Infrastructure sizing related to storage capacity, spillways, rip rap, total maximum daily loads, best management practices, and safety may be required to effectively manage more intense precipitation flows.

Power, Flow, Temperature

An energy-related consideration of climate change impacts is the tradeoff between hydroelectric power generation and water supply. Hydropower production is dependent upon the amount of water available, among other parameters. Changes in precipitation patterns, as described above, result in changes in water availability from surface waters and groundwater aquifers, further altering water deliveries and affecting power generation. Predictions indicate opposite changes are anticipated for hydropower production: increases in winter water supply and decreases in summer (Chung 2006). These changes are adverse to the increased energy demand patterns for summertime and lower demands in winter. Central Valley Project power generation decreases in all months for one "dry" scenario compared to the base, except for the month of March where it increases slightly (Figure 23-1). Additionally, under the same dry scenario, the largest decrease in hydropower generation is predicted to occur in September and October at the end of summer where storage levels are at the lowest (Figure 23-1). In areas that largely rely on hydropower as an energy source, such as Southern California, this issue is just an indication of the difficult tradeoffs water utilities will need to evaluate when faced with climate change.

Rain Versus Snow

Another impact resulting from climate change is reduced snowpack. Snow cover losses constitute a concern for water utilities in Southern California, as this region's water supply heavily relies on basins fed by snowmelt. These observed changes are mainly due to a shift in the ratio of precipitation that falls as rain rather than snow during the winter season. These observed decreases in snowpack are projected to continue. The DWR report (Chung 2006) predicted warmer temperatures will raise the snowline elevations and reduce the area where snowpack accumulates in California. Snow cover losses and rising average snow elevations constitute a threat to water supply in Southern California. As a result of these changes, the runoff quantity from snowmelt has been observed to decrease and the duration of snowmelt and associated timing of spring runoff have been observed to be shorter and arrive as much as three weeks earlier than they did in the middle of the 20th century (Nelson et al. 2007). The shift in the timing of (1) peak flow and (2) runoff to earlier in the year results in a greater fraction of peak flow now occurring in March rather than between April and June,

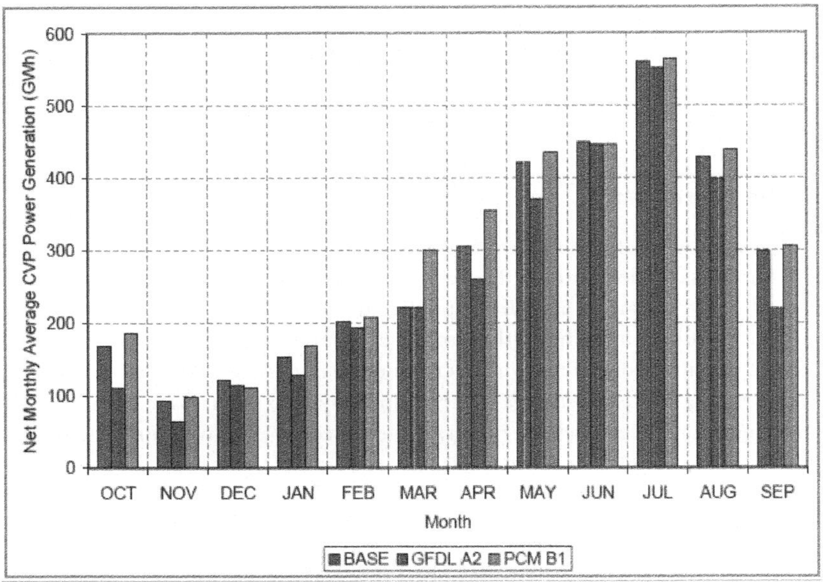

Figure 23-1. Monthly average Central Valley Project power generation
Source: Easton (2006).

which results in reduced summer runoff and stream flows (Palmer and Wiley 2005, Nelson et al. 2007, ICPP 2008). These changes in runoff volumes and timing may cause additional flood control challenges, in turn causing water-yield loss for water utilities.

Sea-Level Rise

Sea-level rise is another of the anticipated impacts of higher temperatures with climate change. Sea-level rise is primarily the result of thermal expansion of seawater as oceans warm and melting of polar ice (Nelson et al. 2007). Sea-level rise along the California coast may be in the range of 0.3 of a foot to 2.9 feet from 1900 to 2100 (Chung 2006). According to the USEPA (2008b), "Josselyn and Callaway examined the impacts of a 1-, 2-, and 3-meter (40-, 80-, and 120-inch) sea-level rise by the year 2100. Of the three scenarios, a 3-meter rise by the year 2100 is regarded as the most probable."* Increases in sea levels may pose a significant burden to coastal communities, as sea level rise of 3 meters would potentially inundate 7,000 square miles of dry land and 50 to 80% of US wetlands, resulting in costs of over $100 billion in the US (USEPA 2008c).

Sea-level rise would potentially increase saltwater intrusion in deltas and coastal aquifers, subsequently increasing the total dissolved solids in the source water and causing subsequent operational challenges for drinking water providers. In response, utilities would have to install barriers for groundwater wells, utilize desalination technologies for water treatment, and diversify their water sources. In addition to the potential impacts to coastal aquifers, sea-level rise could result in the inundation of coastal infrastructure, local flooding near mouths of rivers and streams due to increased potential for levee failure, adverse impacts on flow controls, saltwater intrusion into sewers, increased erosion, impacts on salt wedge/wetlands, impacts on development patterns in bay and estuaries, and effects on water intakes subject to tidal influence (brominated disinfection by-product formation, chloride, total dissolved solids).

Water Quality

Contaminant Concentration

Similar to the water quantity impacts previously described, water quality is also affected by climate change. One major concern is changes in soil moisture and subsequent erosion impacts due to warmer temperatures, which can burden a utility's source water and affect water treatment. Projections indicate an overall decrease in soil moisture in the arid regions of the western US, such as Southern California (Lettenmaier and Miles 2007). Recent simulations of erosion in the US indicate that when increased precipitation was projected, the estimated amount of erosion rose by 15 to 100%. Erosion, particularly as a result of more intense storms and runoff, increases the occurrence of organics, bacteria, and other pathogen and nutrient loadings.

Nutrient loading to source water is more likely with more frequent runoff, particularly in human-dominated landscapes where increases in phosphorus, pesticides, or acids may occur (Gleick and Kiparsky 2003). Water utilities may need to adapt their treatment processes to address new contaminants entering their source water. Increased erosion may also limit the assimilative capacity of wastewater discharges to receiving waters, further degrading mixing zone impacts. Additionally, increased erosion may increase impacts on sensitive aquatic species, which are highly vulnerable to climate variables and to changes in water parameter concentrations. Ecosystems are particularly vulnerable where climate change will act with other anthropogenic stressors such as large water withdrawals or wastewater returns (Gleick and Kiparsky 2003), thus impacting the assimilative capacity of wastewater discharges to receiving waters. Erosion farming operations will also be impacted by reduced nutrient levels in their soils as a consequence of soil moisture reduction.

Chemistry

"A key impact of warmer air temperature is warmer water temperature" (USEPA 2008a), which can pose additional burdens to water utilities. Climate models for California rivers predicted increasing temperatures under two climate change scenarios, and more substantially in one (Figure 23-2). Warmer water temperatures contribute to changes in water quality such as lower dissolved oxygen concentrations, which can release more inorganics into the water stream and cause increased acidification of water sources.

"Increases in water temperature can also make some contaminants, such as ammonia and pentachlorophenol, more toxic for some species and foster the growth of microbial pathogens in sources of drinking water" (USEPA 2008a), which can lead to water quality issues, such as nitrification in drinking water distribution systems. Additional challenges come into play with warmer water temperatures and aquatic species. "Higher temperatures will decrease salmon, trout, and other fish habitat, thereby increasing conflicts over water resources" (Nelson et al. 2007), due to lower dissolved oxygen concentrations. Anadromous fish such as salmon, which primarily live in the sea but breed in fresh water, are highly dependent on the hydrologic processes that influence streamflow timing, volume, and stream temperature; all are highly sensitive to climate variations (Lettenmaier and Miles 2007).

Higher nutrient loading from erosion and high runoff rates can result in eutrophication and more rapid algae growth in source waters, which reduces dissolved oxygen concentrations, impairs aquatic life, and creates conditions for invasive species, such as Quagga mussels, to thrive. Pollutant loading may also be exacerbated by anticipated changes to wetlands, which have natural mechanisms that improve water quality (Gleick and Kiparsky 2003). When higher water temperatures promote the growth of algae, this can further deplete the

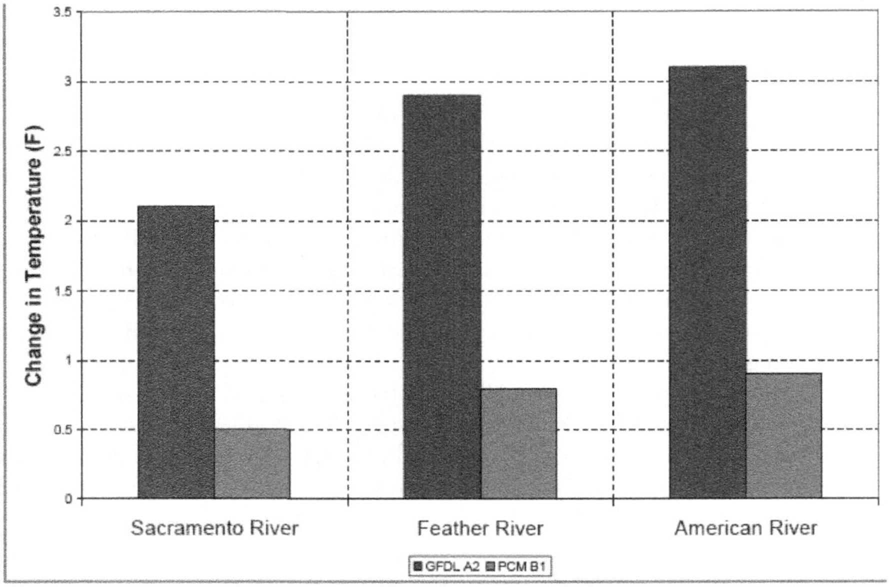

Figure 23-2. Changes in water temperature under two climate change scenarios
Source: Easton (2006).

amount of dissolved oxygen in the water, creating stressful or fatal conditions for fish (Palmer and Wiley 2005), in turn causing taste and odor as well as other water treatment problems for utilities.

Higher water temperatures will significantly affect natural aquatic systems and the ability for indigenous species to thrive. This is already being seen in some areas of the West where "many stream reaches...have water temperatures that exceed water quality standards, and the incidence of violations, especially in summer, will increase with warmer summer temperatures and reduced low flows due to earlier snowmelt" (Lettenmaier and Miles 2007). Source water quality issues are not only of concern for aquatic life but also for water utilities. As source water quality worsens, there is greater potential for fish kills and impaired source waters, requiring more advanced treatment of drinking water. An additional challenge will lie with water utilities to balance the needs of minimum stream flows to sustain aquatic life, while maximizing the use of water rights to meet potable demands.

Climate change will have other indirect impacts for water utility operations. For example, higher water temperatures can lead to adverse effects such as disinfection by-products, regrowth, and difficulty maintain chlorine residuals in water treatment and distributions systems (Cromwell et al. 2007). Changes to water quality can pose other impacts on infrastructure, reducing its useful life through corrosion and other damage.

Sediment Transport

Another water quality impact of higher atmospheric temperature is increased sediment transport, which may constitute a burden not only to natural systems but also to surrounding communities. Climate change predictions may cause reduced stream flows and potentially more frequent wildfires, both of which contribute to sedimentation of water supplies with debris and increased erosion. In general, these impacts result in increased turbidity in source water, which can create potable water treatment challenges for water utilities. Source water protection and additional treatment technologies may be required to effectively treat water to potable standards. Further, increased sedimentation may lead to increased sluicing frequency, basin maintenance, and loss of reservoir storage capacity.

Predictions for decreased precipitation may cause reduced stream flows, which can exacerbate temperature increases and increase sedimentation. In addition, less precipitation could reduce dissolved oxygen concentrations, increase the concentration of pollutants, and increase salinity. In contrast, higher water flows can dilute point-source pollutants, increase loadings from non–point–source pollutants, decrease chemical reactions in streams and lakes, reduce the flushing time for contaminants, and increase export of pollutants to coastal wetlands and deltas. Where surface flows decline, erosion rates and sediment transport may decline, which may improve lake clarity while increasing the concentration of pollutants (Nelson et al. 2007). Thus, the water quality impacts associated with climate change are diverse and will pose significant challenges to water utility operations.

Habitat Destruction

Changes to the volume and timing of stream flow and stream temperature as well as the indirect climatic influences of human use of water resources will burden natural habitats. For example, the sensitivity of freshwater salmon habitat is reliant on "watershed geomorphology, vegetative cover, the relative importance of groundwater hydrology in the stream reach of interest, water resources infrastructure (dams and diversions), the amount and timing of stream flow diverted to out-of-stream uses, and the degree to which key hydrologic processes have been impaired by changes in watersheds" (Lettenmaier and Miles 2007). Projected climate variables also influence physiological processes for trees, such as photosynthesis and respiration, as well as "hydrological variables that control tree growth (e.g., soil moisture limitations); the biophysical environment (e.g., fuel moisture); and disturbance (principally fire and insect outbreaks) (Lettenmaier and Miles 2007). There are additional habitat concerns due to sea-level rise such as wetland displacement, accelerated coastal erosion, limited ability for horizontal migration, and higher salinity levels (USEPA 2008a). Thus, habitat protection is an important consideration in a utility's use of water resources.

Consequences to Water System

Water Demand

Projected impacts to California's water supply from climate change modeling show that future urban water demand will increase by approximately 8 million acre-ft/year between 2020 and 2100 (Tanaka et al. 2006). Thus, there is already a growing need for more water, which will be further exacerbated by climate change. With climate change, water utilities will be subject to increases in water demand associated with warmer temperatures and changes in precipitation, such as drought.

Generally speaking, there is a correlation between high temperatures and low precipitation and the increased demand for water (Gutzler and Nims 2005). Thus, with climate change there may be added "emphasis on planning and demand management rather than construction of new facilities (which) marks an important change in traditional water-management approaches (that) in the past have relied on the construction of large and expensive infrastructure" (Gleick and Kiparsky 2003). Until the application of new strategies, such as reuse water, become more prevalent, residential, commercial, and industrial sectors will continue to rely on potable water to address their water needs. This trend sets the stage for greater application of conservation rate structures to manage potable water demands. In addition, water conservation measures that include "education, industrial and commercial reuse, modern plumbing standards, and pricing policies can be extremely effective at mitigating the effects of climate change on regional water supplies" (Gleick and Kiparsky 2003). These types of demand management strategies are an effective means of reducing stress on water supplies in light of climate change predictions.

Irrigated agriculture is a large holder of water rights that is extremely susceptible to water shortages. With climate change, farmers may be subject to reduced crop production, soil losses due to dust storms, and higher water costs during drought (Gleick and Kiparsky 2003). Further, with higher temperatures and water variability there may be a need to change farming practices, cultivate different crops, or even stop farming specific areas due to fallow lands. Warmer temperatures may also create environments that are ideal for invasive species, pests, or disease, which could necessitate the use of pesticides or herbicides that may alter source water quality. A possible benefit of climate change is that changes to irrigation and farming practices may free water to be transferred to potable use.

In addition to irrigation, process water systems are another significant source of water demand for water utilities. Process water is vital to the operation of power generation as well as manufacturing facilities, such as biotechnology. There are growing trends to encourage use of recycled water for these purposes (CEC 2005). Actions by water utilities to reduce demand and/or slow demand growth can play a major role in reducing the impacts of climate change that are felt by the utility.

Infrastructure

Another impact of climate change on water systems will be in the planning and design of infrastructure. Current planning and design criteria are typically based on observed historical climate patterns. However, evidence shows that climate change would likely result in greater divergence from historical records, possibly leading to longer periods of drought and greater risk of flooding, which might pose a burden to water utilities. Although there are many uncertainties, it is likely that current planning criteria and design standards may be not be adequate to fully respond to climate change. For example, spillway design may need to be upgraded to accommodate higher flows and prevent flooding. Adjustments to planning criteria may be needed to accommodate increased changes in demand patterns, highlighting the infrastructural and operational challenges facing water utilities in the future. Further, other destabilizing factors associated with climate change could shorten the useful life of millions of dollars worth of existing, buried assets (Cromwell et al. 2007). Thus, utilities will need to reevaluate their planning criteria, design standards, and operational practices to cope with impacts related to climate change.

UTILITY ENGAGEMENT

Faced by important challenges to maintain a safe and reliable water supply, water managers have already demonstrated strong leadership and engagement in addressing impacts related to climate change using a variety of options. Some options undertaken by utilities to address such impacts include mitigating greenhouse gases (GHG), adapting planning to incorporate climate change, and tracking regulations. This section discusses these three options.

Mitigating GHG Contribution

Since the beginning of the 21st century, the scientific community and the general public have shown strong interest in climate change and increased global awareness in carbon footprints and GHG. Through the assessment of GHG contribution, a utility can infer its carbon footprint, and thus its impacts on climate change. As beacons in sustainability, numerous utilities have been active in determining their carbon footprint.

The main contributions to GHGs by water industry utilities come from their operational activities, such as energy use for conveyance, treatment, and distribution of water, and to a lesser extent, fuel use for utility's fleet vehicles to support operations and perform maintenance, and nitrous oxide and methane production from wastewater secondary treatment. Thus, an effective way utilities can mitigate their GHG contribution is by reducing their energy use for their operations through the implementation of energy management programs. For example, water utilities can install operationally-efficient equipment to significantly reduce the energy used to pump, transport, treat, and distribute water. To minimize energy use at treatment plants, utilities can incorporate energy-efficient equipment, such as motors, pumps, variable frequency drives, and lighting; install adjustable speed drives or variable frequency drives for chemical feed pumps, cooling water pumps, ozone destruct motors; install power monitoring systems, and replace light fixtures, occupancy sensors, HVAC control devices. Additionally, reconfiguring pump stations to be more energy-efficient could save up to 50% in energy use, possibly providing energy cost savings over $1 million in energy cost savings and gaining $500,000 in energy rebate.

Another way to minimize energy use for utilities is to promote water conservation on the consumers' end. Water conservation would reduce water demand, ultimately resulting in energy savings and reduced GHG contribution. Water-efficiency through conservation or demand management strategies can be achieved by installing ultralow-flow plumbing fixtures and other low-flow showerheads and toilets, using Energy Star certified appliances and supplies, increasing recycled water use, promoting landscape irrigation conservation, using leak detection systems, and developing public awareness, among others. Developing public awareness is crucial for utilities, not only to promote water conservation, but also to shift employee and global community behaviors, which can further mitigate GHG contribution for utilities. For example, utilities can support and encourage ridesharing among employees, the installation of LCD monitors (with about 50% electricity savings), and the purchase of hybrid vehicles.

To further minimize their energy use and reduce their GHG contribution, many water providers have supported the development and use of alternative energy sources. Utilities can reduce their GHG contribution through the use of renewable energy strategies (especially wind or solar), the use of biofuels, and/or hydropower generation. Using alternative energies is of utmost importance in today's world where energy prices are rising. While alternative energy systems can have high initial costs, they provide subsequent cost savings over their design life.

Adaptation

Traditional water planning has relied on the assumption that future climate data would have the same characteristics and variability as historical climate data (Gleick and Kiparsky 2003). However, because of climate change, the reliance on past data is risky. Water utilities need to adapt to address population growth and increased demand in the context of new climatic conditions.

The *Handbook on Methods for Climate Change Impact Assessment and Adaptation Strategies* (Feenstra 1998) described eight main adaptation strategies, namely, bear losses, share losses, modify the threat prevent effects, change use–planning buffers, change location, research–assessing potential impacts, educate, inform and encourage behavioral change. Bear losses, change use–planning buffers, and research–assessing potential impacts are the most relevant adaptation strategies for utilities and are described as follows.

Adaptation to climate change would start by overcoming a first set of problems while potentially accepting losses. Impacts related to climate change have already been observed and assessed; however, there are a lot of uncertainties in projected impacts. High uncertainties are associated with general circulation models (GCMs) as well as regional downscaled models. Indeed, results can vary highly depending on the assumptions chosen and the type of model used, potentially leaving parameters outside of consideration and resulting in a lack of transparency. Additionally, modeling projected impacts related to climate change can be costly and pose a financial burden to municipal utilities, who struggle to determine how to respond to these predictions. Adaptation to impacts may involve high-stakes decisions, potentially leading to expensive investments and costs.

Change in use with planning buffers is another adaptive strategy that water utilities can apply by adjusting demand projections for temperature effects, diversifying supplies, building buffers into supply targets, increasing storage capacity, and evaluating energy (hydro) impacts. Another adaptive strategy is to conduct in-depth research assessing potential impacts by evaluating physical facility vulnerability to flooding, evaluating potential low water flow impacts on receiving waters discharges/influent flows, encouraging conservation (e.g., evapotranspiration controllers), developing tiered pricing, and engaging stakeholders.

Monitoring Regulations

Global GHG regulations are governed by the Kyoto Protocol. However, this agreement has not been signed by the US, making voluntary actions to reduce GHG more common at the local, regional, and state levels. Water utilities have often led the way in instigating such actions, in which they continue to be active. The most important action for water utilities in Southern California is at the state level, with California's Global Warming Solutions Act (AB32) from 2006. AB32 is the first act regulating GHG emissions in the US, which requires cutting the state global GHG emissions 25% below 1990 levels by 2020 and 80% below 1990 by 2050. The water sector is not considered a major emitter sector and thus is not subjected to GHG mandatory reporting. However, the water sector is included in the State Scoping Plan, which recommends GHG reduction measures of 5 million metric tons of equivalent carbon dioxide for the water sector. Consequently, water utilities will have to mitigate their GHG emissions to achieve the goals set by the AB32 Scoping Plan. More information on GHG contribution mitigation opportunities for water utilities is described above.

Other actions in which water utilities can be involved exist at the regional and national/federal levels. For example, action at the regional level where Southern California is involved includes the Western Climate Initiative (WCI) formed by the states of Arizona, California, New Mexico, and Washington in 2007. This initiative set a goal to reduce participants' emissions 15% below 2005 levels by the year 2020. At the national level, the USEPA set a national goal to reduce GHG emissions 18% over a 10-year period between 2002 and 2012. The USEPA is also in the process of creating a federal mandatory reporting rule, which will be ready for approval in the second half of 2009.

---

*Editors Note:

The IPCC projected that sea level rise by 2100 will be 0.2 to 0.6 meters above circa 1990 levels (Solomon et al., 2007). Recent research suggests that when potential for melting from major ice sheets such as Greenland is considered, sea level by 2100 may be as high as 1.5 to 2 meters (Rahmstorf, 2006; Pfeffer et al., 2008). Nonetheless, the analysis is useful to show potential long-term risk from sea-level rise.

REFERENCES

CEC (California Energy Commission). 2005. California's Water–Energy Relationship. CEC-700-2005-011-SF.

Cayan, D.R., A.L. Luers, G. Franco, M. Hanemann, B. Croes, and E. Vine. 2007. Overview of the California climate change scenarios project. *Climatic Change.* DOI 10.1007/s10584-007-9352-2.

Chung, F. 2006. Progress on Incorporating Climate Change into Planning and Management of California's Water Resources. Technical Memorandum Report. July.

Cromwell, J., J. Smith, and R. Raucher. 2007. Implications of Climate Change for Urban Water Utilities. Association of Metropolitan Water Agencies. December.

Easton, D. 2006. Preliminary Climate Change Impacts Assessmnet for SWP and CVP Operations. Calfed Science Conference. October 25.

Feenstra, J.F. 1998. *Handbook on Methods for Climate Change Impact Assessment and Adaptation Strategies.*

Florsheim, L.D., and M.D. Dettinger. 2007. Climate and floods still govern CA levee breaks. *Geophysical Research Letters.* 34, L22403, doi:10.1029/2007GL031702.

Gleick, P., and Kiparsky, M. 2003. *Climate Change and California Water Resources: A Survey and Summary of the Literature.* Pacific Institute for Studies in Development, Environment, and Security.

Groves, D.G., R.J. Lempert, D. Knopman, and S.S. Berry. 2007. Preparing for an Uncertain Future Climate in Inland Empire: Identifying Robust Water-Management Strategies. Rand Corporation. September, 17.

Gutzler, D.S., and J.S. Nims. 2005. Interannual Variability of Water Demand and Summer Climate in Albuquerque, New Mexico. *Journal of Applied Meteorology.* 44, 1777–1787.

IPCC. 2008. Technical Paper on Climate Change and Water. April.

Kim, J., J.-E. Lee, and N.L. Miller. 2001. A multi-year regional climate simulation for the Western United States using the Mesocale Atmospheric Simulation Model. *J. Geophys. Res.–Atmosphere,* LBNL-44548.

Lettenmaier, D., and Miles, E. 2007. HB 1303 Interim Report: A Comprehensive Assessment of the Impacts of Climate Change on the State of Washington. University of Washington JISAO CSES Climate Impacts Group.

Madsen, T., and E. Figdor. 2007. *When It Rains It Pours: Global Warming and the Rising Frequency of Extreme Precipitation in the United States.* Texas Environment Research and Policy Center. December.

Nelson, B. et al. 2007. *In Hot Water: Water Management Strategies to Weather the Effects of Global Warming.* Natural Resource Defense Council.

Palmer, R., and Wiley, M. 2005. *Impact of Climate Change on Municipal Water Supplies in Washington.* University of Washington Climate Impacts Group.

Pfeffer, W. T., J. T. Harper, and S. O'Neel. 2008. "Kinematic Constraints on Glacier Contributions to 21st-Century Sea-Level Rise." *Science.* 321: 1340-1343.

Rahmstorf. S. 2006. "A Semi-Empirical Approach to Projecting Future Sea-Level Rise." *Science.* 315: 368-370.

Solomon, S. D. Qin, M. Manning, Z. Chen, M. Marquis, K. B. Averyt, M. Tignor, and H. L. Miller (eds.). 2007. (IPCC). Intergovernmental Panel on Climate Change. 2007. *Climate Change 2007: The Physical Science Basis. Contribution of Working Group I to the Fourth Assessment Report of the Intergovernmental Panel on Climate Change.* New York: Cambridge University Press.

Tanaka,S.K., T. Zhu, J.R. Lund, R.E. Howitt, M.W. Jenkins, M.A. Pulido, M. Tauber, R.S. Ritzema, and I.C. Ferreira. 2006. Climate Warming and Water Management Adaptation for California. DOI: 10.1007/s10584-006-9079-5.

USEPA. 2008a. *National Water Program Strategy: Response to Climate Change, Public Review Draft.* Office of Water, U.S. Environmental Protection Agency.

USEPA. 2008b. http://epa.gov/climatechange/effects/coastal/SLRLandUse.html. *The Potential Effects on Global Climate Change in the United States. Report to Congress.* Chapter 14. California by G.A. King, R.L. DeVelice, R.P. Neilson, and R.C. Worrest. Originally published in 1989 by the USEPA Office of Policy, Planning, and Evaluation.

USEPA. 2008c. http://epa.gov/climatechange/effects/coastal/SLRLandUse.html Chapter 1. Sea Level Rise and Wetland Loss: An Overview by J.G. Titus. USEPA Office of Policy.

Wallis, M., M.R. Ambrose, and C.C. Chan, 2008. Climate change: Charting a water course in an uncertain future. *AWWA Journal.* 100:6, 70–111.

– 24 –

Melbourne Water Climate Change Study
Impact and Adaptation of Melbourne's Water Supply, Wastewater, and Stormwater Systems

Carol Howe and Bruce Rhodes

INTRODUCTION

Since 1997, southeastern Australia has had an extended period of below-average rainfall and higher average temperatures. This has resulted in average annual streamflows into Melbourne's main water storages for the past 12 years being around 39% below the long-term average recorded to 1996 (see Figure 24-1) and storage volumes falling from capacity levels in 1996 to levels (December 2008) of around 34%. The reduction in mean annual streamflow and low storage volumes has meant water restrictions have been imposed at various times since 2002, with severe water restrictions in place since 2007 when the lowest annual streamflow on record occurred. Climate research suggests *The combination of record heat and widespread drought during the past five to ten years over large parts of southern and eastern Australia is without historical precedent and is, at least partly, a result of climate change* (Australian Government Bureau of Meterology, Drought Statement, 3 Dec 2008).

In 2002, the Water Resource Strategy for the Melbourne Area Committee released its final report, *Water Resource Strategy for the Melbourne Area*. The report recognised the increasing body of scientific evidence that gives a collective picture of a warming world and other climate changes and the potential of significant implications for our water resource systems. The committee recommended that Melbourne Water and the Melbourne retail water companies continue ongoing, active evaluation of climate change impacts on water supply and demand measures. In 2005, Melbourne Water engaged the Australian Scientific and Industrial Research Organisation (CSIRO) to undertake a study to investigate the implications of climate change for Melbourne Water's water, sewerage, and drainage systems. The study was among the first globally to examine these systems in a combined manner. This paper provides a summary of the results of the 2005 study. The paper also provides an overview of the situation and adaptation actions as of the end 2008 due to many actions being bought forward as a result of the observed rapid shift in climate conditions across southeast Australia.

Key features of the 2005 study included:

- Development of Melbourne area climate change scenarios for 2020 and 2050 based on "downscaling" of the Intergovernmental Panel on Climate Change (IPCC) scenarios.
- An overview of the potential implications of climate change coupled with potential water demand scenarios for Melbourne's water, sewerage, and drainage systems including identification of major risk areas.

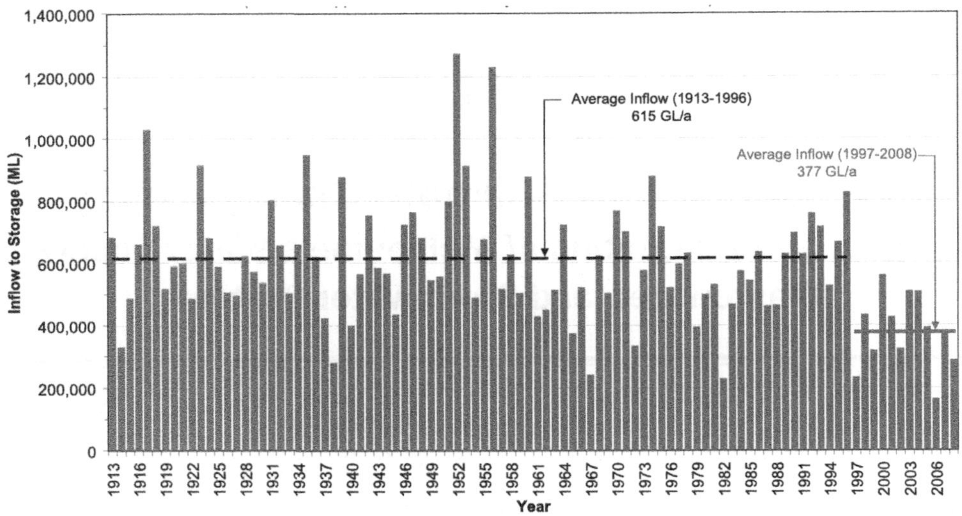

Figure 24-1. Annual streamflow at Melbourne's major harvesting reservoirs (Thomson, Upper Yarra, O'Shannassy, and Maroondah reservoirs)

- A detailed quantitative case study for Melbourne's water supply system and potential adaptation options.
- Semiquantitative case studies of the potential for increased sewerage overflows and local flooding.
- Identification of potential adaptation strategies to mitigate the impacts of climate change given the range of uncertainties identified during the study.

CLIMATE CHANGE PROJECTIONS FOR THE MELBOURNE AREA

In spite of uncertainties associated with climate change predictions, consistent trends for Melbourne could be interpreted from the study. These trends include:

Increased Temperature

Annual temperature increase is 0.3°C to 1.0°C, with a mean of 0.5°C in 2020, and 0.6°C to 2.5°C, with a mean of 1.4°C in 2050. There is a slight tendency for the warming to be higher in summer and less in the other seasons.

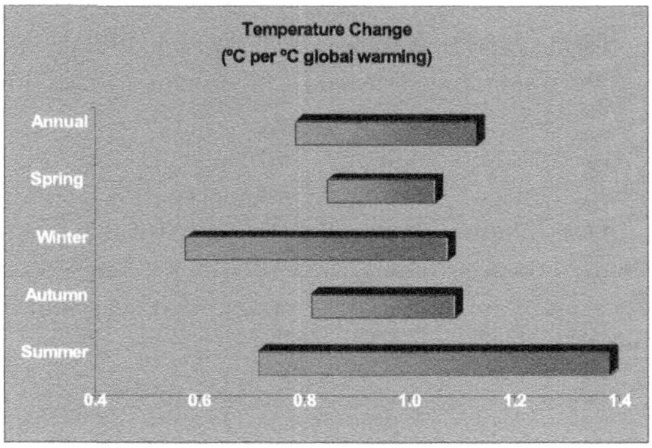

Reduced Rainfall

Annual precipitation change is 5% to 0%, with a mean of –2% in 2020, and –13% to 1%, with a mean of –4% in 2050. Two patterns of change became apparent. One pattern is typical of southern Australia and tends toward decreased rainfall in spring and an increase in summer and autumn. The other pattern is typical of Tasmania and tends toward increases in winter and decreases in summer.

More extreme events with more hot days (above 35°C), more dry days, and increased rainfall intensity during storm events are anticipated.

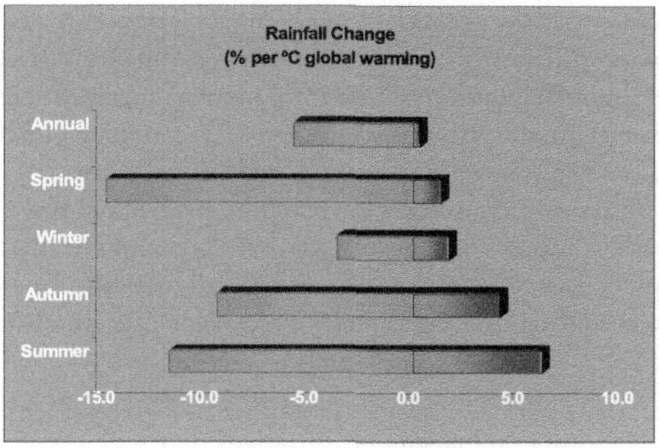

AREAS OF POTENTIAL RISK

Workshops and specific case studies were undertaken to identify the range of implications of climate change scenarios for the water supply system and water availability, the sewerage system, the drainage network, and Melbourne's waterways. The workshops with staff and external experts highlighted a range of potential risks to Melbourne Water's systems. The major risks identified in the 2005 study included:

Water Supply
- Reduced water supply due to decreased stream flows
- Decreased stream flows due to increased risk of bushfires in catchment areas
- Reduced environmental condition of streams with associated implications for water harvesting in regulated and unregulated streams

Sewerage System
- Increased potential for corrosion and odours caused in the sewerage network as a result of increased sewage concentrations associated with water conservation, increasing ambient and seasonal temperatures, and longer travel times within the sewer network.
- Increased incidence of sewer overflows due to increased rainfall intensity during storms
- Increased risk of pipe failure and collapse due to dry soil conditions
- Increased salinity levels in recycled water due to rising seawater levels resulting in increased infiltration to sewerage network and at wastewater treatment plants

Drainage
- Increased flooding risk and property damage due to increased rainfall intensity during storms
- Increased risk of damage to stormwater infrastructure and facilities (e.g., underground drains, levee banks, pump stations, etc.) due to higher peak flows

Receiving Waters

- Reduced health of waterways due to changes in base flows
- Potential for negative water quality impacts in Port Phillip Bay due to increased concentration of pollutants entering the bay (longer periods between runoff events and then high intensity events leads to concentrated pollutant runoff) and higher ambient bay water temperatures

POTENTIAL ADAPTATION STRATEGIES

Climate change scenarios and the potential implications for water service provision, sewerage treatment and disposal, and health of receiving waters highlight increased uncertainty for future resources planning. Adaptation will require ongoing review of climate change, population growth, land-use changes, water use, effectiveness of implemented water demand and supply-side programs, and the capacity of current systems to cope.

The Melbourne 2005 study identified a range of initiatives above those planned that should be investigated further, including changes to water, sewerage, and drainage system design criteria (particularly in greenfield or redevelopment areas) and investigation of additional demand or supply-side options including nontraditional sources such as groundwater and desalination.

Water Supplies

There are many water resource management initiatives that will aid adaptability or provide a buffer for climate change. Many of these initiatives have been investigated for Melbourne or in other locations throughout Australia and the world. For water supply, many of the initiatives are described in detail in the Victorian Government's "21st Century Melbourne: a WaterSmart City" (October 2002) and in "Securing Our Water Future Together: Our Water, Our Future" (June 2004). Information on adaptation actions considered after the release of the 2005 study are detailed in "Sustainable Water Strategy Central Region, Action to 2055" (October 2006) and "Our Water, Our Future, The Next Steps in the Government's Water Plan" (June 2007). The range of options identified in the 2005 study for Melbourne that might be considered in adapting to climate change for water supply management included:

Maintaining focus on efficient domestic water use

- Long-term behaviour change achieved through a range of levers including education and regulation including restrictions, incentives, pricing, and rating schemes
- Water efficient fittings, showerheads, appliances
- Improved gardening and outdoor water-use practice
- Installation of rainwater tanks, greywater, effluent, and stormwater recycling systems and water-sensitive urban design, particularly in new developments
- Closed-loop water systems in new commercial and residential high rises

Industrial and commercial water use efficiency

- Water audits and management plans
- Improved water efficiency and increased use of recycled water

Open-space management

- Audits and water management plans
- Improved water-use practices and recycling schemes

Water authorities

- Leakage reduction
- System optimisation

Existing surface water supply augmentation and efficiency options
- Reconnection of the Tarago Reservoir to supply
- Improvements in system operation to improve water harvesting efficiency
- Additional pumping into Sugarloaf reservoirs subject to environmental considerations

Catchment and reservoir management
- Manage forested catchments to minimise the water yield impacts from disturbances such as bushfires or logging
- Evaporation reduction or rainfall-enhancement measures

Investigation of possible new water sources
- Desalination
- Major recycling schemes to supplement or replace water currently supplied from drinking water sources including releases for environmental flows
- Aquifer storage and recovery of recycled water
- The potential for water sharing and/or trading between sectors (e.g., irrigation and urban) and basins

Sewerage System

Investigation of the potential implication of climate change scenarios overlaid on expected long-term changes in population and demographics and subsequent sewage flows suggest that gradual climate changes could be more readily adapted to than rapid shifts in climate. Possible strategies identified in the 2005 study to assist the sewerage system in adapting to climate change are consistent with normal planning practices and might include:

Measures to reduce peak flows in wet weather
- Increased indoor demand management activities that lower sewage flows and reduce overflow risk
- Ongoing and periodic review of sewerage system strategies and operations
- Local treatment plants to reduce network loading during peak flows
- Ongoing review of strategies to address hydraulic constraints and overflow risks
- Increase storage capacity within system
- Limit growth expansion and/or connections to parts of the system when there are potential capacity constraints
- Review of strategies for emergency relief structure operations
- Monitoring and maintenance of sewer systems to reduce infiltration
- System rehabilitation
- Encouraging sewer construction methods which have much lower infiltration rates in new developments

Measures to manage sewage quality
- Lining of sewers to avoid "corrosion" by sulfuric acid–producing microbes
- Ongoing monitoring of trends in sewage quality, quantity, and temperature, and treatment processes
- Maintain sewer rehabilitation and cleaning regimes
- Examine potential for oxygen injection and odour control to manage changes in temperature

Other initiatives
- Increase levee bank height along lagoons to provide buffer against sea-level rises
- Encourage cluster systems and local treatment to aid recycling options
- Seek long-term improvements to sewage water quality to reduce salts, nutrients and other pollutants which will assist recycled water quality

- Industrial waste minimisation
- Domestic sewage quality, e.g., soaps, detergents, etc.
- Auditing of new installations
- Increase community education on sewers, effluent quality, garden plants, etc. to assist in improving sewerage quality and infiltration
- Ongoing development of recycling schemes, including new developments, sewer mining, and opportunities for retrofits

Drainage and Waterways

Potential drainage management adaptation strategies identified in the 2005 study included:

- Ongoing and periodic review of waterways and strategies, flood modelling, and/or drainage schemes to take account of potential climate change scenarios
- Ongoing monitoring, including water quality, environmental indications and groundwater levels, particularly in wetlands
- Water-sensitive design in new developments to reduce runoff rates and improve water quality
- Retain flood plains and floodways in new urban developments
- Identify options for upstream flow attenuation for flood mitigation including increased local storage
- Education to assist local waterway management and maintenance
- Maintain programs for stream rehabilitation and stream frontage management
- Ongoing review of planning and operational strategies for:
  - Drainage network at a local and system level to take into account current knowledge of potential climate change scenarios
  - Floodgates, levees, storage, and pump stations
  - Combined sea level, surge, and storm frequencies in specific areas
- Periodic review of environmental flows, drought plans, streamflow management, and water allocations and water allocation principles for managing diversions from waterways for climate change
- Develop incentives to aid water use and allocation efficiencies

Planning

The 2005 study identified that climate change scenarios should be included in periodic reviews of major strategies across the business, including water supply plans, drought plans, sewerage strategies, and environmental flow assessments.

CASE STUDIES

A key part of the 2005 study was the combination of qualitative and quantitative assessments of the potential impacts of climate change. Three case studies were selected to further explore higher-risk areas and to undertake quantitative assessments of the extent of these risks. These included a review of the water system to identify potential reductions in water availability and the sewerage and drainage networks for increases in sewer overflows and flooding.

Case Study 1—Water Supplies

The 2005 study undertook detailed quantitative analysis of the impact of reduced streamflows and found that average long-term streamflows potentially would be reduced between 3% and 11% by 2020, and as much as 7% to 35% by 2050. This assessment was based on detailed downscaling of climate change projections to local catchment conditions and then using rainfall runoff models to assess changes to catchment runoff conditions. Water supply systems headworks analysis was then undertaken to assess potential supply availability against a range of population and water demand scenarios. Although any of the projected climate and population scenarios could eventuate, the greatest confidence is in the mid-range area.

Results of the mid-range climate change scenario project a 4% reduction in water supply by 2020, rising to 9% by 2050. Mid-range estimates are quite conservative of the impacts on Melbourne's water supplies. The estimates did not account for the cumulative impact of climate effects beyond parameters evaluated in the case study, assume that the 15% per-capita demand reduction target outlined in strategies at the time would be met by 2010 and that further per-capita demand reductions and system augmentations to Tarago treatment plant, O'Shannassy pipeline duplication, Winneke treatment plant upgrade, and East–West transfer upgrades will occur by 2050.

At the time of the 2005 study, it was expected that planned demand- and supply-side actions would compensate for decreased supply through 2020, although some actions may need to be implemented sooner than currently planned. Some of the system augmentations outlined in the water resources strategy by 2050 may be required by 2020 to maintain existing levels of service. Additional measures to those outlined in the water resources strategy (such as desalination or system augmentation) were identified as potential options between 2020 and 2050 to maintain existing levels of service, especially if anything other than mild climate change and growth projections occur. Contingency planning is required to ensure appropriate water supply adaptation techniques if major-step changes to climate should occur.

As noted in the introduction, since 1997 streamflow conditions into the major reservoirs have been significantly lower than the long-term average and similar to projections expected under high climate change scenarios by 2050. As a result, many actions to respond to the shortfalls have been implemented which are outlined later in the paper. From the 2005 study and based on water-supply yield estimates, population and climate change projections illustrated in Figure 24-2 show the extent of volumetric surplus and deficits expected for the various scenarios.

Case Study 2—Sewer Overflows

The 2005 study identified that the intensity of storm events was the most important variable for the drainage and sewerage systems. Under climate change scenarios, changes in rainfall totals, temporal patterns, and intensities of storm events are expected to increase the risk of sewer overflows within both the transfer system and local reticulated sewage collection system.

A desktop study was undertaken as part of the study in 2005 to analyse the impact on the sewerage and drainage network of 5, 10, and 20 percent increases in rainfall storm event totals per degree of climate change. The flooding analysis was done on "representative" small catchment areas.

Conclusions were that sufficient capacity exists within most of the sewerage and drainage system to accommodate moderate changes (5–20%) in storm rainfall totals with minimal increased overflows or flooding. However, the magnitude of changes due to changed storm event intensity are not well understood, are very site specific, and cannot easily be interpreted from existing climate change information requiring further research.

Melbourne's sewerage system is separate to the stormwater systems and is designed to operate independently. However, water may enter the sewerage system as infiltration into pipelines from rainfall, or through illegal property connections (mainly stormwater runoff from roofs) going directly into the sewerage system. Rainfall does result in extra volumes of water entering the sewers. During extreme storm events, high volumes of infiltration can

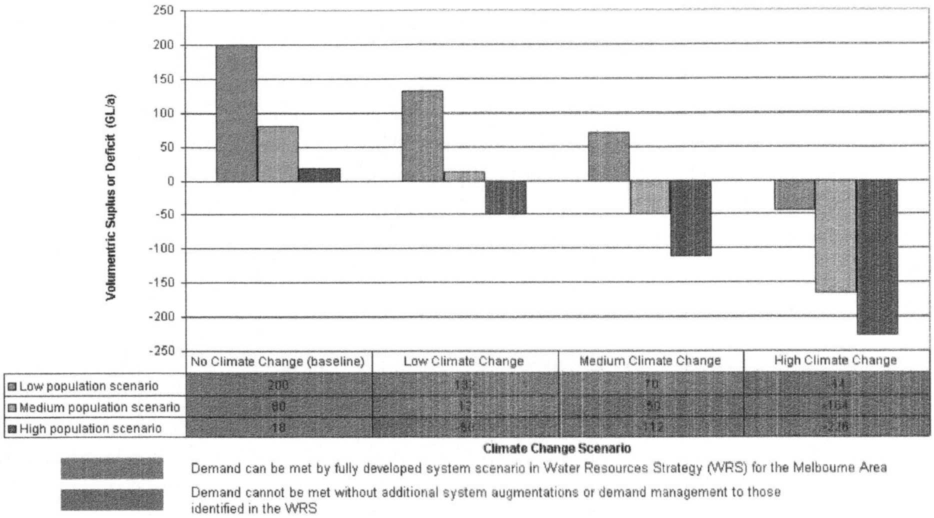

Figure 24-2. Potential buffer (positive value) or shortfall (negative value) of system yield for climate change and projected average annual demand (billion litres) scenarios in 2050

result in increased flows in sewers and in local areas this can cause sewers to overflow from emergency relief structures to the environment. Melbourne Water maintains approximately 43 emergency relief structures thorough the transfer system.

Under climate change scenarios, rainfall totals and intensity are expected to increase. However, the extent of these changes and the implications for storm events and the design of sewerage systems was uncertain. Detailed modelling and analysis of the hydraulic behaviour of sewer flows in the main transfer system operated by Melbourne Water for the 1-in-5-year storm event enabled assessment of the potential for sewer overflows under future development, flow, and storm scenarios.

Rainfall temporal patterns and storm event rainfall totals are key factors in assessing the changes in expected flow rates and the risk of sewer overflows. CSIRO Division of Atmospheric Research suggested a preliminary estimate of a 5% increase in total storm event rainfall per degree of climate warming to assess the sensitivity and implications for the serage system. These estimates were preliminary and provided a useful basis for assessing the sensitivity of the sewerage system to such variations in rainfall totals. Assessment of the potential implications of change to rainfall totals for the sewerage transfer network suggests that changes in rainfall totals of up to 10% for a given recurrence interval event may not impact on system performance, but the extent of overflows is affected, indicatively, by larger shifts in rainfall totals on current events.

While the climate change models in the 2005 study suggested the potential for increased storm intensity, the translation of the climate change scenarios to local storm and catchment events is highly uncertain given the range of influences on local conditions. The sewerage transfer system model was therefore run, and system behaviour assessed for:

- Design event behaviour for 2020
- System behaviour in 2020 with a 5% and 10% increase in rainfall totals per degree of climate change

The study highlighted that the 24-hour rainfall duration events were critical events for the design of Melbourne Water's sewerage transfer network. Shorter duration events are generally more critical for the design of smaller catchments and pipes.

Under the above-mentioned 24-hour scenario, the total rainfall increases were 3% over current design totals. Preliminary assessment in the 2005 study showed that this small increase in rainfall total over the 24-hour duration event was expected to only result in minimal increase in the modelled rainfall-induced infiltration to the sewerage transfer network and to the risk of overflows in the sewerage network. The modeling study showed that the sewerage transfer network in the year 2022/23, under current augmentation plans, would be compliant for the 1-in-5-year event with the small shift in additional rainfall anticipated under climate change. This preliminary model result suggested if the shifts in rainfall totals were to increase for a given ARI, for example rainfall events of the 1-in-10-year ARI were to become a 1-in-5-year event, there may be some limited increases in sewer overflows. This increase in rainfall total from a 1-in-5-year ARI event to 1-in-10-year ARI event represents a 10.6 mm or 14% increase in rainfall totals. This suggested a potential increase in infrastructure investment to meet standards should storm event totals change significantly under climate change.

Storm events are the key factor contributing to sewer overflows from emergency relief structures in the sewerage transfer networks. Under climate change scenarios, rainfall totals and intensity are expected to increase. However, the extent of these changes and the implications for recurrence intervals for storm events is highly uncertain. Under climate change scenarios, changes to rainfall totals, temporal patterns, and intensities in storm events could increase the risk of sewer overflows both within the transfer system and local reticulated sewage collection system. This risk will also be dependent on other factors such as urban development patterns and sewage system repair and renewal rates.

Case Study 3—Flooding

The same climate patterns as described for sewer overflows could also lead to increased local flooding. Factors affecting the frequency, depth, and duration of flooding in urban areas are very complex, and may be influenced by a range of interacting influences. Local property impacts of flooding, often the most visible and costly impact of flooding, will also be influenced by local factors, including:

- Local backwater influences and wave action
- Design standards used for property siting. This includes the return period for the flood event and the freeboard above the flood level to provide for local uncertainties such as backwater and wave action
- Extent of hail and rubbish in runoff and extent of flow blockage during an event
- Prior warning of storm, flood, or potential flooding conditions

Melbourne's drainage system has been in place for more than 70 years. Design standards, which have been in place in for new development areas for more than 25 years provide for a "100-year" ARI design standard for flooding. This ARI is used to define the extent of the overland flow along the drainage system or flooding from overtopping of river banks which would occur on average once in every 100 years. In other words, it means that there is a one-percent chance in any one year of such an event occurring.

In many parts of the greater Melbourne area, often in older established areas, the drainage system was built to the standards of flood frequency and severity that existed when they were designed, which may be less than that provided under current standards in newer areas. In areas where drainage systems were developed prior to the 100-year ARI standard and there is local redevelopment, Melbourne Water has a capital works program to improve infrastructure and work with local councils to address flooding issues.

Rainfall temporal patterns and storm event rainfall totals are key factors in assessing the changes in expected flood frequency, depth, extent, and duration over current estimates for a given recurrence interval flood event. Similar to the sewerage system case study above, the 2005 study was based on a preliminary 5% increase in total storm event rainfall per degree of climate warming, as suggested by CSIRO. The translation of the climate change scenarios to local storm and catchment events is highly uncertain given the range of influences on catchment flooding described above.

A desk study was undertaken of a small catchment in the outer eastern suburbs of Melbourne for which detailed flood studies for flood mitigation works for the 100-year ARI event had been undertaken. The drainage infrastructure in this catchment is typical of that in many urban areas of Melbourne.

The use of an area for which flood studies had been undertaken enabled existing catchment models to be used for quantitative assessment of the impact of climate change scenarios using calibrated catchment models on the expected flow rates associated with flooding and flood levels.

Modelling work for the 100-year ARI event showed that the two-hour duration event produced critical flows for the case study catchment. Rainfall totals for the 100-year ARI, two-hour storm event were factored in accordance with a 5% increase in totals per degree of climate change for years 2020 and 2050 and were based on a mean temperature increase for the years 2020 and 2050 of 0.6°C and 1.5°C, respectively. Flood flow rates and their impact were seen for each scenario.

Table 24-1 shows the results of the study for a 5% increase per degree of climate change and shows the changes in flow rates, and flood levels with a 5% increase in rainfall per degree of climate change by 2020.

The table indicates that for the small catchment in the case study, there are limited incremental shifts in rainfall totals, peak flow rates, and flood levels associated with a 5% increase in rainfall total. In particular, flood level changes were well within the 300 mm freeboard criteria adopted for developments in potential flood areas, and well within the expected local on-site variability from model error, backwater effects and wave action, and local variability. In order to test the sensitivity of the results to the increase in rainfall total, further assessment were undertaken assuming a 10% and 20% increase in rainfall per degree temperature increase for the 100-year ARI. Table 24-2 shows that for the 10% and 20% case studies, the average increase in the flood levels are 30 to 50 mm in 2020, respectively. Freeboard allowances included in existing and proposed development provide a significant buffer for climate change. Even in older areas, many properties, particularly along waterways, currently have a freeboard of up to 600 mm.

Climate change is projected to reduce the interval between intense events. As shown in the table, a 5% increase in rainfall intensity will see the current 130-year event become a more frequent 100-year event by 2020. With a 20% increase in rainfall intensity, the current 210-year event could be expected to occur every 100 years.

There would be considerable cost to the community to upgrade the existing drainage system to the 100-year ARI in all areas. In these circumstances, the most practical and equitable solution is to adapt development to suit the existing drainage system.

The 2005 study suggested further work to assess the implications of climate change on rainfall totals as well as temporal patterns. At the time of the study, it was recognised that it was difficult to make a direct interpretation of climate change impacts for flooding other than in general terms. Flood impacts are likely to be site-specific and will need to be reviewed for local conditions and in the context of local flood management controls and investment. Further work on climate change has been included in the Flood Management and Drainage Strategy, which was completed in 2007 and is discussed later.

Table 24-1. Case study results: 5, 10, and 20% increase in rainfall totals per degree of climate change

Change by 2020	5%	10%	20%
Total rainfall (mm)	60	61.7	65.2
Increase in total rainfall from current (mm)	0.8	3.5	7
Increase in peak flow rate (cumecs)	0.5 (4%)	1.0 (8%)	2.0 (17%)
Average increase in flood levels (mm)	10	30	50

Table 24-2. Results of ARI analysis

Case Study	Scenario	Rainfall Total (mm)	Average Rainfall Intensity (mm/h)	Average Recurrence Interval (years)
5%	Year 2020	60.0	30.00	130
	Year 2050	62.6	31.30	160
10%	Year 2020	61.7	30.85	150
	Year 2050	66.9	33.45	240
20%	Year 2020	65.2	32.60	210
	Year 2050	75.7	37.85	>500

The drainage case study was done for only a single catchment, and did not reflect the effects of increased rainfall on other catchments as the impacts of rainfall events and flooding will be site dependent. Other factors, such as the permeability of soils under different climatic conditions, are also poorly understood.

RECOMMENDATIONS FROM THE 2005 STUDY

Historically, Melbourne has depended on large surface water storages for its water supplies. This resulted in large buffer capacities within the system to counter impacts such as climate variability and population increases. The move to holistic demand and supply-side planning, along with the need for improved efficiencies, has meant that system capacities are moving toward closer alignment with expected demand, resulting in reduced buffer capacity in the system to cope with change.

The 2005 study found that climate change for Melbourne is a measurable variable with defined sensitivities and made it possible to consider climate change both qualitatively and define the quantitative impacts. The study showed that climate change is likely to have significant and measurable impacts on water resource management, including significant decreases to water supplies. The study resulted in a better knowledge of climate change risks across the system and raised some questions about the adequacy and reliance that can be placed on historical climatic and hydrologic data in planning for human-induced climate change.

The 2005 study identified a number of actions to assist ongoing adaptation, including:

- Ongoing monitoring, including periodic review of climate change scenarios and population projections and implications for natural and water resources systems
- Continual review of best practices adopted throughout the world for management for climate change and variability

- Incorporation of climate change implications and assessment to the extent possible in the design, planning, and operation of major resource management systems—in effect, a "no-regrets" policy and undertaking of contingency planning for climate change
- Exploration of nontraditional (desalination, recycling, aquifer storage recovery, catchment management, water sharing and pricing) and other resources management options
- Development of a strategy for climate change that looks at adaptation and prioritisation of planned activities to maximise resilience against climate change
- Further evaluation of the implications of climate change on sewerage and flood planning; in particular, on the potential implications of climate change on storm event temporal patterns and return intervals, and implications on sea-level changes and surges in local areas
- Review of engineering design criteria adopted for planning purposes, including for water supply design, drought security of supply, and yield estimation practice given the potential implications of climate variability and change
- Risk assessment of multiple, potentially cumulative factors such as climate change impact to water supplies, coupled with other water supply factors such as increased risk of catchment bushfires and environmental flow requirements
- Additional case studies of high-risk areas to better quantify the potential extent of impact

THREE YEARS LATER—THE SITUATION BY END 2008

Ongoing below-average streamflow conditions, and the 2006 drought year, which saw the lowest recorded annual streamflow in Melbourne's history, has resulted in increased and ongoing attention to Melbourne's water systems.

In 2006, a major review of Melbourne and the central region water requirements was undertaken through the Victorian Government's Central Region Sustainable Water Strategy (see www.ourwater.vic.gov.au/programs/sws/central/documents) and supported by the Water Supply Demand Strategy. These strategies saw the inclusion of climate change projections in regional water resources supply and demand planning, based on the 2005 study projections and further detailed assessments of climate change impacts on major basin streamflows undertaken by the Department of Sustainability and Environment. The projections used for Melbourne provided for the high, medium, and low climate change projections from the 2005 CSIRO/Melbourne Water study but also assumes a "climate shift" scenario based on the continuation of the reduction in streamflow recorded across Melbourne's main water supply catchments to that time.

The severity of the 2006 drought year, in conjunction with the longer-term observed shift in streamflow conditions, resulted in a series of major actions being announced by the Victorian Government in June 2007 (Our Water, Our Future—The Next Stages of the Government's Water Plan—see: www.ourwater.vic.gov.au/programs/next-stage). This $AUS4.9-billion plan is currently being implemented and will provide water security for Victoria's growing population. Specific actions to respond to the observed shift in hydrologic conditions include:

- The reconnection of Tarago Reservoir (storage capacity 37,500 ML) by mid-2009.
- A 150,000 ML/a desalination plant to augment water supplies from nonclimate-dependent sources by late 2011
- Modernisation of irrigation infrastructure in the Goulburn Murray region of Victoria and access for Melbourne to some of the water savings (the pipeline will transfer an average of 75,000 ML/a) through an interbasin transfer from the first half of 2010

- Further studies to investigate opportunities for increased recycling, including upgrading treatment to tertiary standard from Melbourne's Eastern Treatment Plants (ETP) to enable it to be reused. Options being investigated include for power system cooling and substituting recycled water for high seasonal flows in the Yarra River.
- Supporting new and ongoing water conservation programs which in 2006/07 saw a 28% reduction in per capita water consumption from the average levels of water use in the 1990s
- Expansion of Victoria's water grid to improve interconnectivity between systems, including between Melbourne and Geelong

While much attention has been given to water supplies in the region, in 2007 Melbourne Water completed the Flood Management and Drainage Strategy. This strategy outlined that the change in rainfall patterns is more likely to result in an increase in overland flows rather than riverine flooding because the short, sharp nature of these intense storms results in sudden large volumes of runoff water. A sensitivity analysis undertaken by Melbourne Water of increasing rainfall intensity by 5% per degree of temperature increase suggested a similar increase in runoff volumes of 4% by 2020 to 25% by 2100.

Climate change considerations are one of the key elements being addressed by the current review of the sewerage strategy for Melbourne. This study is being undertaken as a joint study between Melbourne's retail water companies and Melbourne Water and will include a range of climate change scenarios to assess the robustness of the sewerage system and treatment processes to climate change risks such as increased storm intensity and storm surge/sea-level rise.

Melbourne Water also continues with its own internal processes for climate change adaptation. The Melbourne Water Climate Change Adaptation Program is building on work from the 2005 study and subsequent risk assessments which provided an overview of business risks and will seek to identify risks within business-level activities and at an operational level. The risks are identified at a detailed level and prioritized through a risk assessment process considering the likelihood of impacts and consequences. Following risk identification, appropriate actions and responses can be developed throughout the business, including the consideration of appropriate triggers for action including further studies, reviews of planning criteria or policies, or climate and weather trends and events. The combination of these approaches results in consideration of adaptation at the strategic level, through relevant major plans and strategies and at the operational level.

Climate change and variability presents a key challenge for sustainable water management in southeastern Australia and for Melbourne's water, sewerage, and drainage systems. Continuing low levels of rainfall and record low inflows to Melbourne's water storages since 1997 and the completion of the climate change study in 2005 have had significant implications, and actions have been implemented by the Victorian Government to secure water supplies for the next 50 years. The 2005 collaborative climate change study between CSIRO and Melbourne Water represented one of the first globally to examine water systems in a combined manner and provided key information on the risks and possible extent of impacts to support ongoing adaptation actions for responding to climate change.

There are significant challenges and uncertainties in translating climate change projections to a local scale; however, the Melbourne case study has shown the considerable learnings and benefits in working through such studies.

ACKNOWLEDGMENTS

The authors would like to acknowledge the contribution of CSIRO (Roger Jones, Paul Durack, Shiroma Maheepala, Grace Mitchell, and Penny Whetton) and Melbourne Water (Robert Yurisich, Udaya Kalarthna, Simone Esler, Geoff Gardiner, Rod Watkinson, Merran Price, Keith Boniface, Michelle Wotten, Tony Corbett, Alex Walton, Eamonn Kelly, and Peter Scott) to the original 2005 CSIRO–Melbourne Water Study, as well as Carolyn Tsioulos, Dr. K.S. Tan, and Vicki Brown for their ongoing work in this area.

REFERENCES

Howe, C., Jones, R.N., Maheepala, S., and Rhodes, B. 2005. Melbourne Water Climate Change Study. Implications of Potential Climate Change for Melbourne's Water Resources. A collaborative project between Melbourne Water and CSIRO Urban Water and Climate Impact Groups, CMIT-2005-106 March 2005, CSIRO, 36 pp.

Jones, R., Durack, P., Maheepala, S., Howe, C., Rhodes, B., Yurisich, R., Kularathna, U., and Esler, S., 2005. Melbourne Water Climate Change Study. Case Studies: Water Supply System, Drainage and Urban Waterways and Sewerage System. A collaborative project between Melbourne Water and CSIRO Urban Water and Climate Impact Groups, Technical Report 5 CMIT-2005-105 March 2005, CSIRO, 76 pp.

BOM. 2008. 2008 BOM Drought Statement http://www.bom.gov.au/announcements/media_releases/climate/drought/20081203.shtml.

Water Resources Strategy. http://www.watersmart.vic.gov.au/downloads/final_report_-_summary.pdf.

Our Water, Our Future. http://www.ourwater.vic.gov.au/programs/owof.

Our Water, Our Future: The Next Stage. http://www.ourwater.vic.gov.au/programs/next-stage.

The Central Region Sustainable Water Strategy. http://www.ourwater.vic.gov.au/programs/sws/central.

– 25 –

Climate Change Impacts on Urban Drainage Systems in Scandinavia

O. Mark, G. Svensson, A. König, and J.J. Linde

STATUS ON ACTIONS IN DENMARK, SWEDEN, AND NORWAY

Predicted Climate Changes in Denmark

In Denmark, three climate change scenarios have been selected for the analyses related to climate changes, i.e., A2, B2, and EU2C. The likely occurrences of three scenarios are considered to have an equal probability. The Danish EPA has initiated a proactive investigation on how the Danish municipalities, in a cost-efficient way, can mitigate the impact of climate changes on sewer systems. The management and mitigation of the impacts of climate changes demand a holistic and integrated planning process across departments in the municipalities, as more types of legislative procedures will be affected compared to today's work flow. Analyses of the expected climate change scenarios have revealed that the number of high-intensity rainfall events will increase during the summer and that the mean sea level in the Danish waters will increase as well. Both of these phenomena may have an influence on the performance of the sewer systems and hence their impacts must be clarified through analyses. Estimates of impacts from climate changes are proposed to be carried out on three levels. After the impact assessment, the problems should be prioritized, e.g., by use of risk assessment tools. Based on the findings, the municipalities must develop a plan for timely management and mitigation of the impacts from climate changes. The plan should contain descriptions of how and when climate changes are analysed and managed for:

1. Planning and design of new sewer systems.
2. Existing sewers where maintenance and reconstruction are already planned.
3. Existing sewers where no maintenance and reconstruction are scheduled today.

Status on National Guidelines and Recommendations for Urban Drainage and Sewer Systems

The Danish Water and Wastewater Association (DANVA) has taken the strategic decision to carry out research directly applicable for the Danish utilities and municipalities within adaptation to climate change. So far, the outcome of the DANVA research programme is:

- Description of new extreme "climate proof" design rainfall (DANVA 2007a).
- A set of guidelines for the Danish municipalities concerning how to analyse the impact of climate changes (DANVA 2007b).

These research projects provide the Danish municipalities with the consistent methodologies to develop a timely and cost-efficient strategy for coping with climate changes and their impacts on sewer systems. The guideline outlines how Danish municipalities in a consistent framework can analyze the impact for future climate changes. The guideline contains:

1. Three levels of analyses depending on the complexity of the sewer system.
2. A framework for analyses of: if, where, and when the climate change statistically will violate today's standards. Will it be in 10, 25, 50, 100 years, or is today's design climate proof ?

Predicted Climate Changes in Sweden

The Commission on Climate and Vulnerability (The Swedish Government 2007) came to the conclusion that it is important that measures are taken now. The water and wastewater systems are vulnerable and the municipalities must start taking climate change seriously and address the increase in probability of extreme weather events in the physical planning. Municipalities must make sure that increasing risks due to climate change are not built into society. The Swedish Association for Water and Wastewater has prepared two reports on how climate change may affect the drinking water systems as well as the wastewater systems. The two reports, M134 and M135, are a part of the report from the Commission on Climate and Vulnerability. Costs for adaptation of the wastewater systems have been calculated to 25 billon SEK for the coming 25 years on top of the regular rehabilitation, which is estimated to be 50 billion SEK for the same period (Svenskt Vatten 2007). The drainage systems will be affected by increasing rainfall volumes as well as more frequent long, wet periods that give rise to increased runoff volumes. The Rossby Center at SMHI predicts from the global climate models that the autumn and winter periods will get a 50% increase in rainfall volume all over Sweden. The rise of sea levels will affect all municipalities close to the sea.

Status on National Guidelines and Recommendations for Urban Drainage and Sewer Systems

All municipalities need to analyze the highest known level in rivers, lakes, and in the sea close to built-up areas. To these levels, an addition according to the climate scenarios has to be made and also a safety margin has to be added. This information has to be included in all planning activities and also used for risk analysis of the existing sewer systems (The Swedish Government 2007). New areas must be planned according to the information given on future water levels according to climate change scenarios. Existing areas where water levels already today cause flooding at extreme events have to be secured by walls and pumping of drainage from inside the walls. Stormwater pipes have to be secured against backwater from the receiving water (The Swedish Government 2007).

The climate change scenarios for Sweden show less rainfall during summer and more rainfall during autumn and winter. All Sweden will get an increase in rainfall volume with at least 40–50% during winter. The challenge will be to adapt the existing sewer systems to increased rainfall volumes parallel to the ongoing renewal of the sewer systems. Climate change will be another issue to take into account when planning renewal activities. An important measure will be to plan for secondary drainage pathways to be used when the primary pathway, the sewer, has insufficient capacity. These secondary pathways will have to convey the runoff to areas that may be flooded without causing costly damages (Svenskt Vatten 2007).

Predicted Climate Changes in Norway

Norway has concentrated its efforts on mapping regional climate change scenarios. The Norwegian Research Council started in 1997 the support of a coordinated research project called RegClim. The overall aim was to produce scenarios for regional climate change suitable for impact assessments in Northern Europe, bordering sea areas, and major parts of the Arctic, and to quantify uncertainties due to the choice of methods, global scenarios, and due to poorly understood processes influencing our region's climate.

Status on National Guidelines and Recommendations for Urban Drainage and Sewer Systems

The development of mitigation and adaptation strategies to meet the consequences of climate change is carried out at the respective ministries. Those activities are coordinated and aggregated on a national level to ensure a horizontal and vertical exchange of knowledge and strategies. In the report on vulnerability and adaptation strategies to climate change in different sectors in Norway, issued by the Ministry of Environment in April 2006, only general conclusions are drawn. It is confirmed that Norway has to expect higher frequencies in intensity and volume for stormwater. This will lead to more frequent surcharge situations, increased combined sewer overflow events and volume, increased stormwater overflows, and increased load to the treatment plant. It is also pointed out that rising sea levels may have a backwater effect on the discharge pipes. As a consequence, pollution of receiving waters increases, resulting in the complete range of negative health and environmental effects. Direct damages on properties and infrastructure will increase in frequency and extension. Accordingly, the overall socio-economic costs will grow. Currently, the key for mitigation measures is put into the planning and building law on the municipal level. Municipalities are responsible for addressing climate change effects in their master plans for sanitation, development, and area planning. They have to carry out a consequence analysis whenever the planning might come in conflict with climate change effects. Focus is laid on avoiding damages when area planning is done and building permits are given. The report (Miljøverndepartementet 2006) does not explicitly address an adaptation for sanitary systems. It refers to the new sanitary guidelines, effective from January 2008, where the duration of CSO events has to be registered for all overflows. Volume, pollution, and stormwater discharge are, however, not covered in that regulation. Otherwise, Norway has adapted the EN-752 that regulates return intervals for urban flooding caused by sewer systems. The report (Miljøverndepartementet 2006) also includes some suggestions for immediate measures issued by the Ministry of Environment. They see a solution in separation and alternative stormwater treatment as local infiltration and retention, and controlled surface floodways. The municipalities already have the means to regulate these measures through the planning and building law.

The tasks of the Ministry of Environment are to ensure that:

1. Municipalities as local planning authorities have the necessary warranties to prevent unwanted building development.
2. Long-term area planning includes a more restrictive approach in area protection.
3. Procedures are developed that integrate climate change effects in area planning.
4. Municipalities have access to knowledge around climate change effects.
5. The interdisciplinary planning of several sectors is coordinated.

Complimentary to that report, the coordinated research on climate change effects in the RegClim project resulted in concrete quantifications of climate change effects for the different regions in Norway over the next 50 years. In summary, these effects will be:

1. The yearly precipitation volume increases between 5 and 20%. The major increase will be in coastal areas in the southwest and in the northern parts.
2. On a seasonal level, precipitation increases mainly during autumn. In the western parts, middle, and northern Norway, the increase exceeds 20%.
3. During the summer season, the eastern and southern parts of Norway will receive up to 15% less precipitation.

4. The western parts of Norway will experience an additional 15 days of rain events with over 20 mm, which is an increase of 20%.
5. Rain will appear more regularly apart from southern and eastern parts, which will have up to 10% more dry days.
6. Heavy rain events will happen more often. Current events with a yearly return interval will, for example, appear up to three times a year.

IMPACTS ON THE PRECIPITATION DUE TO CLIMATE CHANGE

Transformation of the Rainfall from the Climate Model

In engineering of urban drainage, extreme precipitation is based on rain gauge measurements at a single point. Hence, it is desirable to express the effect of climate change on urban drainage in terms of point measurement extreme statistics. A three-step methodology for transformation of the rainfall from a grid of 25 km × 25 km to a point rainfall has been developed (Grum et al. 2005). The method is demonstrated for a cell covering Copenhagen with 16 point rainfall gauges within a 25 km × 25 km area. Further, this method is applied for Danish cities and the rainfall intensities and return periods have been found for the future climate scenarios. The three step approach comprises (see also Figure 25-1):

1. **Present–Point => Present–Surface.** Measured data (1979–1996) from each of the 16 rain gauges is first converted from minute intensities into hour intensities. A single hourly time series is then created from a weighted average of all of the 16 hourly rainfall intensities. The weights have been chosen based on Thiessen polygons constructed around the rain gauges for the entire chosen 25 km × 25 km area.
2. **Present–Surface => Future–Surface.** Extreme statistics of the gauge based surface is transformed from the *present* to the *future*. This was done under the assumption that the ratio between *present* and *future* extreme statistics of the climate model would be the same as the ratio between *present* and *future* of the 25 km × 25 km surface series based on the 16 point gauges.
3. **Future–Surface => Future–Point.** Finally, the *future* extreme statistics of the 25 km × 25 km surface series based on the 16 point gauges is transformed back into a corresponding *future* **point** extreme statistic. This is based on an assumption that the ratio between surface and point statistics is the same in the *future* as it has been found to be in the *present*.

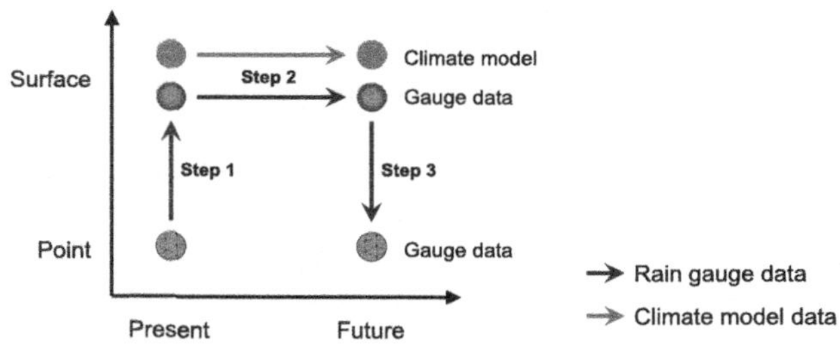

Figure 25-1. Three-step methodology for transforming rainfall from a grid to a point rainfall

Illustration of the three-step analysis: (1) point rain gauge measurements are converted to a surface measurement, (2) relationship between is climate models *present* and *future* is used to project statistics of the surface rainfall into the future, and (3) surface rainfall statistics of the future are converted to point rainfall statistics of the future.

The computed increase in one-hour rainfall intensities has been carried out for a number of Danish cities. The increase for Scenarios A2 is in the magnitude of 10–90% compared to today's rainfall statistics. However, insufficient data are available to make a proper statistical analysis of the computation of the 10-year return rainfall, as only 30 years of rainfall are available. Further, it is not certain what exactly causes the variation in calculated future rainfall over Denmark. It may be statistical uncertainty or local geographical variations, but limited information is available to support analyses of this phenomenon. Hence, it has been chosen to adopt an average increase of 20–50% in rainfall intensity for all of Denmark, where 50% may be applied at exposed locations.

CLIMATE MANAGEMENT STRATEGIES

Problem Identification

In Denmark, sewer systems are typically analyzed by CDS rains or historical rain series. To investigate the effect of climate change, the rains used today are multiplied with a factor corresponding to the expected growth in rain intensity (1.2–1.5). By use of these rains, critical locations or areas in the catchment can be identified, and it can be decided how and when problems have to be solved. It is proposed that clarification of the impacts of climate changes are carried out on three levels:

- Level 1: Check if the impacts of climate changes will result in a violation of the standards.
- Level 2: If the standard is violated, then an estimate of the extent of the violation is required. i.e., estimate the flood depth, the extent of the flood, and the flood damage.
- Level 3: If the extent of the flood, the flood depth, and flood damages are significant and are due to climate changes, then this assessment must feed into a prioritized planning scheme on how to mitigate the expected problems.

When a complete problem identification has been carried out, then the problems should be prioritized, e.g., by use of risk assessment analysis.

Timely Management of Climate Changes

The assessment of the impact of climate changes on sewer systems will be a continuous process in the future. Information about the development of the climate will, in the future, continue to be improved and refined. Hence, it is expected that the evaluation of the impacts from climate changes on sewers will be an ongoing process where new and improved information concerning climate scenarios and their associated uncertainty will be updated and made available to the municipalities on a regular basis. It is expected that new information concerning climate changes will be minor adjustments of existing scenarios, e.g., in the form of a better geographical resolution of the variation of precipitation over local areas.

As mentioned above, the increase in a 10-year rainfall event for Scenario A2 will be 20–50%. Scenario A2 is the worst of the three scenarios adopted by the Danish EPA and futures scenarios different to those predicted in A2 may occur. If all Danish sewers were to be upgraded today to comply with the rainfall foreseen in scenario A2, then the additional yearly investment in Danish sewers would be on the order of magnitude of 20% (a first estimate by the authors) of the yearly maintenance budgets. As the future is uncertain, the main idea behind the recommendation to the municipalities is:

- Not to sit and wait until climate changes are 100% documented based on historical records.
- Not to invest heavily today in upgrading the entire infrastructure as soon as possible.
- To comply with the worst-case future climate scenarios.
- To get an overview of the possible consequences of the likely impact of the climate changes and then to incorporate that knowledge into their maintenance plans.

Based on these ideas, the municipalities are strongly recommended to develop a plan for timely management with constant care to handle the impacts arising from climate changes. The plan must contain the following:

- Planning and design of new sewer systems under the influence of climate changes.
- Guidelines for sewers, for which maintenance and reconstruction projects are already planned, taking into consideration the impact of climate changes.
- Guidelines for analyses of the impact of climate changes on sewer systems, for which no maintenance or reconstruction plans are scheduled.

CASE STUDIES ON ADAPTATION OF URBAN SEWER SYSTEMS TO CLIMATE CHANGES

Helsingborg: Lussebäcken Catchment

Lussebäcken catchment in Helsingborg (Semadeni-Davies et al. 2006a, 2006b) was tested with a number of cases, including both new development and climate change effects as more rainfall over the next few decades according to the A2 scenario. The simulated development will add a further 364 ha (115 ha impervious) to the area contributing to stormwater flow in the stream. Source control to reduce stormwater generation and ponds to regulate flow into the stream will be a central part of any development. For this study there are three cases:

1. Current situation: This case was used for comparison with the other cases in which nothing but climate is changed.
2. Current situation with increased population: This case assumes that the city size increases. The new development will have a separate system; all impervious surfaces will lead to the stormwater system (i.e., the Lussebäcken Creek).
3. Continuation of current trends: This is the most likely case for developments given current trends in Swedish urban water management. Ponds and source control are installed, as currently planned, in the new subdivision; thus, the impervious area will effectively be reduced to 25% and more storage will be available.

Specific findings were:

- Without further city development (case 1), climate change projected for Helsingborg will increase the current drainage problems associated with urbanisation.
- City growth (case 2) could slightly increase flood risk in the upper reaches of Lussebäcken catchment.
- The combination of city growth (case 2) and the A2 climate scenario have the potential to cause the worst drainage problems by increasing peak flow volumes.
- Source control and increased storage capacity (case 3) are able to mitigate the impact of urbanization but may not be enough to eliminate the combined impacts of urbanization and climate change.

Kalmar: Lindsdal Catchment

Future urban drainage impacts have been thoroughly investigated (Olofsson 2007, Berggren 2007) in two Licentiate works from Luleå Technical University. The study area was a separated urban catchment in the city of Kalmar (Olofsson et al. 2007). The size of the catchment is 54 ha with 20 ha impervious surfaces. The total number of inhabitants is 3,000. The stormwater system contains 410 nodes and is designed according to the current design standards, which means that the system should manage rains of at least a 10-year return period without surface flooding. The objective was to investigate the effects on flooding for the climate change projection on this part of Sweden. The method used is outlined in Figure 25-2.

The regional climate models produce temperature and precipitation data for the future with a spatial resolution of 50 km × 50 km and a temporal resolution of 30 min. To be able to use the climate data in urban drainage models, temporal and spatial resolution must be improved. A modification of the so-called Delta change method, where the changes are related to the rainfall intensity level, is presented to transfer the changes in rain characteristics from different future time periods to an observed series. Effects of climate change due to higher rainfall intensities show that higher water flow–ratios in pipes, longer durations of floods, and more frequent floods can be expected. Since the renewal rate of pipes in the existing urban drainage system is relatively slow, emphasis must not lie only on city development but also on future climate change. Design criteria, therefore, need to be changed according to changes in precipitation. Climate change also affects urban drainage in different ways, depending on where in Sweden the city lies. In northern Sweden, problems can arise with changing snowmelt patterns, for example.

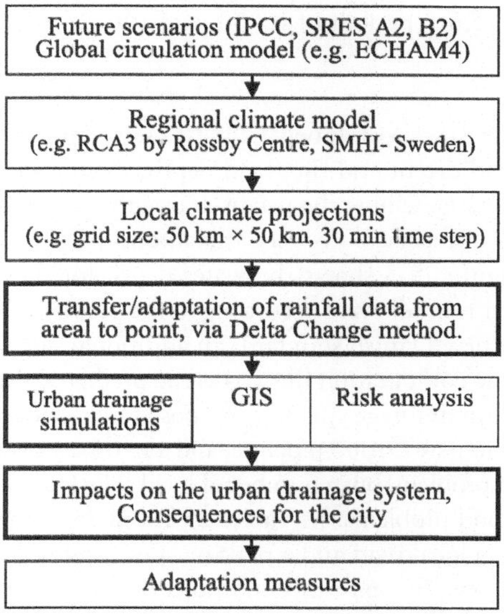

Figure 25-2. Main strategy for assessing impacts on sewer systems (Olofsson et al. 2007)

Odense, Denmark—Urban Flooding

This study is a case where the municipality is experiencing flooding and wants to develop a plan for the system, which will both mitigate the current flood problems and the impact from future climate changes. The future design rainfall is a local CDS rain with a 10-year return period for Odense based on climate scenario A2. The catchment for the study has both combined and separate sewers. The upstream area has only separate sewers and an area of 273 ha (87 ha reduced) with a load of 10,200 PE. The combined sewer system covers an area of 178 ha (50 ha reduced) with a load of 6,100 PE. Flooding occurs during heavy rainfalls.

Modeling Approach

The modeling of the potential impact due to climate changes is carried out combining a two-dimensional (2-D) surface flow model explicitly coupled with a one-dimensional (1-D) sewer system model. Hence, the model illustrates the sewer system and the surface flow simultaneously. Flow exchange may occur from the surface to the sewer system or vice versa depending on the flow conditions. Initially, water starts to enter manholes as surface runoff. At the beginning of the runoff event, there is still free capacity in the pipe network. If there is surcharging of the sewer system, the water surcharges to the surface and causes surface flooding. At the end of the event, the pipe network drains the flooded water from the surface. To calculate the flow exchange between the surface and the pipe system, the systems may be considered as two linked systems and hence an energy equation is applied to calculate the flow from the surface to the pipe system through the manhole and vice versa. The flow is driven by the energy head available between the two basins. The computations are carried out using the combined hydrodynamic sewer and surface flow model. A digital terrain model with a grid size of 2 m is the basis of the 2-D hydrodynamic surface flow model. Computations are carried out with CDS rainfall events with return periods of 10, 50, and 100 years.

Project Results

The results below illustrate how a combined 1-D–2-D sewer and surface model in combination can compute the flood extent and the flood depth. According to the Danish standard, the municipality must investigate how the urban drainage system responds to situation with an overload. This kind of information is very valuable in preparation of upgrade measures and emergency plans. Figure 25-3 shows the water depth for the existing conditions. The rainfall applied is a 10-year rainfall event according to the predictions by climate scenario A2. Further, according to the new Danish standard, an additional safety factor of 20% has been applied on the rainfall. The computation illustrates the predicted flood problems.

An upgrade of the main drainage system is proposed by the local consultant to be 200 m new Ø1200 pipeline, 650 m new Ø1600 pipeline, and a new basin with a volume of 4,000 m^3. This solves the main flood problem, but a number of local solutions must still be found in order to completely solve the flood problems. In Tables 25-1 and 25-2, flood depths at two important locations in the area (a local road and a parking spot at a kindergarten) and the number of flooded buildings are shown for: existing conditions/the upgraded system; today's rainfall/a future climate scenario (A2). The impacts from the climate change are clearly demonstrated, e.g., will the current conditions for a rainfall event with a 50-year return period almost correspond to a future 10-years (Scenario A2) rain—including a safety factor of 20%?

A sensitivity analysis has been carried out on the dominant physical parameters, i.e., the Manning number in the pipe model, the roughness in the surface model, and the rainfall. The sensitivity analysis shows the model is not very sensitive to changes in the roughness in the pipe and surface model. As expected, the change of rainfall in the order of 20–44% has

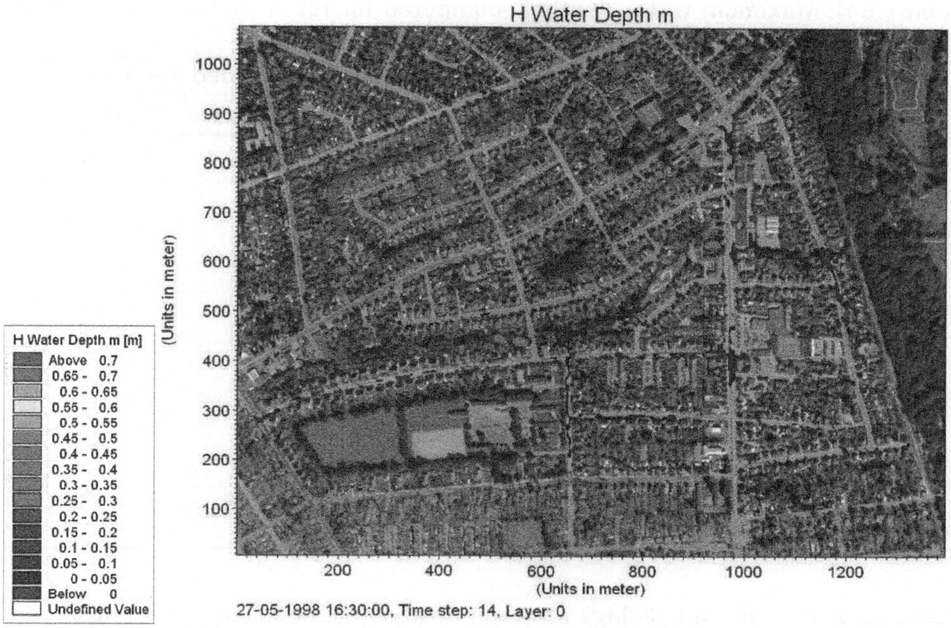

Figure 25-3. Flood map computed for a rain with a 10-year return period taking impact from potential climate changes into account (climate scenario A2)

an impact on both the flooded area and the flood depth. Hence, the model is demonstrated to be robust and to reflect the dynamics of climate change.

SUMMARY AND CONCLUSION

The most important national findings are:

- Denmark has come up with new rainfall design and produced a consistent procedure for analyses of impact on sewer systems from climate change.
- Norway has concentrated its efforts on mapping regional climate change scenarios and now elaborates the consequences for different sectors through concerted research actions.
- Sweden has recently presented a roadmap for actions concerning climate change with respect to urban areas.

Estimates of impacts from climate changes are proposed to be carried out on three levels. After the impact assessment, the problems should be prioritized, for example, by use of risk assessment tools. Based on the findings, the municipalities must develop a plan for timely management and mitigation of the impacts from climate changes. The plan should contain descriptions of how and when climate changes are analysed and managed for:

- Planning and design of new sewer systems.
- Existing sewers where maintenance and reconstruction are already planned.
- Existing sewers where no maintenance and reconstruction are scheduled today.

Case studies on the mitigation of the climate change impacts have been carried out for catchment areas in Sweden and Denmark. The overflow conditions and the flood conditions with respect to the extent of the flood, flood depth, and the number of affected houses under the current conditions in the present catchment area have been analyzed.

Table 25-4. Maximum water depths—computed for return periods and climate impacts

Return Period/ Water Depth	Existing System (cm)		Upgraded System (cm)	
	Kindergarten	Dalumvej	Kindergarten	Dalumvej
T = 10 years	58	0	0	0
T = 10 years + climate change	71	17	0	0
T = 10 years + climate change + safety factor	84	27	0	0
T = 50 years	91	29	5	0
T = 100 years	99	35	7	9
T = 50 years + climate change + safety factor	122	56	17	15
T = 100 years + climate change + safety factor	124	59	26	21

Table 25-5. Number of buildings affected by flooding—computed for return periods and climate impacts

Return Period and Potential Climate Impact	Number of Flooded Buildings	
	Existing System	Upgraded System
T = 10 years	20	8
T = 10 years + climate change + safety factor	36	13
T = 50 years	37	15
T = 50 years + climate change + safety factor	53	30
T = 100 years	50	22
T = 100 years + climate change + safety factor	92	37

The development of a flood mitigation strategy for sewer systems under the impact of climate changes has resulted in a unique set of guidelines for municipalities. It is believed that this set of guidelines will provide the municipalities with a timely and cost-efficient strategy for coping with climate changes and their impacts on sewer systems.

REFERENCES

Berggren, K. (2007) Licentiate Thesis, Luleå Technical University, Luleå, Sweden.

DANVA. (2007a) Klimaændringernes betydning for ændringer i kraftig nedbør i høj opløsning i tid og sted.

DANVA. (2007b). En kogebog for analyser af klimaændringers effekter på afløbssystemer – med fokus på oversvømmelser. F& U rapport nr. 5. ISBN: 87-90455-74-6.
Grum, M., Jørgensen A.T., Johansen, R.M., and Linde, J.J. (2005). *The Effect of Climate Change on Urban Drainage: An Evaluation Based on Regional Climate Model Simulations,* 10th International Conference on Urban Drainage, Copenhagen, Denmark, August 2005.
Miljøverndepartementet, Norway. (2006). Rapport om sårbarhet for og tilpasning til klimaendringer i sektorer i Norge. En oppfølging av interdepartementalt seminar 31. August 2005 om tilpasning til klimaendringer.
Olofsson, M. (2007). Licentiate Thesis, Luleå Technical University, Luleå, Sweden.
Olofsson, M., Berggren, K., Viklander, M., and Svensson, G. (2007) *Hydraulic impact on urban drainage systems due to climate change. J. Hydrology* (submitted).
Semadeni-Davies A., Hernebring, C., Svensson, G., and Gustafsson, L.G. (2006a). The impacts of climate change and urbanisation on drainage in Helsingborg, Sweden: 1. combined sewer system. *J. Hydrology* (submitted).
Semadeni-Davies, A., Hernebring, C., Svensson, G., and Gustafsson, L.G. (2006b). The impacts of climate change and urbanisation on drainage in Helsingborg, Sweden: 2. suburban stormwater. *J. Hydrology* (submitted).
Svenskt Vatten (2007). Klimatförändringarnas inverkan på allmänna avloppsnät, Meddelande M135, Svenskt vatten, Stockholm, Sweden.
The Swedish Government. (2007). Sweden and Climate Change—Threats and Opportunities. Report M2007:60, Stockholm, Sweden.

Index

Note: *f.* indicates a figure; *t.* indicates a table.

ABC Waters. *See* Active, Beautiful, Clean Waters Programme
ACM0010—Consolidated Baseline Methodology for GHG Emission Reductions from Manure Management Systems, 166–168
ACM0014—Mitigation of Greenhouse Gas Emissions from Treatment of Industrial Wastewater, 165–166
Active, Beautiful, Clean Waters (ABC) Programme, 246, 248
Adaptation, 7–8, 55, 174, 176, 179, 182, 191, 206–207, 249, 252, 259–260, 266, 275
case studies on, 282–285
climate change and, 1–2, 3, 41–42, 42*f.*, 149, 170, 203, 222
  focus on, 155
  identification of, 242
  innovative, 239
  low regrets, 169, 203
  no-regrets, 169, 203
  water and, 42, 188–190
  by water utilities, 202–203
Adaptation strategies, 8, 9, 11, 176, 180, 215, 238, 248, 259, 266–268, 279
  developing, 191, 217, 218, 221
  drainage management, 268
  flexible, 239
  global warming challenges and, 10
  overview of, 230*t.*
  potential for, 175, 230–234, 264
Adelaide
  energy/water use in, 113*t.*
  rainfall in, 112
Aeration, 76, 162, 165
Aerobic systems, 159, 160, 161, 162, 164
Agriculture, 157, 189, 199, 227
AgriVerde Ltd., 160
Agrosuper, 167
Air quality, 161, 165
AiRO process, 236
Algae, 60, 82, 228
  growth of, 14, 68, 220, 255–256
Alliance to Save Energy, 97, 123, 126, 135
Allocations, 146, 183, 188, 190, 268
Alternative sources, 45, 150, 178, 231–232, 231*f.*, 239, 259
American National Standards Institute (ANSI), 152
American Water Works Association (AWWA), 1, 2, 104, 206, 213
America's Climate Security Act (2007), 145
AMS III.F—Avoidance of Methane Emissions through Controlled Biological Treatment of Biomass, 162–164
AMS III.H—Methane Recovery in Wastewater Treatment, 160–162
AMS III.I—Avoidance of Methane Production in Wastewater Treatment through Replacement of Anaerobic Systems by Aerobic Systems, 164–165
Amsterdam
  drinking water resources for, 231*f.*
  municipal water chain, climate footprint of, 66–67, 66f.
Anaerobic digesters, 138, 150, 162, 163
Anaerobic treatment system, 160, 167
Analysis, 80, 138, 221, 269, 270
  discriminant, 50
  end-use, 104
  financial, 119
  results, 273*t.*
  statistical, 281
  supply chain, 58
  three-step, 281
  threshold, 8

vulnerability, 15–16
See also Life Cycle Analysis
Anatolian development project (GAP), 34
Andes, water resources in, 31–32, 37
Animal waste management systems (AWMSs), 159, 166–167
ANSI. *See* American National Standards Institute
ANUBIX–B reactor, 166
Approved Methodology, 158, 159, 160
Aquatic environment, 81, 180, 256
Aqueducts, 17, 18, 219, 251, 265
Aquifer storage, 154, 195, 234, 267, 274
Aquifer Storage and Recovery (ASR), 234
Aquifer Storage transport and Recovery (AStR), 228
Aquifers, 146, 206, 241, 253
brackish, 234
coastal, 1, 8, 202
saline, 8, 153, 154, 202
AR4. *See* Fourth Assessment Report
ARI, 271, 272
analysis results, 273*t.*
ASR. *See* Aquifer Storage and Recovery
Assessment Reports (IPCC), 5, 6, 10, 205, 207, 209, 215, 222, 241, 248
Assessment tools, development of, 52–53
AStR. *See* Aquifer Storage transport and Recovery
Auckland
energy/water use in, 113*t.*
rainfall in, 112
Audits, 128, 129, 130, 150, 266, 268
Augmentation, 100–101, 102, 267, 269, 271
Austin Energy, 140
Australia, map of, 112*f.*
Australian Scientific and Industrial Research Organisation (CSIRO), 111, 263, 272, 274, 275
Automation, importance of, 131–132
AWMSs. *See* Animal waste management systems
AWWA. *See* American Water Works Association
Awwa Research Foundation, 1

Back flow, 209
Base flow, 8
Basins, 191
Behavior
changing, 71*f.*, 259, 266
hydraulic, 270
Benchmarking, 79, 132
with climate footprint, 69, 70*f.*
Best efficiency point (BEP), 131
Bethune polder, water from, 64
BIG/CC. *See* Biomass-fired Integrated Gasifier Combined Cycle
Bio-capacity, 56, 57f.
Bio-electricity, 82, 92
WF of, 88, 90–91, 93
Bio-energy, 94
production of, 81, 93
replacement with, 82
WF of, 83, 92, 93
Bio-ethanol, 81, 82, 85
WF of, 90, 93
Biochemical oxygen demand (BOD), 76
Biodiesel, 81, 84, 86, 87
food crops and, 93
oil crops and, 82
WF of, 90, 91*f.*, 93
Biofouling, 236
Biofuels, 4, 97, 153
biomass and, 87, 92
energy/crop unit, 88, 90
first-generation, 85–87, 88, 90
next-generation, 82, 87–88, 91
production of, 92, 93, 155
replacement with, 81
WF of, 85–87, 87–88, 90–91
Biogas, 66, 73, 82, 114, 223
capture and flaring system, 161, 162, 164, 168
collecting, 166
conversion of, 78
power from, 160
production of, 118, 167, 208
Biomass, 4, 146, 162
biofuels and, 87, 92
heat/electricity from, 82, 85, 88, 92
production of, 81, 94
WF of, 85, 88, 89*t.*, 93
Biomass-fired Integrated Gasifier Combined Cycle (BIG/CC), 85
Biorefineries, 155
Biosolids, 138, 223
Black-water systems, 79
BOD. *See* Biochemical oxygen demand
Bolivar Wastewater Treatment Plant, 114
Brabant Water, 55, 68, 73, 233
climate footprint of, 60, 61, 62*f.*, 63, 63*f.*
climate neutrality for, 60–63
employee involvement at, 62–63, 70
map of, 56*f.*

planned approach of, 61–62
water cycle and, 57
Brackish water, 195, 231, 232, 233, 233*f.*, 234, 235
Brisbane
energy/water use in, 113*t.*
rainfall in, 112
Brisbane Water, energy use profile of, 114
Bromate formation, 197, 254

Cachuma Operations and Maintenance Board (COMB), 47
California
climate change for, 48
electricity use in, 208*f.*
population growth in, 251
See also Southern California
California Assembly Bill 1058, 18
California Climate Action Registry, 224
California Climate Change Center, 220
California Department of Fish and Game, 220
California Department of Water Resources, 218
California Energy Commission, 139–140, 208, 222
California State Water Project, 265
California Urban Water Conservation Council (CUWCC), 104
Camanche Reservoir, 13, 15, 16, 217, 218
Cap-and-trade system, 137, 138, 145–147, 206, 222, 224
Carbon capture, 146
Carbon credits, 158, 225
Carbon cycle, 148, 223
Carbon dioxide, 50, 199, 205, 206, 207, 215, 222, 223
concentration of, 200, 200*f.*
emissions, 73, 123, 146, 225
injection of, 153, 154
mandatory reporting of, 137
release of, 86, 138
sequestration of, 149, 153, 154
short-cycle, 67, 75
treatment process and, 75
underground injection of, 146
Carbon dioxide equivalents, 58, 59*f.*, 68, 73
Carbon footprints, 56, 220, 236, 238, 258–259
Carbon monoxide, 154
Carbon sinks, 68, 148
Catchments, 22, 25, 26, 27, 111, 114, 115, 185, 241, 242, 265, 269, 271, 273, 282, 283
expanding, 245
in Latin America, 190
local, 243, 244
management of, 267, 274
CCM3. *See* Community Climate Model
CCSP. *See* U.S. Climate Change Science Program
CCX. *See* Chicago Climate Exchange
CDM. *See* Clean Development Mechanism
Ceará, water demand management in, 188
Central Region Sustainable Water Strategy, 274
Central Valley Project, 251, 253
power generation by, 254f.
Certified emissions reduction credits (CERs), 97–98, 158, 159, 165
CFCs. *See* Chlorofluorocarbons
Challenge, 247
Chemical oxygen demand (COD), 164, 165, 166
Chemicals, truck transport of, 208
Chemistry, 154, 255–256
Chicago Climate Exchange (CCX), 224
Chlorination plants, 114
Chlorofluorocarbons (CFCs), 199, 205
CHR. *See* International Commission for the Hydrology of the Rhine Basin
City Datum, 209, 210*f.*
City West Water, 111
Clean and Diversified Energy Initiative of, 140
Clean Development Mechanism (CDM), 98, 157, 158–159, 159*f.*
Clean Water Act, 148
Climate change
addressing, 2, 155, 170, 222
coping with, 44, 45, 73, 189, 191, 202–203, 273, 277–278, 286
evaluating, 18, 42, 43, 45, 53, 281, 282
future of, 199–201
human-induced, 5, 148, 273
impact of, 1–2, 8, 18, 41, 43, 44, 45, 103, 159, 173, 176, 183, 187, 190, 206, 213, 217, 239, 242, 243, 244–245, 249, 252, 253, 260, 268, 269, 274, 277, 278, 279, 281, 284, 285, 286
managing, 3, 17, 19, 149, 281–282
parameters of, 181, 285
preparing for, 17–18, 215–216

projections for, 26, 43, 50, 174, 176, 203, 258, 269, 272
responding to, 148–149, 154, 169, 181, 225, 275
understanding, 3–4, 217
vicious circle reinforcing, 100*f.*
"Climate Change 2007" (IPCC), 5
Climate Change and Adaptation (IWA), 1–2
Climate Change and Water Resources: A Primer for Municipal Water Providers (Awwa Research Foundation), 1
Climate footprint, 58–59, 69, 74, 234–236
distribution of, 65*f.*
measuring, 4, 56, 58, 64, 71
methodology of, 55–56
reducing, 55, 60, 68, 80, 238
renewable electricity and, 63*f.*
Climate management, strategies for, 281–282
Climate model projections, 51*t.*, 176
Climate neutrality, 18, 60–63, 69
Climate Partners/PW Advies, 55, 65
Climate trends, 251, 252
Climate variables, 9, 48, 111, 148, 182
changes in, 175
establishing, 174
short-term impacts of, 51–52
Climate zones, map of, 33*f.*
Coastal habitats, 7
pollutants and, 257
salinity of, 1, 202
COD. *See* Chemical oxygen demand
Cold storage projects, 61
Color, 51, 53
Colorado River (California), 21, 31, 100
drought and, 196
Lake Powell and, 194*f.*
salinity in, 193, 195
SNWA and, 196, 197
Southern California and, 251
wastewater to, 197
water from, 37
Colorado River (Texas), 179, 180, 181
Colorado River Aqueduct, 265
COMB. *See* Cachuma Operations and Maintenance Board
Combined Heat and Power Partnership, 140
Combined sewer overflow (CSO), 209, 211, 279
Commission on Climate and Vulnerability (Sweden), 278
Community Climate Model (CCM3), 48, 50
Congress, 145
mitigation and, 155
water efficiency and, 147
Conservation, 97, 146, 218, 249
consumer, 259
demand-side, 126
encouraging, 147, 191, 196, 246, 260
energy, 152–153, 237
groundwater, 180
improving, 186
programs, 191, 196, 275
water, 149–152, 189, 190, 195, 223, 238
Contra Costa Water District, 195
Cooling, 62, 103, 160
with groundwater, 61*f.*
Cooperative Programme on Water and Climate, 55
Corona del Mar Water Treatment Plant, 47
Corrosion, 256, 265
Council on Environmental Quality, 147
Crop coverage, 83–84
Crops
characteristics of, 86*t.*
drought-resistant/early-maturing, 189
water for, 88, 92
CROPWAT, 84, 92
CSIRO. *See* Australian Scientific and Industrial Research Organisation
CSO. *See* Combined sewer overflow
Culture, changing, 71*f.*
CUWCC. *See* California Urban Water Conservation Council
Cyanobacteria, 228

DAF. *See* Dissolved air floatation
Dam construction, 184–185, 186, 189
Danish EPA, 277, 281
Danish Water and Wastewater Association (DANVA), 277
Danube River, 21, 24, 25
DANVA. *See* Danish Water and Wastewater Association
DBOO. *See* Design, build, own, and operate agreement
DCW. *See* Digital Chart of the World
Deforestation, 199
Degassing, 73, 235, 237
Demand-side strategies, 184*t.*, 190, 230, 239, 269
Demand simulation methods, 43–44
Demographics, 267

Denitrification, 75, 76
Denmark, climate change in, 277–278
Department of Sustainability and Environment (Victorian Government), 274
Dephosphorization plants, 67
Desalination, 10, 17, 81, 99, 111, 115, 116, 146, 177, 195, 220, 231, 242, 244, 245, 247, 248, 266, 267, 269, 274
 benefits of, 186
 criticism of, 186
 energy-intensive, 237
 partial, 188
 in Perth, 185–186
 systems, 189, 190
 technologies, 254
Desalination plants, 101, 102, 114, 186, 190
Design, build, own, and operate (DBOO) agreement, 244
Design storms, 210
 comparison between, 213*t*.
Diesel fuel, 222
Digital Chart of the World (DCW), 24, 36
Dissolved air floatation (DAF), 164, 167
Dissolved oxygen, decrease in, 255, 257
Distillation, 190, 237
Distribution systems, 101–102, 147, 206
 augmentation of, 102
 energy intensity of, 102
 leaky, 124
Division of Atmospheric Research (CSIRO), 270
DIY water-saving kits, 246
DOE. *See* U.S. Department of Energy
Domestic water, GWP of, 79
Drainage, 170, 248, 265, 268
 guidelines for, 277–278, 279–280
Drakens Mountains, water from, 33, 34
Drinking water, 63, 73, 187, 235
 alternative sources of, 231*f*.
 availability of, 153
 carbon footprint of, 236
 climate change and, 146, 234, 238
 distribution of, 56, 74, 237, 237*t*., 238
 energy-efficient technology for, 236–237
 energy for, 222
GHG and, 64, 79
 groundwater and, 233
GWP of, 235*f*., 236, 236*f*.
 production of, 56, 73, 74, 227–230, 232, 234, 236–237, 236*f*., 237*t*., 238, 247
 rainfall and, 234
 sea water and, 237
 standards for, 14, 244
 sustainable supply of, 234
 treatment of, 65, 150, 256
 underground sources of, 154
Drinking water utilities, 148, 148, 150
 adaptation by, 155
 conservation/management for, 151
GHG and, 145
 policies for, 155
 See also Water utilities
Drought, 6, 37, 55, 101, 184, 188, 221, 257, 263, 274
 climate change and, 99
 extended, 227, 228
 planning for, 189, 193, 194, 196–197, 268
 resistance, 10
 river quality and, 228
 severe, 170, 196
 summer, 30
 vicious circle reinforcing, 100*f*.
 water service and, 100
 in western U.S., 193–194
Dutch water companies, map of, 56*f*.

East African Highlands, water from, 33
East Bay Municipal Utility District (EBMUD), 3, 14, 15, 170, 215, 219, 220
 adaptation strategies by, 218
 analysis by, 221
 CCX and, 224
 climate change and, 15, 217, 221, 225
 energy use by, 224
 flood control and, 16
 GHG and, 18
 inventory by, 224
 management approach by, 217
 mitigation by, 224–225
 project investigations by, 17
 renewable resources and, 223
 Sacramento County Water Agency and, 221
 sea levels and, 219
 storage capacity of, 15
 wastewater treatment plant of, 216*f*.
 water demand for, 13, 16
 water from, 217, 218
 wet repair by, 18
 WSMP and, 220
Eastern Climate Registry, 137, 140
Eastern Treatment Plants (ETP), 275

EBMUD. *See* East Bay Municipal Utility District
Economic growth, 42, 199
Ecosystems, 14, 148, 255
Education, 246, 268
Efficiency, 104, 118, 146, 152, 178, 189, 238, 249, 267
  appliance, 266
  cost-benefit arguments for, 125
  design, 133–135
  end use, 106–107
  energy, 77–78, 134, 135*f.*, 137, 138, 150–151, 208, 247
  financing, 125
  guidelines for, 134
  improving, 10, 58–59, 69, 79, 97, 111, 112, 123, 125, 126, 126*t.*, 137, 138, 142, 142*t.*, 147, 186, 208, 273
  measures for, 130–133
  monitoring, 142*f.*
  motor, 134, 135, 259
  neglecting, 124
  pump, 128, 130, 134, 135, 139, 140–141, 208, 259
  public services and, 123
  safety margins and, 134
  solutions to, 141–143
  water, 106–107, 195, 266
  water supply, 97, 126*t.*
  water utility, 123, 133–135, 150–151
Effluent, 160, 164, 228, 232, 234
  anaerobic treatment for, 165
El Niño/Southern Oscillation (ENSO), 3, 16, 48, 211, 218, 241
  indicators, 51, 52, 53
Electric Power Research Institute, 208
Electrical systems, monitoring, 129, 131, 133
Electricity, 87, 138
  bio-, 82, 88, 90–91, 92, 93
  biomass and, 85, 88, 92
  cost of, 9
  food crops and, 93
  green, 102
  nonrenewable, 61
  reducing use of, 141, 161
  renewable, 62, 63, 63*f.*, 67
  saving on, 132, 259
  theft, 125
  use of, 208, 208*f.*
  WF of, 85, 88, 89*t.*, 93
Emission reductions (ERs), 157, 159, 166
Emissions trading, 157
EN-752, 279
Energy
  conservation of, 152–153
  production, 119
  saving, 104, 133, 134, 135*f.*, 142, 143, 237
  sustainable, 146
  water and, 99, 123–124, 145, 153
  See also Renewable energy
Energy consumption, 74, 130, 134, 159, 208, 246, 297
  national/residential, 106
  per capita, 275
  reducing, 116, 190, 238
Energy demand, 106, 220
  increasing, 99
  local conditions and, 118
  for water, 117*t.*
"Energy Down the Drain: The Hidden Costs of California's Water Supply" (Pacific Institute), 100
Energy intensity, 101, 106, 107
Energy management programs, 223, 259
Energy Policy and Conservation Act, 146
Energy saving devices, 134, 135*f.*
Energy Star, 146, 150, 259
Energy systems
  alternative, 259
  inter-relationship with, 106*f.*
  sustainability of, 99
Energy use, 97, 119–120, 219
  estimating future, 116
  public reporting of, 118
  reducing, 97, 153, 159, 170, 238
  water supply and, 119
ENSO. *See* El Niño/Southern Oscillation
ENSO SOI Monitoring Stations, map of, 49*f.*
ENSO Teleconnection Model, 53*f.*
  water quality and, 53*t.*
Environment and Water Research Programme, 247
Environmental flows, 180, 268
Erosion, 155
  coastal, 7, 257
  nutrient loadings and, 255
ERs. *See* Emission reductions
Estate-scale systems, 101, 102
Estuaries, 206
  salinization of, 229, 230
ETBE. *See* Ethyl tert-butyl ether
Ethanol, 90
  energy from, 89*t.*

food crops and, 93
HHV of, 88
production of, 82, 86, 87, 93
WF of, 91*f.*
Ethiopia, dam construction in, 184–185
Ethyl tert-butyl ether (ETBE), 228
ETP. *See* Eastern Treatment Plants
European Alps, 22, 24, 27
contribution of, 25*f.*
glaciers in, 28
Rhine river basin and, 32
rivers originating in, 24*f.*
runoff from, 285
temperature/precipitation changes and, 26
water from, 32, 37
Evaporation, 1, 47, 202, 227, 267
Evapotranspiration, 25, 37, 84, 173, 219
controllers, 260
increase in, 28
Executive Order 1342 (2008), 152

FAO, 81, 82, 84
FAO Penman-Monteith method, 84
Federal agencies, 145, 147–153
Fermentation, 85, 86
Ferrous sulfate, 77, 235
Fertilizers, production of, 93
First Assessment Report (IPCC), 5
First National Communications, 184
Fish kills, 256
Fishery management, 14, 15
Flex Water concept, 234
Flexibility, 8, 134–135, 173, 191, 239
Flexible sources concept, 234, 238
Flood control, 15, 173, 177, 217, 248, 252, 272
challenges of, 254
changes for, 8
impact on, 14, 16
water supply and, 218
Flood Management and Drainage Strategy, 272, 275
Flood map, 285*f.*
Flooding, 14, 37, 103, 188, 228, 241, 247–248, 252, 271–272, 273, 275
analysis, 269
buildings affected by, 286*t.*
climate change and, 274, 283
damage from, 8, 281
extreme, 227
frequency of, 272
increase in, 7
local, 264
planning for, 189, 284
rainfall and, 248, 284
storm surges and, 219, 248
urban, 284–285
Flow conditions, 253, 284
Flow monitoring, advances in, 142*f.*
Flow rates, 142, 272
Flow regulators, 246
Flushing, 246
rainwater and, 78, 231
Food crops, energy from, 93, 94
Food-processing waste, energy from, 224
Food production, water for, 187
Forecasting, 41, 42–43, 44, 45, 49, 51, 188
Forests, global warming and, 7
Fossil fuels, 138, 145, 160, 162, 163
bio-capacity and, 57*f.*
burning, 199
displacing, 81, 82
reduced use of, 225
vicious circle reinforcing, 100*f.*
Four National Taps, 243, 244, 248–249
Fourth Assessment Report (AR4) (IPCC), 5, 6, 207, 215, 222, 241
global warming and, 205
hope in, 10
projection in, 248
"Fresh keeper" concept, 233, 233*f.*
Freshwater, 4, 33, 183, 227, 233, 241
Funding Database (USEPA), 140
Future-Surface-Future-Point, 280

Ganges–Brahmaputra, 37
water from, 35, 38
GAP. *See* Anatolian development project
General circulation models (GCMs), 48, 260
Generators, 224
GHG emissions at, 164
microturbine, 18
See also Power generation
Geographic information system (GIS), 128
Geothermal, 152
GHG. *See* Greenhouse gas
Gianyar Regency, 163, 164
GIS. *See* Geographic information system
Glaciers, 23, 24, 26, 27
map of, 24*f.*
mass balance, 29*f.*
melting of, 3, 28, 29, 30, 31, 32, 36, 37, 199
in South America, 31*f.*

Global warming, 9, 13, 30, 32, 205, 209
GHG and, 137–138
impact of, 7–8, 22, 55, 191
water resource management and, 6, 28, 31
Global warming potentials (GWP), 73, 74, 75, 77, 78, 234
black-water systems and, 79
distribution of, 74*f.*
of domestic water, 78*t.*, 79
of drinking water, 235*f.*, 236, 236*f.*
GHG and, 235
of wastewater, 236
in water sector, 74*f.*
Global Warming Solutions Act (AB32) (2006), 206, 221–222, 260
Gold Coast
energy/water use in, 113*t.*
rainfall in, 112
Goleta Water District, 47, 52, 53
Granja Tres Arroyos Methane Avoidance in Slaughterhouse Effluent Project, 164
Great Plains Institute, 140
"Green building" design, 152–153
Green roofs, 152
Green water, 84
Greenhouse gas (GHG), 151, 155, 168, 213, 215, 234
concentration of, 199, 200
emission allowance transfer system, 145–146
emissions targets for, 137
excess, 141
human-caused, 220
increase in, 199, 222
inventorying, 222, 224
nature's contribution of, 205
production of, 190
synthetic, 205
unpredictability of, 209
Greenhouse gas (GHG) emissions, 4, 55, 66, 69, 103, 106
addressing, 10
anthropogenic, 137
avoiding, 94
carbon dioxide equivalents of, 58
concentration of, 10
direct, 67, 75, 76*t.*
indirect, 66*f.*, 67, 77, 235
net difference in, 108*f.*
per capita, 59
reducing, 1, 2, 9, 13, 18, 65, 68, 70, 79, 97–98, 107, 112, 119, 137, 138, 140, 141–143, 146, 149, 153, 157, 158, 159, 162, 164, 166, 168, 169, 170, 206, 207, 208, 213, 221–222, 222, 225, 258–259
stabilizing, 6, 18, 68
vicious circle reinforcing, 100*f.*
Greenhouse Gas Protocol, 69
Greenpower purchase, 112
Greywater reuse systems, 102, 147, 188
Groundwater, 63, 154, 177, 232, 241, 266
allocation of, 190
banking, 17
brackish, 233, 233*f.*, 235
cooling with, 61*f.*
demineralization of, 197
development of, 180, 233*f.*
drinking water and, 233
dynamics of, 228
heating with, 61*f.*
infiltration, 47
management, 56
methane and, 78, 237
protection of, 190
pumping, 81
salinization of, 228, 233
sequestration and, 146
storage, 154, 222
GWP. *See* Global warming potentials

HadCM2 climate model, 48, 50, 53
HadCM3 climate model, 34
Handbook on Methods for Climate Change Impact Assessment and Adaptation Strategies (Feenstra), 259
HCFCs. *See* Hydrochlorofluorocarbons
HDPE. *See* High-density polyethylene
Heat, 238, 263
GHG from, 66, 138
with groundwater, 61*f.*
WF of, 85, 88, 93
Heat waves, 7, 28, 202, 215, 216
Helsingborg, climate change in, 282
HFCs. *See* Hydrofluorocarbons
HHV. *See* Higher heating value
High Atlas Mountains, water from, 33
High-density polyethylene (HDPE), 160
Higher heating value (HHV), 85, 88
Highland Lakes reservoir system, 179–180
Himalayas, 21, 27
water from, 25

Hindu-Kush-Himalayas mountain range (HKH), 21
glaciers in, 36*f.*, 37
map of, 36*f.*
water from, 35–37
Hot water systems, 97, 116, 117, 118
solar, 120
volume/energy, 117*t.*
House Bill 631 (2009), 147
Hurricane Katrina, 217
Hurricanes, intense, 202
Hybrids, 18, 120, 225
Hydraulic behavior, 267, 270
Hydrochlorofluorocarbons (HCFCs), 146
Hydroclimate, 48–49, 52
lake, 47–48
predictors, 49, 50, 51
water quality and, 49, 50
Hydroelectric plants, 34
Hydroelectricity, 190, 225
Hydrofluorocarbons (HFCs), 146, 206
Hydrogeology, 3
Hydrologic cycles, 13, 184, 252, 273
changes in, 6, 7, 27, 202, 257
snowpack and, 23
Hydrological variables, 13, 188, 189, 257
Hydrology, 8, 25, 33, 221
catchment, 22
climate change and, 26–30, 41, 220
groundwater, 257
heatwaves and, 28
models, 215
mountain, 26–30
Hydropower, 13, 31, 32, 37, 38, 217, 219–220, 223, 225
cheap, 186
decrease in, 220
as energy source, 253
generation of, 28
in-conduit, 208
water for, 187
Hyflux Ltd., 244

Ice sheets, 23, 201, 209, 216
IDF curves. *See* Intensity-duration-frequency curves
IEUA. *See* Inland Empire Utilities Agency
Impacts, 178, 252–258
direct/indirect, 205
measuring, 4
Impellers, 131, 132, 139, 141
Imported water, 244
Independent Science Board, 207
Indus River, 3, 21
water from, 35, 37
Industrial water, 187
conservation of, 151–152
Industrialization, 35, 38, 117–119, 268
Infiltration, 267, 268
Infiltration pond, 229*f.*
Infrastructure, 99, 100, 208, 211, 252, 253, 279
agricultural, 189
changes in, 175, 206
collection system, 213
construction of, 104, 183, 215
costs of, 102
green, 152–153
investment in, 102, 125, 220
leakage rates and, 124
planning/design of, 206, 258
sewer, 207
stormwater, 205, 265
upgrading, 282
vulnerabilities with, 217
water quality and, 256
Injection wells, 154
Inland Empire Utilities Agency (IEUA), 139, 251, 252
Inorganics, 163, 255
Institute for Sustainable Futures (ISF), 101, 102, 104
Integrated resource planning (IRP), 104, 106, 107, 190
framework for, 105*f.*
Integrated water resources management (IWRM), 190–191
Intensity-duration-frequency (IDF) curves, 170, 205, 209, 211, 213, 248
compared, 212*f.*
Intercatchment transfers, 101, 102, 107
Intergovernmental Panel on Climate Change (IPCC), 34, 58, 184, 199, 210, 215, 241, 248
adaptation and, 7
downscaling of, 263
hope from, 10–11
nitrous oxide and, 77
planning and, 8
projections by, 7, 200, 207
reports by, 189, 201, 205, 207, 242
role of, 5, 6
International Association of Cryospheric Sciences, 28

International Commission for the Hydrology of the Rhine Basin (CHR), 28
International Water Association (IWA), 1, 2
Invasive species, 255
IPCC. *See* Intergovernmental Panel on Climate Change
IRP. *See* Integrated resource planning
Irrawaddy River, water from, 35
Irrigation, 34, 38, 187, 190, 258
- in Asia, 36*f.*
- efficiency in, 189
- glacier meltwater for, 37
- modernization of, 274
- requirements for, 37, 88
- supplementing, 188

ISF. *See* Institute for Sustainable Futures
IWA. *See* International Water Association
IWRM. *See* Integrated water resources management

Kalmar, case study on, 283
Köppen-Geiger system, 33, 34, 35, 36
Kwinana Desalination Plant, 186
KWR-Grontmik project, 73
Kyoto Protocol, 97–98, 158, 260
- emissions reductions and, 157
- GHG and, 222
- mitigation and, 168

Lagoons
- aerobic, 160, 161
- anaerobic, 159, 160, 161, 164, 165, 166, 167

Lake Cachuma, 3, 47, 51
- climate change for, 48, 49
- water balance for, 48
- water quality for, 49*t.*, 50
- water supply from, 52, 53

Lake Mead, 193
- surface levels at, 196, 196*f.*
- water rights to, 197

Lake Powell, 193
- Colorado River and, 194*f.*

Land use, 42, 148, 186
Lawrence Berkeley National Laboratory, 213
Lawrence Livermore National Laboratory, 13–14, 213
LCA. *See* Life Cycle Analysis
LCRA. *See* Lower Colorado River Authority
Leadership for Energy and Environmental Design (LEED), 152
Leakage, 189, 245
- inefficiencies from, 125
- infrastructure and, 124
- reducing, 130, 151, 188, 190, 266

LEED. *See* Leadership for Energy and Environmental Design
Levees, 17–18, 219
LHV. *See* Lower heating value
Licentiate works, 283
Lieberman-Warner Climate Security Act (2008), 145
Life Cycle Analysis (LCA), 58, 65, 107
Lift stations, 138, 141, 142
Lindsdal catchment, 283
Lobith, 30
- discharge at, 29*f.*

Lower Colorado River Authority (LCRA), 179, 180, 181
- map of, 180*f.*

Lower Colorado River Authority–San Antonio Water System Water Project, 179–181, 181*f.*
Lower heating value (LHV), 85
Lussebäcken catchment, 282

Madison Center for Sustainability and the Global Environment, 84
Mandatory renewable energy target (MRET), 102
Manure management projects, 166
Marina Barrage, 248
Marina Reservoir, 243
Master thermal units (MTUs), 132
Mechanical degradation, 140–141
Mekong River, water from, 35
Melbourne
- climate change in, 264–265, 273
- energy/water use in, 113*t.*
- rainfall in, 112
- reservoirs, streamflows in, 264*f.*

Melbourne Water, 111, 114, 263, 275
- climate change and, 265
- emergency relief structures of, 270
- sewerage transfer network of, 271
- study by, 274

Melbourne Water Climate Change Adaptation Program, 275
Melbourne Water Flood Management and Drainage Strategy, 275
Membrane Bioreactor Plant, 247
Membrane concentrate disposal, 233

Membrane filtration, 197, 236, 237, 246
Memstill, 237
Metering, 130, 132–133
Methane, 69, 73, 140, 157, 199, 205, 206, 215, 216, 222, 235, 259
anaerobic lagoons and, 161
avoidance, 158
burning, 162, 167, 223
digesters, 138
emissions of, 66, 68, 75, 76, 77, 137, 138, 164, 165, 166
energy from, 159, 224
groundwater and, 237
leaking of, 163
recovering, 62, 78, 79, 160, 163, 164, 165, 237
reducing, 168
Methane capture, 158, 159, 160, 162, 167
Methane Capture and Combustion from Swine Manure Treatment for Peralillo study, 167
Methyl tert-butyl ether (MTBE), 228, 229*f.*
Metropolitan Water District of Southern California, 18
Meuse River
Evides and, 228
MTBE in, 228, 229*f.*
salinization of, 229, 230
MF. *See* Microfiltration
MHHW. *See* Monthly mean higher-high water
Microfiltration (MF), 244, 246
Midwestern Regional GHG Reduction Accord, 222
Milieucentraal, 58
Ministry of Environment (Netherlands), 73
Ministry of Environment (Norway), 245, 279
Ministry of Environment and Water Resources (Singapore), 245
Ministry of National Development (Singapore), 248
Mississippi River, 22
Mitigation, 71, 73, 77–79, 80, 152, 158, 170
anticipating, 149
climate change and, 106, 107, 145, 151, 168, 213, 221–223, 252, 277, 285
consequences of, 153
flood, 268, 272, 286
GHG, 10, 11, 98, 112, 207–208, 221, 222, 223*f.*, 224–225, 258–259, 260
implications of, 9–10
measures for, 4, 279
potential for, 234–238, 237*t.*
strategies for, 1, 97, 206, 215, 217, 239, 279
technologies for, 153, 154, 155
in water cycles, 78, 79*t.*
water utilities and, 223, 234
Mixing zones, 255
Models, 191, 257–258
climate, 34, 48, 50, 53, 169, 174, 176, 203, 213, 253, 270, 280–281, 283
discriminant analysis, 50
econometric, 43
end-use breakdown, 111
first principles, 111
general circulation, 48, 260
hydrologic, 176, 215
lake water quality, 49–50, 50*f.*
pipe, 284
predictive, 43, 45, 215
sewer, 284
surface flow, 284
Mokelumne Aqueducts, 17, 18
Mokelumne Hill, runoff from, 16*f.*
Mokelumne River, 13, 14, 15, 16, 17
changes for, 217
fish hatchery on, 220
storm-runoff development/flooding potential in, 17*f.*
Monitoring, 49*f.*, 129, 130, 131, 142*f.*, 247, 268
improving, 118
metering and, 132–133
regulations for, 260
technology for, 141–143
Monthly mean higher-high water (MHHW), 209, 210*f.*
Motors
auditing, 129
efficient, 134, 135, 259
Mount Kenya, glacier remains of, 26
Mountain ranges
African, 33–34, 33*f.*
map of, 22*f.*, 33*f.*
Mountain types, 21–22
map of, 22*f.*
MRET. *See* Mandatory renewable energy target
MTBE. *See* Methyl tert-butyl ether
MTUs. *See* Master thermal units
Multiple sources concept, 232–233, 238

MultiTrode Inc., 141–142
Murray River, 101, 114

Nanofiltration, 228
National Adaptation Taskforce, 248
National Center for Atmospheric Research, 1, 48, 213
National Climate Change Strategy (NCCS), 170, 248
National Climate Data Center (NOAA), 47, 213
National Economic Council, 147
National Geodetic Vertical Datum, 209
National Oceanographic and Atmospheric Administration (NOAA), 47, 209, 213
National Research Foundation (NRF), 247
National Resource Council (NRC), 155
"National Taps," 170
National Water Commission, 104
National Water Program (NWP), 150–151, 152–153
National Water Program Climate Change Workgroup, 148, 149
National Water Program Strategy (USEPA), 148–149
National Water Program Strategy: Response to Climate Change, 149
National Weather Service (NWS), 209, 210, 211
Natural gas, 66, 222, 224
NCCS. *See* National Climate Change Strategy
NEWater, 170, 242, 244, 245, 246, 247, 248–249
Niger River, water from, 33
Nile River, 3, 21
  Egypt and, 33–34
  water from, 33, 37
  water shortage in, 38
Nino SST Monitoring Regions, map of, 49*f.*
Nitrates, 60, 76
Nitrification, 75, 76, 157
Nitrites, 76
Nitrogen, 76, 154
Nitrous oxide, 73, 157, 168, 199, 205, 206, 222, 259
  emissions of, 66, 68, 75, 76, 77
  reducing, 78
No-regrets solutions, 18, 187, 221, 274
NOAA. *See* National Oceanographic and Atmospheric Administration
Nonpoint pollution impacts, 153
North Head Sewage Treatment Plant (STP), 114
North Pine Treatment Plant, 114
Norway
  climate change in, 278–280
  rainfall in, 280
Norwegian Research Council, 278
NRC. *See* National Resource Council
NRF. *See* National Research Foundation
Nutrient loadings, 255
NWP. *See* National Water Program
NWS. *See* National Weather Service

Odor, 47, 161, 165, 256, 265
Office of Air and Radiation (USEPA), 149, 150
Office of Management and Budget, 147
Office of Research and Development (ORD) (USEPA), 147, 149
Office of Science and Technology Policy, 147
Offset programs, 61, 97, 102, 146
Oil crops, 83, 84, 86
  biodiesel and, 82
  energy from, 89*t.*, 90
  HHV from, 90
  WF of, 91*f.*
Operational energy-efficiency programs, 118
Operational scopes, 58
  greenhouse gas protocol and, 59*f.*
ORD. *See* Office of Research and Development
Organics, 81, 82, 163, 167, 255
Organizational scope, 56–57
O'Shannassy pipeline, 269
Ouderkerkerplas, 68
"Our Water, Our Future, The Next Steps in the Government's Water Plan," 266, 274
Outside world, organizing relations with, 70–71, 71*f.*
Overflow tanks, 101
Overflows, sewer, 8, 209, 211, 264, 265, 269–271, 279
Oxidation, 236
Ozonation, 197, 234
Ozone, 228

Pacific Decadal Oscillation (PDO), 209
Pacific Gas & Electric Company, 15

Pacific Institute for Studies in Development, Environment, and Security, 100, 213
Palm oil mill effluent (POME), 160
Pardee Reservoir, 13, 15, 217
Payback periods, 126*t.*, 128
PCM, 48, 50, 53
PDD. *See* Project Design Document
PDO. *See* Pacific Decadal Oscillation
Peat meadows, 68
Pentachlorophenol, 255
Perfluorocarbons, 206
Perth, 112
  desalination in, 185–186
  energy/water use in, 113*t.*
  rainfall in, 112
  water shortage in, 185–186
Pesticides, 93, 255
Pew Center, 173
Phosphates, 60
Phosphorous, 120, 255
Photovoltaic-energy-generation projects, 225
Pipes, 284
  cast-iron, 245
  design, 135
  stormwater, 278
  transport capacity of, 237
PKS Milano Pinang Awan palm oil mill, 160
Planning, 103, 189, 196, 197, 210, 249, 252, 260, 284
  adjustments to, 8, 258
  alternative strategies for, 45
  climate change and, 18, 169, 179–181, 206–208, 217, 220, 258, 268
  climate data and, 259
  contingency, 269
  implementation, 152, 157, 179
  land-use, 186
  long-term, 18, 183, 191, 206, 279
  strategic, 221, 268
  supply-side, 273
  TWM, 178
  uncertainty and, 45
  water supply, 206, 220–221
  See also Integrated resource planning
Po River, 21, 24, 25
Point rainfall gauges, 281
Pollutants, 27, 74, 148, 228, 266
  chemical, 238
  coastal wetlands and, 257
  concentration of, 257
  water, 279
POME. *See* Palm oil mill effluent
Population density, in Asia, 36*f.*
Population growth, 7, 38, 197, 209, 267
in California, 251
  climate change and, 173, 252
  demand management and, 10
  impact of, 35, 199, 202
  projections on, 269
  water demand and, 42
Port Phillip Bay, 266
Portfolio organization, 8, 9, 190–191
POTWs. *See* Publicly owned treatment works
Power, 160
  green, 150
  solar, 9, 62, 120, 152, 223
  wind, 9, 223
  See also Hydropower
Power generation, 28, 153, 217, 219–220, 253, 254*f.*
  on-site, 152
  projects, 224–225
PPP. *See* Public–private partnership project
Precipitation, 25, 176, 216, 231
  annual values of, 193
  changes in, 26, 37, 199, 202, 206, 207, 209, 210*f.*, 218, 241, 257
  climate change and, 280–281
  decrease in, 201, 217, 221, 257
  extreme, 253, 281
  frequency of, 202, 251, 252–253
  increase in, 201, 202, 252
  influence on, 41, 252
  intensity of, 202, 251, 253
  minimal, 51
  patterns of, 1, 207, 213, 227, 253
  projections of, 184, 205–206
  temperature and, 173, 257
  See also Rainfall; Snowfall
Present-Point-Present-Surface, 280
Present-Surface-FutureSurface, 280
Pricing, 190, 245, 274
Production, 67, 128
  alternative energy, 150
  chemical, 208
  water-related, 149–150
Project Design Document (PDD), 158
PUB, 245, 247, 248, 249
  ABC Waters and, 246
  catchments and, 243
  conservation and, 246
  tagline of, 242–243

water management and, 242
Public-private partnership (PPP) project, 244
Publicly owned treatment works (POTWs), 206
Pump stations, 67, 142–143, 237
monitoring, 133, 141
reconstruction of, 62
Pumping, 222
off-peak, 224
systems, 62, 135
Pumps, 142, 243
BEP for, 131
efficiency of, 128, 130–131, 134, 135, 139, 140–141, 208, 259
energy for, 138, 139, 140
high-head/low-head, 135
individual household, 101
life cycle costs of, 129*f.*
monitoring, 129, 131, 133, 141
operating, 132, 135
oversized, 135, 141
wastewater, 137, 138–139, 142
Punggol-Serangoon Reservoir Scheme, 243
Pyruvate, 85

Qualitative threshold approach, 176, 177
Quelcaya Glacier, 32

Rainfall
capturing, 186
climate change and, 272, 273*t.*, 278, 283
decrease in, 263, 265
drinking water and, 234
five-year average, 210*f.*
flooding and, 248, 284
frequency of, 210–211, 251
future, 210, 281
grid to point, 280*f.*
increase in, 7–8, 265, 267, 270, 273
intensity of, 207, 265, 272, 277, 280, 281, 283
patterns of, 270, 272
runoff and, 51, 53
sewer systems and, 285
snow versus, 253–254
spatial variation in, 211
storm event, 270
temporal patterns, 270
transformation of, 280–281
See also Precipitation; Snowfall
Rainwater, 230, 232, 246
evapotranspiration of, 84
harvesting, 101, 120
supply systems, 101, 102
tanks, 102, 188, 189
toilet flushing and, 78, 231
using, 147, 189
Raw materials, indirect emissions from, 66*f.*
RCSPs. *See* Regional Carbon Sequestration Partnerships
Receiving waters, 266
RECs. *See* Renewable energy certificates
Recycling, 111, 120, 188, 189, 190, 245, 249, 266, 274
increased, 186, 275
promoting, 151–152, 267, 268
technology, 151–152
RegClim, 278, 279
Regional Carbon Sequestration Partnerships (RCSPs), 154
Regulations, 71, 187, 188, 252, 258, 260
Remote terminal units (RTUs), 132
Renewable energy, 94, 97, 216
portfolio standards, 137
strategies, 259
using, 224–225
vicious circle of, 103*f.*
See also Energy
Renewable energy certificates (RECs), 62, 102
Renewable energy credits, 225
Renewable resources, 208, 223, 224
Report in Brief, A (NRC), 155
Reservoirs, 177, 183, 206, 251
building, 184
capacity reduction of, 228
management of, 267
runoff and, 215
Responsibility, 9–10, 224
Reuse, 10, 116, 120, 177, 189, 190, 195, 238
accelerating, 186
greywater, 102, 147, 188
indirect potable, 101
large-scale, 99
promoting, 151–152
rainwater, 147
Reverse osmosis (RO), 228, 237, 244, 246
Rhine River, 3, 21, 68, 227, 228
case study of, 24–25, 28–30
discharge in, 27*f.*, 29*f.*
European Alps and, 32
hydrology and, 285
map of, 26*f.*

salinization of, 229, 230
subcatchments of, 25
water from, 37, 64, 231
Rhone River, 21, 24
Risk, 44, 45, 173
areas of, 265–266
aversion to, 125
climate change, 170, 174–177, 178, 181, 203, 215, 220, 241, 273
flooding, 18, 30, 183, 252, 253, 282, 284
management, 69, 203
microbiological, 230
overflow, 267
quantitative nature of, 176
reducing, 144, 178
system, 182
Risk assessment, 174, 203, 274, 275, 281
climate change, 178, 180
tools for, 277
River flows, 27, 28
RIVERWARE, 191
RO. *See* Reverse osmosis
Rocky Mountains, 21, 22
water from, 30–31
Rossby Center, 278
RTUs. *See* Remote terminal units
Runoff, 1, 21, 23, 25, 37, 193, 275, 285
capturing, 30
change in, 254
decreased, 179, 268
earlier, 9, 14, 30, 206, 213, 218
fertilizer, 155
flood control and, 16
hydrograph, 15
increased, 278
management, 227
map of, 23*f.*
nutrient loadings and, 255
pollutant, 266
projecting, 34
rainfall and, 51, 53, 269
reservoirs and, 215
seasonality in, 27, 30, 183
snowmelt and, 253
storage and, 15
stormwater, 243
surface, 284
temperature and, 37
urban, 246
wastewater, 162

Sacramento River, 17, 221
Sacramento/San Joaquin River Delta, 14, 217
sea levels and, 219*f.*
water from, 219
Sacramento Valley, 251
Safe Drinking Water Act (2003), 146, 148
Salinity, 1, 174, 179, 193, 265
change in, 195*f.*
coastal aquifers and, 8
controlling, 195
dry-season, 194
increase in, 8, 9, 202, 257
managing, 197
plants, 247
Salinization, 234, 236, 254
estuary, 229, 230
groundwater, 228, 233
preventing, 233
Salmon habitat, 257
San Antonio Water System (SAWS), 179, 180, 181
San Francisco Bay Area
flooding in, 219
water for, 217, 251
San Francisco Hydraulic–Hydrologic Data Acquisition Rain (SFHHDAR), 211
San Francisco Public Utilities Commission (SFPUC), 170
sewer system plan of, 209–211, 213
San Pablo Bay, salinity change in, 195*f.*
Santa Ynez Mountains, 47
Santa Ynez River, 47, 48
SAWS. *See* San Antonio Water System
SCADA system. *See* Supervisory control and data acquisition system
Scarcity, 126, 170, 187, 188, 191, 197, 230, 241
Scenarios, 182, 199, 284
climate change, 8, 263, 266, 269, 277, 282, 285
emission, 176
future, 281
population, 269
using, 176–177
Scripps Institute of Oceanography, 213
Sea level
increase in, 6, 7, 8, 14, 55, 170, 176, 199, 201, 202, 206, 207, 209, 213, 215, 216, 217, 219, 219 (fig.), 241, 247–248, 249, 254, 257, 267, 274, 279
temperature and, 254
water utilities and, 219
Sea water, 231, 236

drinking water from, 237
thermal expansion of, 254
treating, 189, 246
"Securing Our Water Future Together: Our Water, Our Future" (Victorian Government), 266
Sedimentation, 7, 8, 160, 256–257
Sembcorp NEWater Pte., 244
SEQ. *See* Southeast Queensland
Sequestration, 112, 146
carbon dioxide, 149, 153–154
geologic, 97, 153–154
Sewage, 63, 78, 266
climate change and, 274
emissions from, 4
flows, 267
GHG and, 64, 76
quality, 267, 268
strategies for, 268
treatment, 56, 73
Sewer systems, 74, 209, 265, 267–268, 271
assessing impacts on, 283*f.*
climate change and, 274, 275, 278, 281, 282–285, 286
construction of, 267, 285
guidelines/recommendations for, 277–278, 279–280
improving, 210, 281
maintaining, 267, 285
planning/design of, 206, 277, 285
SFHHDAR. *See* San Francisco Hydraulic–Hydrologic Data Acquisition Rain
SFPUC. *See* San Francisco Public Utilities Commission
Shoalhaven River, intercatchment transfer from, 101
Shower devices, 238, 266
Sierra Nevada, 13
precipitation in, 251
snowpack melt in, 195
water from, 30–31, 193, 217
Silvretta Glacier, 28
Singapore
catchments in, 243
climate change in, 241, 242, 248
map of, 242f.
rainfall in, 241
water importing by, 244
Singapore Government, 247, 248
Singapore Public Utilities Board, 170
SingSpring Pte Ltd., 244
Sludge, 166, 167, 216
digestion process, 66, 161, 162
energy from, 224
generating, 164
GHG from, 137
managing, 159
ponds, 160
treatment of, 73, 75, 168
"Smart growth" policies, 152
Snowfall
map of, 23*f.*
rainfall versus, 253–254
See also Precipitation; Rainfall
Snowmelt, 28, 195, 218
capturing, 30
changes in, 283
early, 186, 202, 206, 213
reliance on, 217
runoff and, 253
Snowpack, 16, 22, 195, 215
annual, 218
changes in, 7, 170, 202, 213
decrease in, 9, 30, 36, 173, 186, 199, 202, 206, 218, 253
hydrological cycle and, 23
projected, 193–194, 194f.
recent, 194*f.*
storage by, 14, 218
SNWA. *See* Southern Nevada Water Authority
Socioeconomic factors, 45, 148, 195
Sodium hydroxide, 77, 235
Softening, 238
SOI. *See* Southern Oscillation Index
Soil moisture, reducing, 202, 255
Solar collectors, 62
Solar power, 9, 62, 120, 152, 223
Solids, truck transport of, 208
South Coast Conduit, 47
South Delta, 197
South East Water, 111
Southeast Queensland (SEQ), 101, 102, 107
Southern California
Colorado River and, 251
population of, 252
water demand in, 31
water rights challenges in, 170
water scarcity in, 170
See also California
Southern Nevada Water Authority (SNWA), 196–197
Southern Oscillation Index (SOI), 48
monitoring stations, map of, 49*f.*

SSMP. *See* 30-Year Sewer System Master Plan
Stakeholder involvement, 178, 179, 180, 188, 191, 197, 217, 260
State Scoping Plan, 260
State Water Project, 31
Storage, 25, 147, 190, 206
Storage capacity, 23, 177, 253
  impact on, 14
  increase in, 190, 282
  loss of, 8, 257
  runoff and, 15
Storm events, 271
  changes in, 209, 269, 309
Storm surges, 202, 219, 248, 274
Storm tracks, 216, 218, 221
Stormwater, 78, 103, 106, 205, 238
  collecting, 1, 101, 195, 206
  discharge of, 279
  management of, 153, 177, 190
  permit, 152–153
  reuse of, 147, 195
  segregation of, 243
  systems, 269, 282
  treating, 73
STOWA, 73, 77
STP. *See* North Head Sewage Treatment Plant
Streamflows, 274
  conditions, 269
  management, 268
  at Melbourne reservoirs, 264*f.*
  reduction in, 202, 257, 263
  seasonality/shifts in, 23, 30
Sugarloaf Reservoir, 114, 267
Sulfur hexafluoride, 205, 206
Supervisory control and data acquisition (SCADA) system, 132, 134
Supply-side solutions, 104, 107, 184*t.*, 185–186, 190, 231, 269
Surface water, 177, 197, 232, 234
  availability of, 180
  contamination of, 228
  management, 56, 63
  sources of, 184
  storage of, 273
  supply augmentation with, 267
Suspended matter, 228
Sustainability, 146, 158, 161, 185, 194, 221, 230, 259
  developing, 195, 249
  goals for, 165, 167
  long-term, 245
"Sustainable Water Strategy Central Region, Action to 2055" (Victorian Government), 266
Sweden, climate change in, 278
Swedish Association for Water and Wastewater, 278
Sydney
  energy/water use in, 113*t.*
  rainfall in, 112
  water demand management in, 288
Sydney Catchment Authority, 111, 115
Sydney Water, 114, 115
System yield, potential buffer/shortfall of, 270*f.*

Tarago Reservoir, 267, 274
Tarago treatment plants, 269
Taste, 47, 256
Taurus-Elburs-Zagros mountain ranges, 21, 34–35
Taurus Mountains, water from, 34
TECHNEAU project, 227
Technical management, 128–130
Technology, 147, 246–247
  energy-efficient, 134, 135*f.*
  transfer, 195
Tecolote Tunnel, 47
Temperature, 15, 41, 216, 253, 265
  changes in, 26, 37, 201*f.*, 227, 251, 256*f.*, 263
  increase in, 6, 7, 30, 176, 199, 200–201, 200*f.*, 215, 218, 251, 252, 255, 258, 264
  precipitation and, 173, 257
  runoff and, 37
  sea level and, 254
  water, 30, 202, 215, 255, 256, 256*f.*, 257
10 Litre Challenge, 246
Terschelling, 234
Texas, map of, 180*f.*
Thin-film-condensate (TFC) membranes, 236
Third Assessment Report (IPCC), 5, 209
30-Year Sewer System Master Plan (SSMP) (SFPUC), 205, 209–211, 213, 272
Threshold approach, 8, 177, 178, 182, 203
  framework for, 175*f.*
  using, 174–176
Tibesti Mountains, 21
Tigris-Euphrates river basin, 21
  map of, 35*f.*
  water for, 34
Total dissolved solids, 194*f.*, 254

Total water management (TWM), 174–179, 177*f.*, 181
Traveston dam, 102
Treatment, 4, 56, 65, 73, 75, 118, 150, 168, 197, 236, 255, 256
  anaerobic, 160, 165, 167
  systems, 1
  tuning of, 78
  See also Wastewater treatment
Trias energetica, 69, 70*f.*
Turbidity, 7, 51, 53, 220
  increase in, 8, 14, 202
  removal of, 47
"21st Century Melbourne: a WaterSmart City" (Victorian Government), 266
TWM. *See* Total water management

UASB. *See* Upflow anaerobic sludge blanket
UF. *See* Ultrafiltration
UK Environment Agency (UKEA), 104
Ultrafiltration (UF), 244, 246
Ultraviolet (UV) disinfection, 228, 234, 244, 246
Umgeni Water, strategy of, 185, 185*f.*
Uncertainty, 44*f.*, 45, 175
Underground sources of drinking water (USDWs), 154
UNEP. *See* United Nations Environment Program
UNFCCC. *See* United Nations Framework Convention on Climate Change
United Kingdom Hadley Center, 48
United Nations Environment Program (UNEP), 5
United Nations Framework Convention on Climate Change (UNFCCC), 158, 159, 168, 184
United States Bureau of Reclamation, 47
U.S. Climate Change Science Program (CCSP), 147–148, 155
U.S. Conference of Mayors, 140
U.S. Department of Energy (DOE), 130, 139–140, 154
U.S. Environmental Protection Agency (USEPA), 140, 155, 222, 254, 260
  conservation/management and, 151
  drinking water and, 244
  energy efficiency and, 150
  Energy Star and, 146, 150
  GHG allowance transfer system and, 145–146
  "green buildings/infrastructure" and, 153
  mandatory reporting rule and, 206
  NWP and, 152–153
  water efficiency legislation and, 147
  WaterSense and, 151
U.S. Green Building Council, 152
U.S. Mayors' Climate Protection Agreement, 137, 140
University of California, San Diego, 213
Upflow anaerobic sludge blanket (UASB), 159, 166
Urbanization, 218
  climate change and, 282, 285
  impact of, 35, 38, 227, 282
USDWs. *See* Underground sources of drinking water
USEPA. *See* U.S. Environmental Protection Agency
Utility engagement, 258–260
UV disinfection. *See* Ultraviolet disinfection
UV/H202 treatment, 236

Variable frequency drives (VFDs), 134
Variable Salinity Plant, 247
Variable speed drives (VSDs), 131n, 133, 135
Virtuous cycle, 107–108
Vitens, 233, 234
Volcano Antizana, 32
VSDs. *See* Variable speed drives
Vulnerability, 15–16, 175, 181, 190, 217, 218
  assessing, 8, 178
  defined, 182, 220
  strategies for, 279

Wadden Island, 234
Warm water, conservation of, 78, 79, 80, 238
Warmth content, reusing, 78–79
Warragamba Dam, gravity feed from, 112, 114
Wastewater, 64, 73, 170, 193, 197, 232
  collection of, 1
  conveying, 97
  demand, 106
  discharges of, 116, 255
  domestic, 79, 231
  emissions from, 75
  energy, 106
  flow of, 114
  GHG and, 64, 137, 138–139
  GWP of, 236

methane recovery from, 160
nutrients in, 119
pollution equivalents of, 74
pumping, 114
recycling, 186, 189
returns, 255
reuse of, 10, 101, 189, 190, 195
Wastewater Fact Sheet (USEPA), 140
Wastewater industry, 207
climate change and, 205–206
GHG from, 97, 138–140
resources for, 213
Wastewater Master Plan, IDF curves of, 211, 212*f.*
Wastewater systems, 118
efficiency in design of, 133–135
energy–intensive, 102
energy use for, 115*f.*
reconfiguration of, 120
Wastewater treatment, 56, 63, 73, 100, 116, 118, 126, 132, 162, 210, 222, 259
biogas from, 114
efficiency in, 97, 126*t.*, 128, 141
energy consumption during, 159
GHG from, 4, 64, 76, 97, 137, 138
infrastructure for, 124
methane emissions from, 77
nitrification/denitrification in, 75
optimization studies for, 78
pollutant loads from, 27
warm water from, 79
Wastewater treatment plants (WWTPs), 139, 161, 162, 222
effluent of, 228, 232, 234
electricity use by, 208
energy efficiency at, 150–151
Water Agreements (1961, 1962), 244
Water balance, 25, 48
Water banking, 17, 195
Water chains
climate footprint and, 64, 65, 65*f.*, 66–67, 66*f.*
process steps of, 64*f.*
Water Conservation Tax (WCT), 245
Water Corporation, 111, 186
Water cycles, 57, 73, 115, 148
analysis of, 80
climate change and, 99
climate footprint of, 63–68
energy use through, 111
GHG and, 56, 67*f.*, 74, 112
household measures and, 237–238
managing, 173
mitigation in, 78, 79*t.*
studying, 57*f.*
understanding, 148
Water demand, 31, 150, 173, 193, 217, 218–219, 227, 249, 252
allocation and, 188
charge for, 191
climate change and, 41–42, 42*f.*, 44*f.*, 188–189, 202
forecasting, 41, 42, 43, 44, 45
increase in, 99, 183, 202, 246, 257
long-term, 41, 244–245
managing, 10, 106, 116, 119, 146, 184, 187–188, 196
modeling for, 257–258
reducing, 196
risk-based approach to, 44
weather variables and, 43
Water depths, maximum, 286*t.*
Water Efficiency Implementation Guidance, 152
Water Efficient Homes Programme, 246
Water footprint (WF), 4, 93
calculation of, 84, 85
global overview of, 82–83
Water for All: Conserve, Value, Enjoy, 242–245
Water Framework Directive, 238
Water hammer, 132, 139
Water harvesting, 101, 120, 188, 189, 190, 267
Water loop, closing, 243*f.*
Water management, 151, 174–179, 205, 208, 215, 217, 258, 273, 277, 285
climate change and, 3, 31, 179–181
demand-side, 126, 183–190
effective, 195, 197
flexibility/reliability of, 173
goals of, 232
integrated, 169, 242–246
options for, 118
planned, 206, 266
private participation in, 118
reliability of, 9
supply-side, 183–190
urban, 1
Water meters, replacing, 133
Water policy, broader consideration for, 119–120
Water quality, 1, 183, 189, 190, 196, 217, 227, 228, 246
climate change and, 25, 51, 52*t.*, 146, 170, 202, 220, 252, 255

decline in, 194, 197, 199, 227
diminished flows and, 193
ENSO teleconnection and, 54*t.*
hydroclimate and, 49, 50
impacts on, 14–15, 266
improving, 197, 255–256
lake, 47, 49–50
management of, 193
monitoring, 247, 268
outcome probabilities for, 50
poor, 187, 219
standards for, 120
strategies for, 213
temperature and, 256
Water quantity, 187, 252–254
climate change and, 170, 252, 255
management of, 193
Water Reclamation Study, 244
Water Research Foundation, 1
Water Resource Strategy for the Melbourne Area, 263
Water resources
climate change and, 34, 111, 149
consequences for, 6–7
developing, 183
limited, 125
predictions regarding, 6–7
strategy with, 269
Water rights, 170, 197, 258
Water-saving devices/practices, 78, 194, 196
Water sector, 99
energy use for, 81, 115*f.*
GHG and, 55, 73, 102
GWP of, 74*t.*
Water security, 187, 190
Water Services Association of Australia (WSAA), 104, 111, 118–119
Water shortages, 37, 38, 44, 185–186, 238, 258
adaptation and, 42
climate change and, 35
emergency, 194
Water stress, 31, 92, 227–230
managing, 183–190
Water supply, 97, 126, 170, 183, 217, 218, 264, 266–267
availability of, 13–14
case study about, 268–269
climate change and, 30, 41, 146, 173, 202
demand side of, 185
design, 274
efficiency in, 97, 126*t.*
energy and, 99, 115, 119
flood control and, 218
impact on, 7–8, 14
increasing, 100, 128, 169, 173, 215, 221, 245
infrastructure for, 99, 124
long-term, 194–195
low-quality, 186
managing, 184–186, 206, 207, 221, 258
planning, 180, 206, 220–221
reliability of, 15–16
resilience in, 241
supply side of, 185
sustainable, 174–179, 230
vulnerabilities with, 217, 218
Water Supply Demand Strategy, 274
Water supply management program (WSMP), 221
Water systems, 182, 265, 270
automation of, 130
centralized large-scale, 100–101
climate change and, 13–15, 19, 174–175, 178, 179, 227
consequences for, 257–258
decentralized, 118
emission reductions in, 124*f.*, 127*f.*
energy costs for, 102, 120
inter-relationship with, 106*f.*
modification of, 181, 269
optimization of, 130, 266
over-allocated, 30
risk performance of, 176
supply-side, 102
sustainability of, 99
vulnerability of, 175
Water towers, mountains as, 21, 30–31
Water transfers, 81, 101, 102, 107, 186, 195, 267
Water Use Efficiency and Conservation Research Act (2008), 147
Water utilities
adaptation by, 202–203
climate change and, 4, 203, 217–221, 221–223, 230*t.*, 234, 256–257
climate footprint of, 64, 120
cooperation by, 207
data from, 121*t.*
efficiency by, 123, 133–135, 150–151
energy for, 9, 116, 119, 124, 208, 223, 233*f.*
GHG and, 259
mitigation and, 223, 234

planning by, 196
policy/programs, 97
sea level and, 219
water-yield loss for, 254
See also Drinking water utilities
Water Utility Climate Alliance, 173
Water21 (journal), 1
Waterbedrijf Groningen, 55, 68
climate footprint of, 58–59, 60
efficiency improvements at, 58–59
map of, 56*f.*
opportunities outside/within processes of, 60
water cycle and, 57
Waterboard Amstel, 63
Watergy, 97, 123, 125
monitoring and, 132
technical/managerial approaches to, 126, 128–135
Waternet, 55
alternative resources and, 231
climate footprint of, 63–68
drinking water supply of, 232*t.*
employee involvement at, 70
GHG from, 67*f.*
map of, 56*f.*
results for, 65*f.*
targets for, 65*f.*
water cycle and, 57
WaterSense, 151, 152
Watersheds, 177
geomorphology of, 257
natural storage in, 194*f.*
WCI. *See* Western Climate Initiative
WCT. *See* Water Conservation Tax
WEAP, 191
Weather variables, 43, 45
climate change and, 43*f.*
Western Climate Initiative (WCI), 222, 260
Wetlands, 254, 255–256, 257
WF. *See* Water footprint
WGMS. *See* World Glacier Monitoring Service
WHO. *See* World Health Organization
Wind power, 9, 223
Winneke treatment plant, 269
WML (water utility), 234
World Bank, biofuels and, 81
World Economic Forum, 173
World Glacier Monitoring Service (WGMS), 28
World Health Organization (WHO), 244
World Meteorological Organization, 5
WRI/WBCSD Greenhouse Gas Protocol, 58
WSAA. *See* Water Services Association of Australia
WSMP. *See* Water supply management program
WWF, desalination and, 186
WWF Living Planet Report, 55
WWTPs. *See* Wastewater treatment plants

Yangtze River, 21, 35
Yarra River, 114, 275
Yarra Valley Water, 111

Zagros Mountains, water from, 34
Zone 7 Water Agency, 196, 197

IWA Publishing's authorised EU representative for General Product Safety Regulations is Diane D'Arras, 15 rue Duret, 75116 Paris, France, e-mail: safety@iwap.co.uk.

Printed and bound by CPI Group (UK) Ltd, Croydon, CR0 4YY
12/05/2026
02108909-0005